# Aquatic Oligochaete Biology IX

Developments in Hydrobiology 186

*Series editor*

K. Martens

# Aquatic Oligochaete Biology IX

## Selected Papers from the 9th Symposium on Aquatic Oligochaeta, 6–10 October 2003, Wageningen, The Netherlands


*Edited by*

Piet F.M. Verdonschot[1], Hongzhe Wang[2], Adrian Pinder[3] and Rebi Nijboer[1]

[1] *Alterra, Centre for Ecosystem Studies, Department of Freshwater Ecosystems, P.O. Box 47, 6700 AA Wageningen, The Netherlands*

[2] *State Key Laboratory of Freshwater Ecology and Biotechnology, Institute of Hydrobiology, Chinese Academy of Sciences, Wuhan 430072, China*

[3] *Wildlife Research Centre, Department of Conservation and Land Management, P.O. Box 51, 6946 Wanneroo, Western Australia, Australia*



***Reprinted from Hydrobiologia, Volume 564 (2006)***


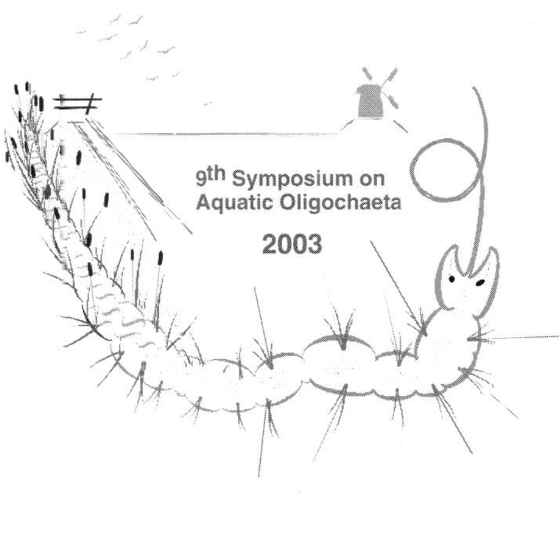



Springer

**Library of Congress Cataloging-in-Publication Data**

A C.I.P. Catalogue record for this book is available from the Library of Congress.

ISBN 1-4020-4781-9

---

Published by Springer,
P.O. Box 17, 3300 AA Dordrecht, The Netherlands

---

**Cover illustration:** Logo of the Symposium

---

***Printed on acid-free paper***



Printed in the Netherlands

## TABLE OF CONTENTS

## PART THREE : ECOLOGY AND TOXICOLOGY



## PART FOUR : SLUDGE STUDY

Hydrobiologia (2006) 564:vii–ix
P.F.M. Verdonschot, H. Wang, A. Pinder & R. Nijboer (eds), Aquatic Oligochaete Biology IX
DOI 10.1007/s10750-005-9701-y


# Preface

The ninth Symposium on Aquatic Oligochaeta held in Wageningen (6th to 10th of October 2003) was organised by Alterra, Green World Research, Centre for Ecosystem Studies and the Wageningen University, Department of Environmental Technology. As all symposia held before, this meeting again provided a unique forum for scientific communication among those who are interested in the various fields of fundamental and applied research on aquatic Oligochaeta. In a stimulating environment the state of the art in aquatic oligochaete research was presented and discussed, along with business and societal interaction.

The Symposium was opened by Bram Klapwijk, head of the Department of Environmental Technology, Wageningen University and Henk Siepel head of the Centre for Ecosystem studies, Alterra. The attendance of 83 scientists from 28 different countries, through a participation of 23 posters contributions and 51 oral communications, set the scope and width of the meeting.

The symposium intended to widen the scientific value of each of the sessions as much as possible. Experts from nine different fields within oligochaete studies presented keynote lectures. Each of the lectures started with a general introduction of the expertise as kind of 'educational' part for all attendees, and secondly gave an overview of the 'state of the art'. Kathy Coates and co-authors keynote dealt with taxonomy and phylogeny ('Taxonomic studies of oligochaete, clitellate, annelids – The Current Condition'). A second keynote on this wide topic was given by Christer Erséus and covered the phylogenetic reclassification of Limnodrilbidinae (Tubificidae). The keynote on ecology with a emphasis on assessment was presented by Trefor Reynoldsen: 'From Aristotle to Advanced Arithmetic Algorithms: the applied use of oligochaetes in ecological assessment'. Guiliano Bonomi and Andrea Pasteris talked about their long life experience with population dynamics studies in aquatic Oligochaeta. The field of physiology and behaviour was covered by Olav Giere who embedded his theme in 'Ecology and biology of marine Oligochaeta – an inventory rather than another review'. Pilar Rodriquez presented a keynote on the role of Oligochaeta in ecotoxicological research. The applied and innovative ideas on sludge reduction by predatory activity of aquatic annelids in wastewater treatment plants, was covered by Christa Ratsak. A special environment for oligochaetes is the unknown world of the groundwater and hyporheic zone. Michel Lafont brought light in this world with his keynote: 'Hyporheic zone and groundwater oligochaetes: key-habitat and key-assemblages for the functional studies of streams and rivers'. Finally, the distribution patterns of freshwater tubificid oligochaetes and the capacities for long-way dispersal of tubificids and other freshwater invertebrates with particular reference to "alien" species and birds as vectors of dispersal were dealt with by Goran Milbrink.

A short course on Oligochaeta and toxicity tests was organised by Doris Vidal together with Pilar Rodriguez and Trefor Reynoldsen. The course provided the attendants with a better understanding of the importance, methods and limitations of the use of oligochaetes as toxicity assessment tools. Furthermore, an exchange of the current understanding and advances in this field of research was established.

A special microscopy discussion session was attended by many of the participants. Worm slides were discussed and knowledge on taxonomy, morphology and ecology was exchanged.

The Symposium was only possible due to the great effort of all co-organisers: Rebi Nijboer.

Tjeerd-Harm van den Hoek (Alterra, Green World Research, Freshwater Ecology) and Christa Ratsak, Bas Buys, and Hellen Elissen (Wageningen University, Sub-department of Environmental Technology.

The symposium was financially supported by WIMEK, STOWA, Alterra Centre for Ecosystem Studies, Wageningen University Sub-department of Environmental Technology, Wageningen University Department of Agrotechnology and Food Sciences, KNAW, LEB foundation, Alterra Green World Research, and Eijkelkamp.

Piet Verdonschot
*Guest Editor*
*April, 2005*

*Group Photograph.* 1 Paula Hartzell, 2 Erica Sjolin, 3 Francesco Zaccanti, 4 Rosanna Falconi, 5 Tetsuya Narita, 6 Chiharu Sato, 7 May Lin Ang, 8 Seray Yildiz, 9 Naime Arslan, 10 Stuart Gelder, 11 Gillian Gelder, 12 Daniel Shain, 13 Enrique Martinez-Ansemil, 14 Wilco Verberk, 15 Carla Bonacina, 16 Ayaki Nakamoto, 17 Akifumi Ohtaka, 18 Takashi Shimizu, 19 Ingrid Verbessem, 20 Brigitte Lods-Crozet, 21 unknown, 22 Sharon Lavoie, 23 Doris Vidal, 24 Francisco Vidal, 25 Beatrice Sambugar, 26 Christa Ratsak, 27 Elzbieta Dumnicka, 28 Philipp Egeler, 29 Marco Ferraguti, 30 Regine Boensch, 31 Ferdinand Sporka, 32 Goran Milbrink, 33 Christer Erséus, 34 Yuansong Wei, 35 Trefor Reynoldson, 36 Jan Soors, 37 Marina Melnik, 38 Patrick Martin, 39 Jaques Juget, 40 Claudia Pavelescu, 41 Brenda Healy, 42 Anne Vivier, 43 Ton Van Haaren, 44 Bastian Buys, 45 Tarmo Timm, 46 Ernst De Bokx, 47 unknown, 48 Hellen Elissen, 49 Kathryn Coates, 50 Jana Schenkova, 51 Tjeerd-Harm Van den Hoiek, 52 Steve Fend, 53 Gert-Jan Cui, 58 Piet Verdonschot, 59 Pilar Rodriguez, 60 Rebi Nijboer, 61 Olav Giere, 62 Gunter Purschke, 63 Mark Wetzel, 64 Michel Lafont, 65 Andrea Pasteris

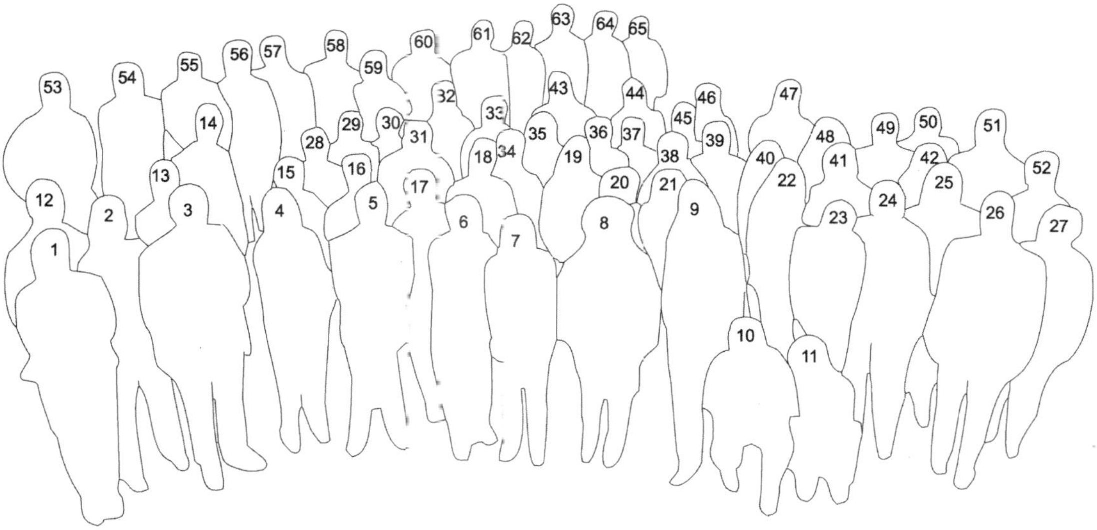
1
2
3
4
5
6
7
8
9
10
11
12
13
14
15
16
17
18
19
20
21
22
23
24
25
26
27
28
29
30
31
32
33
34
35
36
37
38
39
40
41
42
43
44
45
46
47
48
49
50
51
52
53
54
55
56
57
58
59
60
61
62
63
64
65

Hydrobiologia (2006) 564:1–4
P.F.M. Verdonschot, H. Wang, A. Pinder & R. Nijboer (eds), Aquatic Oligochaete Biology IX
DOI 10.1007/s10750-005-9702-x


# In Memoriam: Professor Walter James Harman, PhD (1928–2002)

Mark J. Wetzel
*Illinois Natural History Survey Center for Biodiversity, Champaign, Illinois 61820-6953, USA*
*(E-mail: mjwetzel@uiuc.edu)*

Our dear friend and colleague, Dr Walter James Harman, passed away on 1 April 2002, in Baton Rouge, Louisiana, USA, at the age of 74. He was a long time faculty member at Louisiana State University (LSU) in Baton Rouge. Dr Harman is survived by Dorothy Brewster Harman, his wife of 47 years, two daughters – Dea Harman O'Rourke and Nan Harman Brinson, three grandsons – Colin, Andrew, and Bennett, and his mother – Colvis Lovette.

Dr Harman was born on February 25, 1928 in Strong, Union County, Arkansas. He completed primary and secondary education in his hometown, graduating from Strong High School in 1945. He enrolled as an undergraduate student at Louisiana Polytechnic Institute in Ruston, receiving his B.S. degree in 1948. He was then accepted into graduate school at The University of Arkansas in Fayetteville, receiving his Master of Arts degree in 1950. In January 1950, Dr Harman became an instructor in the Zoology Department of Louisiana Polytechnic Institute (Louisiana Tech). He then served in the U.S. Marine Corps from 1952–1954, primarily as one of the elite Military Police and for a time was assigned with the Secret Service to provide security for President Dwight David Eisenhower.

After his release from military service, Dr Harman returned to Louisiana Tech as an assistant professor of Zoology. In 1955, he enrolled in a doctoral program at the University of Illinois in Urbana-Champaign under the guidance of Dr Max R. Matteson; during his enrollment there, he was elected as a Scholar and Fellow (Southern Fellowship Fund Foundation) in three successive years (1955–1957). Upon completion of class and research requirements for his doctoral degree, Dr Harman returned to Louisiana Tech in 1958, continuing as an assistant professor of Zoology while concurrently writing his PhD thesis on the taxonomy and nomenclature of the earthworms occurring in central Illinois; his doctoral degree from the University of Illinois was awarded on 13 February 1960. He continued as an associate professor at Louisiana Tech until 1961, then took a position as an assistant professor at LSU in Baton Rouge; by 1963 he had attained the rank of Associate Professor. In late 1963, he was promoted to full professor and appointed Chairman of the Department of Zoology and Physiology. He served as chairman of that department until 1977, then continued as a professor until his retirement in 1989.

Dr Harman's research focused primarily on the taxonomy, systematics, and ecology of terrestrial and aquatic oligochaetes. The majority of his publications focused on the Naididae and Opistocystidae of North, Central, and South America, but his research interests also included other annelid groups, nemerteans, mollusks, and cycads. Dr Harman was the author or co-author of 44 peer-reviewed publications, book and book chapters, and popular articles; citations for these publications follow this obituary.

Over the course of his career, he amassed an extensive and diverse collection of oligochaetes. In 1984, Dr Harman donated the earthworm collection resulting from his doctoral research to the Illinois Natural History Survey Annelida Collection, Champaign; after retirement, he transferred the remainder of his extensive collection of primarily aquatic oligochaetes to the Division of Worms at the National Museum of Natural History–Smithsonian Institution in Washington, DC, USA.

Dr Harman was a member of several professional organizations, including the American Association for the Advancement of Science, American Microscopical Society, American Society of Zoologists, Louisiana Academy of Sciences, the Society of Systematic Zoology, and the Cycad Society. Although he thought of himself to be only a regular member of the Cycad Society, his fellow members maintain that he was *the* founding father of that society. He also served as the first editor of the Cycad Society Newsletter. During the preparation of this obituary, conversations with Garrie Landry and other members of the Cycad Society revealed that Dr Harman's interest in oligochaetes was virtually unknown to his fellow botanists; likewise, few of us who interacted with Dr Harman in the world of oligochaetology knew of his botanical interests. The Cycad Society has established the Walter Harman Cycad Research Fund in his honor.

Dr Harman served as editor for the Proceedings of the Louisiana Academy of Sciences (1959–1963), as secretary of the Division of Invertebrate Zoology of the American Society of Zoologists, and as Program Officer (1977) and Vice President (1976–1978) for the American Microscopical Society. He served as a science consultant to the Rijksmuseum (Leyden, The Netherlands) and the Natural Museum of Natural History – Smithsonian Institution, (Washington, DC), and as a member of the Board of Directors for the Greater Baton Rouge Zoo (Baton Rouge, Louisiana).

Dr Harman was a member of our 'Founding Class' of participants at the First International Symposia on Aquatic Oligochaete Biology, hosted by Ralph Brinkhurst and convened at the Institute of Ocean Sciences in Sydney, BC, Canada, in 1979. He also participated in the 1st, 2nd, and 3rd North American Aquatic Oligochaete Symposia convened in a log cabin near Echo Lake and Mount Evans in Colorado, USA, during the early and mid-1980s. Dr Harman served as our host for the 4th International Symposium on Aquatic Oligochaete Biology in Baton Rouge, Louisiana, 14–18 March 1988. Soon after this meeting, on April 8th, Dr Harman suffered a massive stroke that restricted his ability to walk but not his ability to think and communicate effectively – both scientifically, and from his heart.

Although restricted to a wheelchair, Dr Harman's character, inner strength, and his desire as a scientist allowed him to attend and participate in the 5th North American Oligochaete Symposium at the University of Wyoming Biological Station at Jackson Lake, Wyoming, USA, in 1990. At that meeting, Dr Harman talked compassionately about the love he had for science, for the beauty of nature, and for the opportunities he had experienced over the years as an oligochaetologist. He also shared with us that he was [at that time] experiencing the biggest challenge, and disappointment, in his life – 'hosting' intense and lucid mental activity in a failing body; sadly, this was perhaps the last time most of his oligochaete colleagues would be with him.

Dr Harman had a strong and passionate love for teaching, a perpetual calmness in conversation, a dislike for pretension, an occasional desire for a martini so dry that the word vermouth, a two-syllable word, was never spoken out loud in the presence of gin, and a subtle yet humorous wit that was often as dry as his martinis. His gentle charm, love for books and plants, and the worship of God through the uniqueness, diversity, and beauty of the zoological and botanical world in which we live was apparent in each of our interactions with him. First and foremost, however, Dr Harman's colleagues, students, friends, and family will remember him as a courteous southern gentleman (Fig. 1).

During each of the preceding international symposia on aquatic oligochaete biology, and in

*Figure 1.* Dr Walter J. Harman, at his microscope in the Life Sciences building at Louisiana State University in Baton Rouge, ca. 1975; photo courtesy of Dorothy Harman.

many of the published symposia proceedings, we have honored the memory of deceased colleagues and friends – those who have participated in one or more of these symposia, and those who have provided significant influence on the directions of our own scientific endeavors. These include Drs H. Randy Baker, Sergej Hrabe, Clarence J. Goodnight, Andreina Di Chiara, Gilberto Righi, and Perry C. Holt. Therefore, as a participant in each of the nine international aquatic oligochaete symposia, and on behalf of the organizing committee of this present symposium and our fellow oligochaetologists, worldwide, who have known Dr Harman and benefited from both his scientific contributions and his sincere kindness, we hereby dedicate this 9th International Symposium on Aquatic Oligochaete Biology to the memory of Dr Walter J. Harman.

## Acknowledgements

I thank Dorothy Harman, for her friendship for so many years, and for her trust in me to write this dedication for the most important man in, and love of, her life. I extend my appreciation to Dr Earl H. Weidner, a retired professor and former colleague of Dr Harman at LSU, for sharing memories of their friendship and working relationship with me. I also thank Garrie P Landry, a professor at the University of Louisiana – Lafayette, and perennial Cycad Society member and friend of Dr Harman; some of the information presented above was paraphrased with permission from Landry (2002). Dr Michael S. Loden, a former graduate student of and research collaborator with Dr Harman while at LSU, kindly reviewed and contributed to this memoriam. Finally, I thank Joel Dexter, photographer at the Illinois State Geological Survey, Champaign, for his assistance with photographs used in this dedication as presented at the symposium in Wageningen, The Netherlands, 6 October 2003.

## Reference

Landry, G. P., 2002. Dr. Walter Harman (1928–2002). The Cycad Newsletter 25(2): 11.

## Publications of Walter J. Harman (listed chronologically)

Harman, W. J., 1952. A taxonomic survey of the earthworms of Lincoln Parish, Louisiana. Proceedings of the Louisiana Academy of Sciences 15: 19–23.

Harman, W. J., 1954. Some earthworms from southern Oklahoma. Proceedings of the Oklahoma Academy of Science 35: 51–55. [This paper was most likely published in 1955; the proceedings are for the year 1954, and printed information in this volume stated that papers were received in late December, 1954, for publication.]

Harman, W. J., 1955. Earthworms of commercial importance and their effect on distribution. Proceedings of the Louisiana Academy of Sciences 18: 54–57.

Harman, W. J., 1956. Isolation of earthworm setae by use of histophagous protozoa. Transactions of the American Microscopical Society 75: 332–333.

Harman, W. J., 1960. Studies on the taxonomy and musculature of the earthworms of central illinois. PhD. thesis. University of Illinois, Urbana V + 107 pp.

Harman, W. J., 1961a. [1960]. A double-tailed earthworm. Proceedings of the Louisiana Academy of Sciences 23: 75–76. ['1961' is clearly stated as the year of publication on front cover of this volume, even though '1960' is printed as the publication date at the top of the first page of this paper.]

Harman, W. J., 1961b. Plain facts about earthworms. Organic Gardening 8(11): 23–27.

Harman, W. J., 1961c. Seta. In Gray, P (ed.), The encyclopedia of the biological sciences. Reinhold Publishing Corporation, New York, 915.

Harman, W. J. & J. H. Platt, 1961. Notes on some aquatic oligochaetes from Louisiana. Proceedings of the Louisiana Academy of Sciences 24: 90–95.

Harman, W. J., 1962. A fresh water nemertine from Louisiana. Proceedings of the Louisiana Academy of Sciences 25: 32–34.

Dundee, D. S. & W. J. Harman, 1963. *Corbicula fluminea* (Müller) in Louisiana. Nautilus 77(1): 30.

AIBS Subcommittee on Facilities and Standards [members Johnson, W., Balamuth, W., Bell, C. R., Harman, W., McDallan, S. E. A. and Schneiderman, H.], 1963. Standards for staff and facilities in departments of biological science. AIBS Bulletin 13(6): 14–19.

Roberts, J. H., H. B. Boudreaux & W. J. Harman, 1963. Laboratory manual for general zoology. Part two: The invertebrates. Wm. C. Brown Book Co, Dubuque, IA, USA [out of print].

Harman, W. J., 1964. Zoogeography of freshwater Oligochaeta [abstract]. American Zoologist 4(4): 437.

Harman, W. J., 1965a. A key to the genus *Slavina* (Oligochaeta: Naididae) with description of a new species from Costa Rica. Annals & Magazine of Natural History, Series 13–8(93 & 94): 565–568.

Harman, W. J., 1965b. A new species of the genus *Pristina* (Oligochaeta: Naididae) from Louisiana. Proceedings of the Louisiana Academy of Sciences 28: 28–31.

Harman, W. J., 1965c. Life history studies of the earthworm *Sparganophilus eiseni* in Louisiana. The Southwestern Naturalist 10(1): 22–24.

Roberts, J. H., H. B. Boudreaux, W. J. Harman & N. B. Causey, 1965. Laboratory Manual for General Zoology. Directions for use in a two-semester course in introductory zoology. Second edition. Part 2: Second Semester – The Invertebrates. Wm. C. Brown Book Co, Dubuque, IA USA. vi + 79 pp [out of print].

Harman, W. J., 1966. Some aquatic oligochaetes from Mississippi. American Midland Naturalist 76(1): 239–242.

Harman, W. J., 1970. Revision of the family Opistocystidae (Oligochaeta). Transactions of the American Microscopical Society 88(4): 472–478 [Per comment from editor in the successive volume of this journal (Vol. 89, p. 347), the correct year of publication of this issue is 1970, not '1969' – a misprint on the fourth issue of Vol. 88].

Harman, W. J., 1971. A review of the subgenus *Allodero* (Oligochaeta: Naididae: *Dero*) with a description of *D. (A.) floridana* n. sp. from *Bufo terrestris*. Transactions of the American Microscopical Society 90(2): 225–228.

Harman, W. J., 1973a. *Dero (Allodero) hylae* (Oligochaeta: Naididae) in Louisiana anurans. Proceedings of the Louisiana Academy of Sciences 36: 71–76.

Harman, W. J., 1973b. New species of Oligochaeta (Naididae) with additional distributional records from Oklahoma and Texas. The Southwestern Naturalist 18(2): 151–164.

Harman, W. J., 1974. The Naididae (Oligochaeta) of Surinam. Zoologische Verhandelingen Uitgegeven Door Het Rijksmuseum Van Natuurlijke Historie Te Leiden 133: 1–36.

Harman, W. J., 1975. A new species of *Stephensoniana* (Oligochaeta: Naididae) from North America. Proceedings of the Biological Society of Washington 88(1): 1–4.

Harman, W. J. & R. C. Harrel, 1975. *Haemonais waldvogeli* (Naididae: Oligochaeta) now established in North America. Texas Journal of Science 26(3 + 4): 621–623.

Harman, W. J. & A. R. Lawler, 1975. *Dero (Allodero) hylae*, an oligochaete symbiont in hylid frogs in Mississippi. Transactions of the American Microscopical Society 94(1): 38–42.

Harman, W. J. & M. L. McMahan, 1975a. A reevaluation of *Pristina longiseta* (Oligochaeta: Naididae) in North America. Proceedings of the Biological Society of Washington 88(17): 167–178.

Harman, W. J. & M. L. McMahan, 1975b. Scanning electron microscopy as a tool in the study of oligochaetes. [abstract of paper presented at meeting of the Louisiana and Texas Societies for Electron Microscopy, 22–25 February 1975 in New Orleans, Louisiana]. Texas Reports on Biology and Medicine 33(2): 350.

Harman, W. J., 1976. An oligochaete that belongs to two genera. [abstract of contributed paper presented at the 91st annual meeting of the American Microscopical Society, New Orleans, Louisiana, 10–14 November 1975]. Transactions of the American Microscopical Society 95(2): 274.

Harman, W. J., 1977. Three new species of Oligochaeta (Naididae) from the southeastern U.S. Proceedings of the Biological Society of Washington 90(3): 483–490.

Harman, W. J. & M. S. Loden, 1978a. A re-evaluation of the Opistocystidae (Oligochaeta) with descriptions of two new genera. Proceedings of the Biological Society of Washington 91(2): 453–462.

Harman, W. J. & M. S. Loden, 1978b. *Bratislavia unidentata* (Oligochaeta: Naididae), a re-description. The Southwestern Naturalist 23(4): 541–544.

Harman, W. J. & M. S. Loden, 1979. A rediscovery of *Bratislavia bilongata* (Chen, 1944) (Oligochaeta: Naididae) in Florida. Transactions of the American Microscopical Society 98(4): 581–583.

Harman, W. J., M. S. Loden & J. R. Davis, 1979. Aquatic Oligochaeta new to North America with some further records of species from Texas. The Southwestern Naturalist 24(3): 509–525.

Harman, W. J., 1980. Specific and generic criteria in freshwater Oligochaeta, with special emphasis on the Naididae. Pages 1–8, In: Brinkhurst, R. O. & D. G. Cook (eds), Aquatic oligochaete biology. Plenum Press, New York. ix + 529 pp. [Proceedings of the First International Symposium on Aquatic Oligochaete Biology, Institute of Ocean Sciences, Sidney, British Columbia, Canada, 1–4 May 1979].

Loden, M. S. & W. J. Harman, 1980. Ecophenotypic variation in setae of Naididae (Oligochaeta). Pages 33–39, In: Brinkhurst, R. O. & D. G. Cook (eds), Aquatic Oligochaete Biology. Plenum Press, New York. ix + 529 pp. [Proceedings of the First International Symposium on Aquatic Oligochaete Biology, Institute of Ocean Sciences, Sidney, British Columbia, Canada, 1–4 May 1979].

Harman, W. J., 1982a. Oligochaeta. In Hurlbert, S. H. & A. Villalobos-Figueroa (eds), Aquatic Biota of Mexico, Central America and the West Indies. SDSU Foundation, San Diego State University Press, San Diego, California, USA 92182, 162–165.

Harman, W. J., 1982b. The aquatic Oligochaeta (Aeolosomatidae, Opistocystidae, Naididae) of Central America. The Southwestern Naturalist 27(3): 287–298.

Loden, M. S. & W. J. Harman, 1982. *Dero (Aulophorus) intermedia* Nomen Novum for *Aulophorus pectinatus* Stephenson, 1931 (Oligochaeta: Naididae). Freshwater Invertebrate Biology 1(1): 53–54.

Harman, W. J., 1984a. An analysis of the inorganic composition of the setae of *Eisenia foetida* (Oligochaeta, Lumbricidae). Proceedings of the Louisiana Academy of Sciences 47: 19–22.

Harman, W. J., 1984b. Segment numbers in the earthworm, *Bimastos zeteki* (Oligochaeta: Lumbricidae), as a function of age. Proceedings of the Louisiana Academy of Sciences 47: 34–41.

Harman, W. J. & M. S. Loden, 1984. *Capilloventer atlanticus* gen. et sp.n., a member of a new family of marine Oligochaeta from Brazil. Hydrobiologia 115: 51–54.

Harman, W. J., R. O. Brinkhurst & M. Marchese, 1988. A contribution to the taxonomy of the aquatic Oligochaeta (Naididae) of South America. Canadian Journal of Zoology 66(10): 2233–2242.

# Part One
# Taxonomy, Anatomy, Morphology and Physiology

Hydrobiologia (2006) 564:7–17
P.F.M. Verdonschot, H. Wang, A. Pinder & R. Nijboer (eds), Aquatic Oligochaete Biology IX
DOI 10.1007/s10750-005-1703-2



# *Troglodrilus* (Annelida, Oligochaeta, Tubificidae), a new genus from subterranean habitats in southwestern Europe

Jacques Juget[1], Michel C. des Châtelliers[1,*] & Pilar Rodriguez[2]
[1]*UMR CNRS 5023 Ecologie des Hydrosystèmes Fluviaux, Equipe d'Hydrobiologie et Ecologie Souterraines, Université Claude Bernard Lyon 1, Villeurbanne Cedex, F-69622, France*
[2]*Dpt. Zoología y Biología Celular Animal Facultad de Ciencias, Universidad Pais Vasco/Euskal Herriko Unibertsitatea, Apdo. 644, Bilbao, E-48080, Spain*
*(*Author for correspondence: E-mail: michel.deschatelliers@univ-lyon1.fr.)*

*Key words: Troglodrilus* n. gen. *Tubificoides galarzai*, Tubificidae, France, Spain, caves

## Abstract

Over the last 20 years, several specimens of the tubificid oligochaete *Tubificoides galarzai* Giani & Rodriguez, 1988 were collected from subterranean waters of the North-West Iberian Peninsula (cave Mairulegorreta, Spain), the East Hexagon (subterranean gallery of Montgelas near Lyon (département du Rhône, France), and Crotot Cave in the vicinity of Romain (located northeast of Besançon, département du Doubs, France). Subsequent study of specimens from these various sites – and in particular, their genitalia and setal morphology – provided us with the opportunity to measure the variability of characters in specimens from each site, and subsequently assess the intraspecific geographical variation in this species. Examination of serial histological sections of male ducts of French specimens now supports a revaluation of the taxonomic status of *T. galarzai*. The new material confirms that the vas deferens enters the atrium apically, not opposite to the prostate junction (as stated in the original description), thus justifying its exclusion from the genus *Tubificoides*. Although a clear histological separation is not observable in the proximal end of the atrium, two kinds of cells are present in the atrial epithelium near the entrance of prostate gland – similar to the histological structure described in *Lamadrilus sorosi* Timm 1998. The relationship of *T. galarzai* with several species of the genus *Tubifex* is also discussed. A new genus *Troglodrilus* n. gen. is proposed for this tubificine oligochaete that seems to be restricted to subterranean waters.

## Introduction

This work deals with the geographical intraspecific variation of the tubificine *Tubificoides galarzai* Giani & Rodriguez, 1988. Since discovery of four specimens of this species in 1984–1985 from a karst spring and a cave located in the Basque Country of northern Spain, several additional mature specimens have been collected from other sites in southern Europe. Three of them were collected on 8 November 1987, from Mairulegorreta Cave (Alava Province, Basque Country, northern Spain). Between 1998 and 2001, several other specimens were collected from subterranean habitats located in the eastern part of France (subterranean gallery of Montgelas in the 'Mont d'Or' near Lyon, and the Crotot Cave near Romain, northeast of Besançon) (Juget & des Châtelliers, 2001). A comparison of specimens from various localities first induced the authors to study the rate of geographical intraspecific variation of *T. galarzai*. However, the examination of serial histological sections of male ducts of French specimens has subsequently enabled us to revalue the taxonomic status of this species because of divergences from the genus *Tubificoides* Lastockin, 1937 and affinities with the genera *Tubifex* Lamarck, 1816 and *Lamadrilus* Timm, 1998.

## Materials and methods

The cave of Mairulegorreta is located in the Gorbeia Natural Park in Spain (43° 1′ 23″ N, 2° 45′ 29″ W; elevation: ~1000 m). Oligochaetes from the cave of Mairulegorreta were sampled with a hand net (200 $\mu$m mesh size) in a stream one meter wide (Rio Arriba) with gravelly sediments.

Samples from two sites in France – the subterranean gallery of Montgelas (45° 49′ 33″ N, 4° 49′ 08″ E (DMS datum WGS84); elevation: 375 m) and the Crotot Cave (47° 26′ 44.52″ N, 6° 24′ 31.64″ E (DMS datum WGS84); elevation: 411 m), were collected from submerged sediments with prevailing silty sands (median particle size less than150 $\mu$m) using a Surber net (200 $\mu$m mesh size).

Samples were fixed in the field with 4–8% formaldehyde (Spanish specimens) or Bouin's fluid (French specimens), washed at the laboratory through a 200 $\mu$m sieve, then worms were sorted under stereomicroscopes and preserved in 70% ethyl alcohol.

Specimens were stained in Erhlich's hematoxylin (Spanish specimens) or in paracarmin (French specimens), and dissected or whole mounted in Canada balsam. Serial sections (5 $\mu$m) of three mature specimens from the Crotot Cave were stained with trichromatic staining after Masson–Goldner modified (Martojat & Martojat-Pierson, 1967; the acetic aniline blue in place of brilliant green, Dupuy, pers. comm.)

*Examined material*

We have examined the type collection, where the four specimens were Syntypes, and designated Lectotype and Paralectotypes, which are described below with their codes.

*Type material*

*Tubificoides galarzai* Giani & Rodriguez, 1988. Lectotype MNCN16.03/3037, dissected worm collected on 15 December, 1984. Paralectotypes: three specimens, two dissected worms collected on 15 December from type locality, one in MNCN16.03/3038, and other in the collection of Dr P. Rodriguez (University of Basque Country, Bilbao Spain); one dissected worm collected on 23 February, 1985 from Cave Goiketxe (Spain) in the collection of Dr Giani (23-2369).

*Type locality*

Cave Argatxa (North Spain); see description in Giani & Rodriguez (1988).

*Other material*

Two dissected worms from Cave Mairulegorreta (Alava, Spain) collected on 8 November 1987; 1 whole mounted specimen from gallery of Montgelas (France) collected on 1 March, 1998, in the personal collection of Dr P. Rodriguez (University of Basque Country, Bilbao, Spain).

Three sectioned specimens from Crotot Cave (Romain, Doubs, France), collected on 31 January 2001 deposited in the Muséum d'Histoire Naturelle de Lyon, France (MHNL): 1 specimen on 17 plates (MHNL 44002201-1 to 44002201-17) – 1 specimen on 19 plates (MHNL 44002202-1 to 44002202-19) – 1 specimen on 19 plates (MHNL 44002203-1 to 44002203-19). One dissected specimen from gallery of Montgelas (Rhône, France), collected on 3 June 1998; (MHNL 44002204). Four whole mounted specimens from gallery of Montgelas (France) collected on 3 June 1998, (MHNL 44002205-44002208). Eleven whole mounted specimens from Crotot Cave (Romain, France), collected on 31 January 2001 (MHNL 44002209-44002219).

Abbreviations for museums: MNCN: Museo Nacional de Ciencias Naturales (Madrid, Spain); MHNL: Muséum d'Histoire Naturelle de Lyon (France).

*Statistical analysis*

A cluster analysis was done using the Pearson correlation and the unweighted pair-group average linkage to compare the individuals from Spanish and French populations (Statistica 6.0). A first analysis concerned the variability of setae to individuals from Montgelas ($n = 5$) and Crotot ($n = 8$). The maximum length of the hair setae, pectinate setae and ventral bifid setae was measured on bundles II, V and post-clitellar. A second analysis concerned five measures of the penial sheath (Fig. 1) of individuals from Spain ($n = 5$), Montgelas ($n = 12$) and Crotot ($n = 10$).

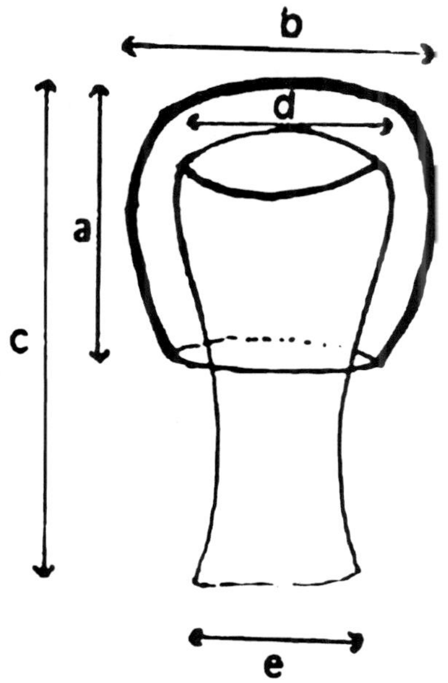


*Figure 1.* Morphological parameters (a, b, c, d, e) of the penial sheath of *T. galarzai* (see Fig. 12). a and b, length and diameter of the collen; c, total length of the penial sheath; d and e, proximal and distal diameter of the stalk.

**Results**

*Comparison between Spanish and French populations*

*External features*

Specimens from France and Spain do not show high morphological differences. Ringed glandular epidermis is more easily discernible in type specimens than in the new material, either Spanish or French (Figs. 2, 3). However, higher heterogeneity, characterizes specimens in the French populations compared to specimens in the Spanish populations as confirmed by the wider range of variation of the characters in specimens from the French populations (see Table 1). Differences between Spanish and French populations were in size and/or number rather than shape. For instance, with regard to the setae, the bent distal part and palmate appearance of the pectinate setae in specimens from the French populations with intermediate teeth, generally longer than lateral teeth, are similar to the original material from Spain. However, the number of intermediate teeth is generally smaller in specimens from the French populations [2–5] than those from the Spanish populations [4–6] (Fig. 4). The upper (distal) tooth of the ventral setae is always longer than the lower (proximal) tooth in specimens from both the Mairulegorreta and French populations (Fig. 5), unlike type specimens that showed this characteristic only in the anterior ventral bundles; the distal and proximal tooth of setae in the mid and posterior bundles of the type specimens are equal in length. In spite of the higher variability of the specimens from France, the length of the setae does not allow to separate both populations of Montgelas and Crotot (Fig. 6). Both populations mix, with maybe a higher variability of the specimens from Montgelas (but only five individuals).

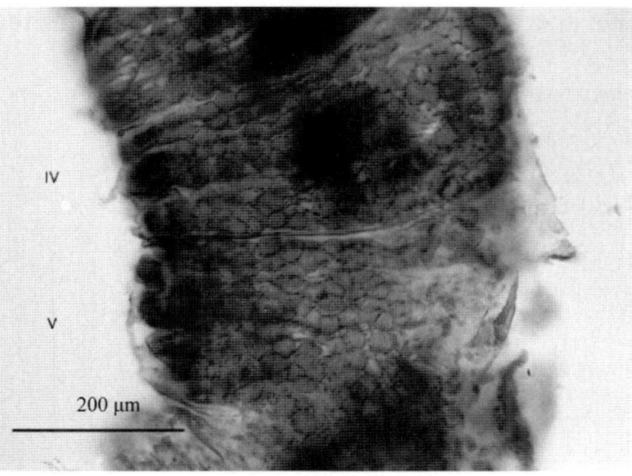


*Figure 2.* Lateral view of the anterior part of the body with rows of the epidermic glands (specimen of *Tubificoides galarzai* from the type locality).

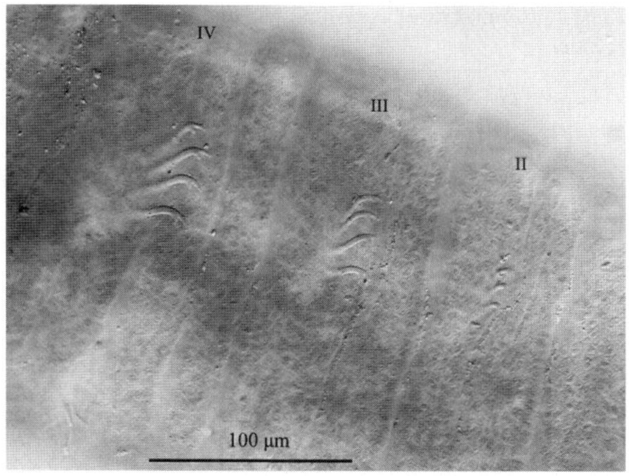


*Figure 3.* Lateral view of the anterior part of the body with rows of the epidermic glands (specimen of *Tubificoides galarzai* from the Crotot Cave).

*Table 1.* Comparative ranges (minima and maxima) of some taxonomic features of Spanish and French populations of *Tubificoides galarzai*

| | Populations | | | |
|---|---|---|---|---|
| | Spanish | | French | |
| N of segments (complete specimens) | 28–38 | ($n=5$) | 25–57 | ($n=13$) |
| Width of segment V ($\mu$m) (mature specimens) | 260–300 | ($n=7$) | 225–300 | ($n=20$) |
| N of pre-clitellar hair setae/bundle | 1–6 | ($n=7$) | 1–6 | ($n=14$) |
| Length of pre-clitellar hair setae ($\mu$m) | 144–224 | ($n=7$) | 84–247 | ($n=14$) |
| N of pre-clitellar pectinate setae/bundle | 3–5 | ($n=7$) | 1–5 | ($n=14$) |
| Length of pre-clitellar pectinate setae ($\mu$m) | 44–82 | ($n=7$) | 28–65 | ($n=14$) |
| N of intermediate teeth of pectinate setae | 4–6 | ($n=7$) | 2–5 | ($n=14$) |
| N of pre-clitellar ventral bifid setae | 2–5 | ($n=7$) | 1–6 | ($n=14$) |
| Length of pre-clitellar ventral bifid setae ($\mu$m) | 55–80 | ($n=7$) | 28–73 | ($n=14$) |
| Width of spermathecal ampulla ($\mu$m) | 112–250 | ($n=7$) | 96–224 | ($n=8$) |
| Width of spermathecal bulb ($\mu$m) | 53–110 | ($n=7$) | 39–62 | ($n=11$) |
| Length of atrium (penial apparatus excluded) ($\mu$m) | 230–250 | ($n=7$) | 224–280 | ($n=8$) |
| Width of atrium (proximal part) ($\mu$m) | 27–50 | ($n=7$) | 45–75 | ($n=8$) |
| Length of penial sac ($\mu$m) | 75–124 | ($n=7$) | 64–97 | ($n=16$) |
| Width of penial sac ($\mu$m) | 61–87 | ($n=7$) | 45–65 | ($n=16$) |
| Length of penis sheath ($\mu$m) | 74–104 | ($n=7$) | 47–94 | ($n=23$) |
| Width of penis sheath ($\mu$m) | 57–76 | ($n=7$) | 32–60 | ($n=23$) |

($n$ = number of individuals measured).

### *Internal anatomy*

Alimentary canal and associated glands have similar characteristics among the five populations investigated: alimentary canal distended in (VIII) IX, chloragogen cells from VI backwards, and pharyngeal glands present in (II) III to VI (Fig. 7).

All studied populations show similar shape of spermathecae with a bulbous duct, spindle-shaped spermatozeugmata, comma-shaped atria without ejaculatory duct, penial sacs, and penial cuticular sheaths (Table 1). However, the spermathecal bulb of specimens from the French populations (Figs. 8 and 9) is smaller – both in absolute value (bulb 39–62 $\mu$m wide instead of 53–110 $\mu$m wide in specimens from the Spanish populations) and relative value (spermathecal bulb less than twice as wide as the spermathecal canal in specimens from the French populations, more than twice as wide as the spermathecal canal in specimens from the

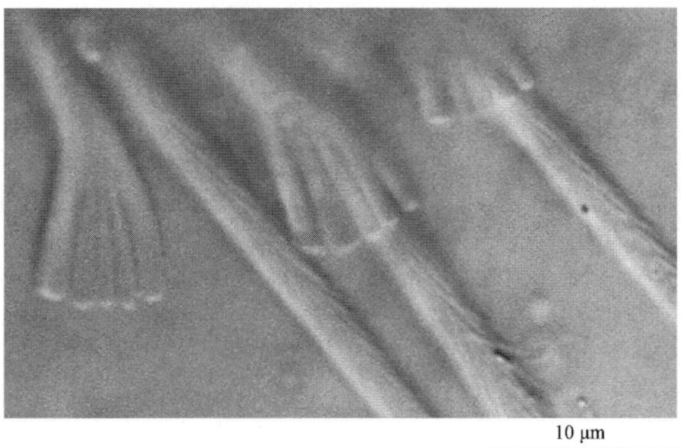


*Figure 4.* Dorsal pectinate setae and serrate hair setae (segment V) (specimens of *Tubificoides galarzai* from the Crotot Cave).

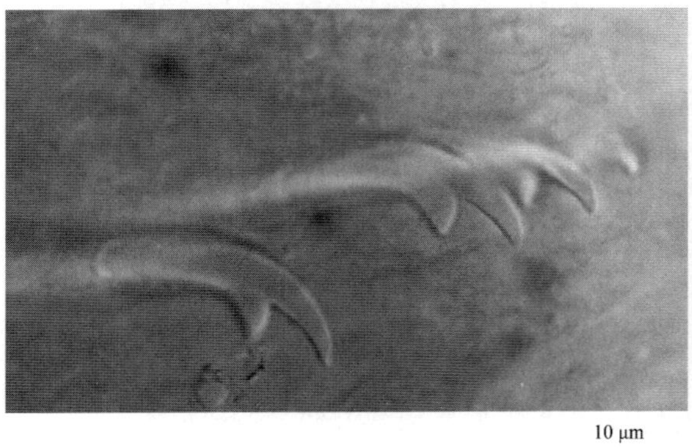


*Figure 5.* Ventral bifid setae (segment V) (specimens of *Tubificoides galarzai* from the Crotot Cave).

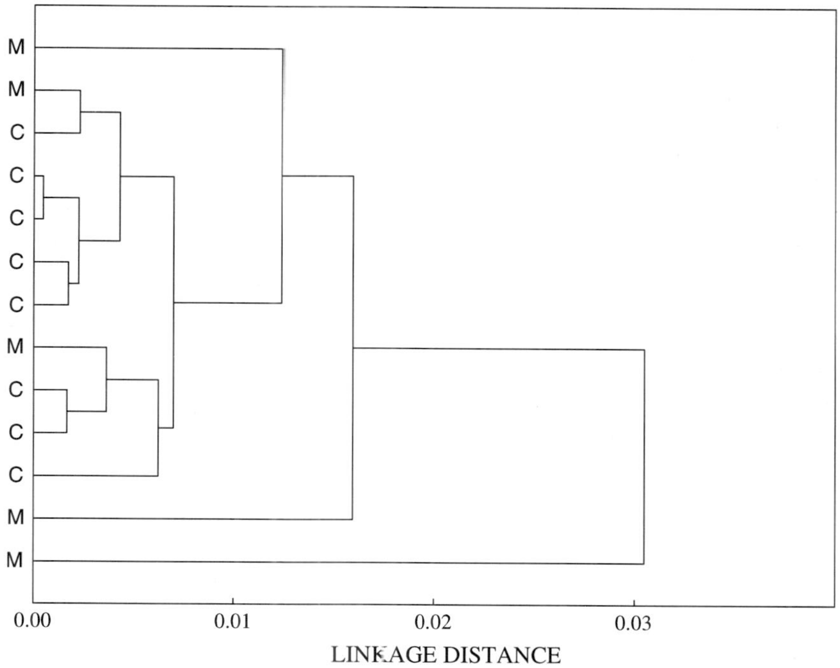


*Figure 6.* Cluster analysis bearing on the maximum length of the hair setae, pectinate setae and ventral bifid setae measured on bundles II, V and post-clitellar: C = specimens from Crotot (France); M = specimens from Montgelas (France).

Spanish populations). Penis free into a cuticular sheath (Figs 1, 10 and 11), characterized by an ental collar with a wall 2 $\mu$m thick and a cylindrical stalk with a wall more or less finely pleated and slightly swollen, (similar to the group E of *Tubificoides*, grouped according to penial sheath type by Helgason & Erséus, 1987). Detailed study of several parameters of the penial sheath (Fig. 1) does not allow to separate specimens resulting from populations of Spain and France (Montgelas and Crotot) which seem very homogeneous for this character (Fig. 12). The male ducts of the

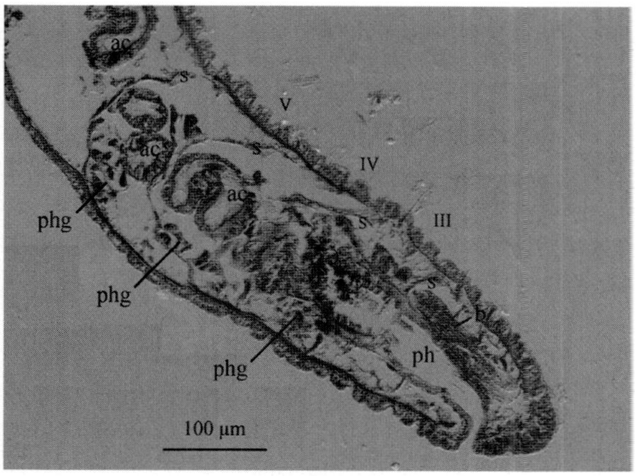


*Figure 7.* Body wall and internal organs of the anterior part of the body in longitudinal section (specimen of *Tubificoides galarzai* from the Crotot Cave). ac, alimentary canal; b, brain; ph, pharynx; phg; pharyngeal gland; s, septum.

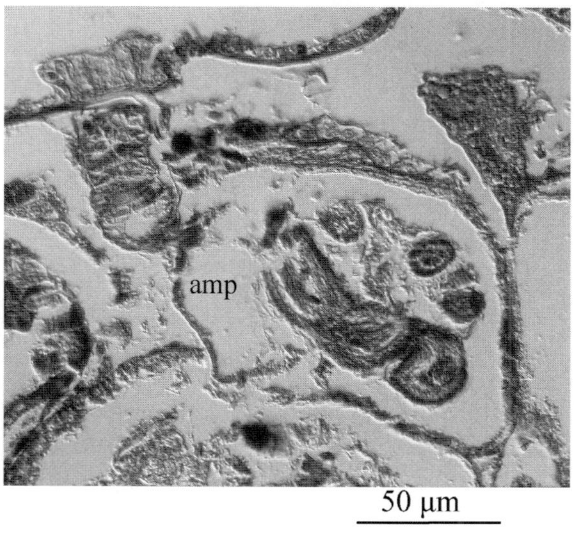


*Figure 8.* Spermatheca with side view of the spermathecal bulb (specimens of *Tubificoides galarzai* from the Crotot Cave). amp, ampulla.

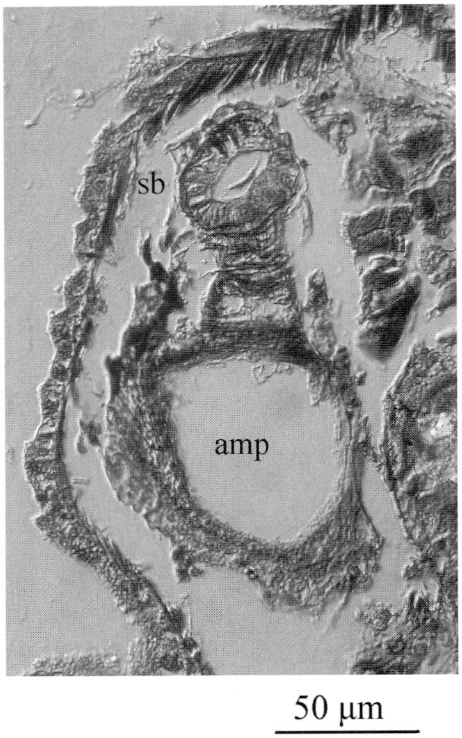


*Figure 9.* Spermatheca with front view of the spermathecal bulb and with distal portion of the ampulla (specimens of *Tubificoides galarzai* from the Crotot Cave). amp, ampulla; sb, spermathecal bulb.

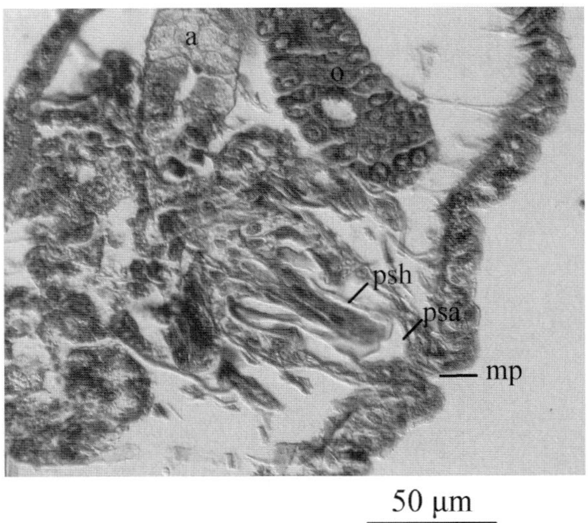


*Figure 10.* Penial apparatus (with penial sac and penial sheath) and distal portion of atrium in sagittal section (specimens of *Tubificoides galarzai* from the Crotot Cave). a, atrium; mp, male pore; psa, penial sac; psh, penial sheath.

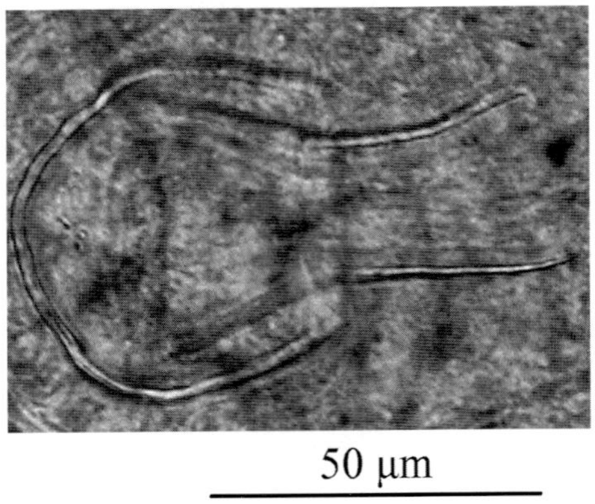


*Figure 11.* Penial sheath (specimens of *Tubificoides galarzai* from the Crotot Cave).

specimens from the Crotot Cave were examined using serial sections that had been prepared using a trichromatic stain (after Masson–Goldner modified, Martojat & Martojat-Pierson, 1967). As suggested by Giani & Rodriguez (1988), from their observations of dissected specimens from Spain, this method has confirmed the entrance of the vas deferens into the apical end of the atrium. Cross and sagittal sections of French material provide this opportunity to present a more detailed description of the structure of the atrial epithelium. Two different types of epithelial lining can be distinguished in the atrium (Figs. 13–15):

- a glandular heavily stained, densely granulated epithelium restricted to the concave side of the ental portion of atrium, both distally and proximally to the junction of the prostate gland, with maximum thickness (about two thirds of the maximum atrium diameter) near the junction of the prostate gland.
- a non-granulated light epithelium, of almost uniform thickness, ciliated near the entrance of vas deferens, and located in the convex side of the proximal portion of atrium and along the distal portion of atrium.

## Discussion

This study has revealed that the taxonomic position of *Tubificoides galarzai* should be revised based on new material collected from subterranean habitats in Spain and France. Giani & Rodriguez (1988) discussed the taxonomy of this species,

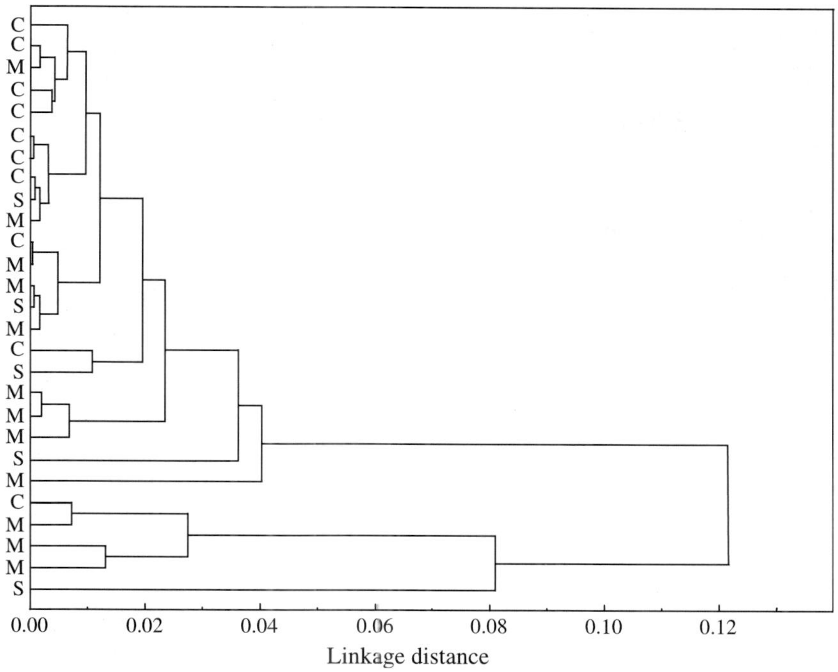


*Figure 12.* Cluster analysis bearing on five measures of the penial sheath (see Fig. 1): C = specimens from Crotot (France); M = specimens from Montgelas (France); S = specimens from Spanish (Argatxa, Goiketxe, Mairulegorreta).

suggesting that it may be intermediate between *Tubificoides* and *Tubifex* (*sensu lato*) due to the variable position of the junction of the vas deferens both apical or subapical, opposed to the junction of the prostate gland. Such differences could be explained as due to different orientations of atria during the setting, but also to differences in the development of the granulated ental part of the atrium. The new material confirms that the position of junction of vas deferens and atrium is apical. However, the presence of several characters (e.g., bulbous spermathecal duct, epidermal glands, and the influence of high tides on the water salinity of the type locality) favoured the initial

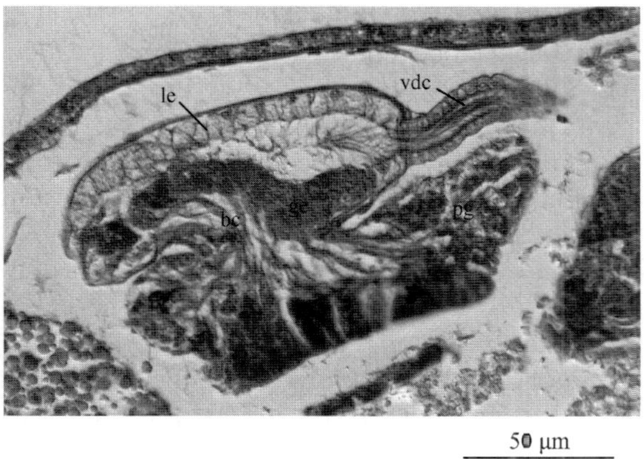


*Figure 13.* Proximal part of atrium in sagittal section showing the two kinds of atrial epithelium and connections respectively with the vas deferens and the prostate gland (specimens of *Tubificoides galarzai* from the Crotot Cave). bc, bundle of canals; ge, granular epithelium; le, light epithelium; pg, prostate gland; vdc, vas deferens ciliated.

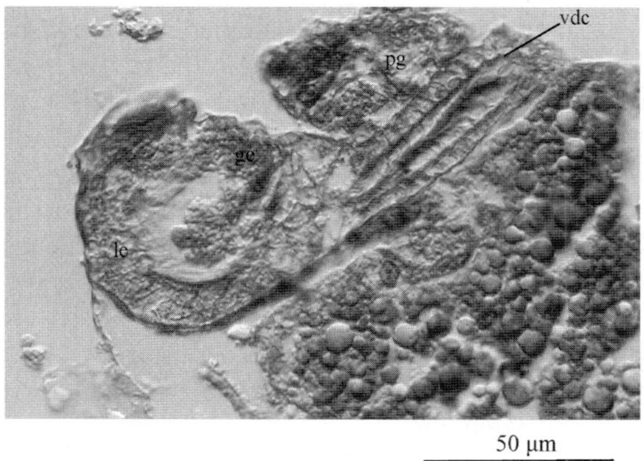


*Figure 14.* Vas deferens (in sagittal section) connected to the apical end of atrium (in cross section) (specimens of *Tubificoides galarzai* from the Crotot Cave). le, light epithelium; ge, granular epithelium; pg, prostate gland; vdc, vas deferens ciliated.

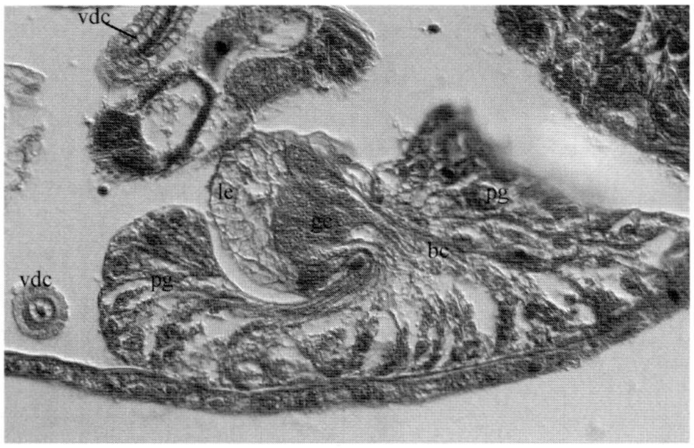


*Figure 15.* Cross section of the atrium near the entrance of the ducts of prostate gland. bc, bundle of canals; ge, granular epithelium; le, ligh epithelium; pg, prostate gland; vdc, vas deferens ciliated (specimens of *Tubificoides galarzai* from the Crotot Cave). bc, bundle of canals; ge, granular epithelium; le, light epithelium; pg, prostate gland; vdc, vas deferens ciliated.

classification of the taxon into the genus *Tubificoides*.

According to the diagnosis based on the contributions of several publications (Brinkhurst & Baker, 1979; Brinkhurst, 1986; Helgason & Erséus, 1987; Erséus, 1989; Milligan, 1996), the genus *Tubificoides* includes more than 50 estuarine and marine species. All these species are characterized by subapical entrance of the vas deferens, generally opposite to the junction of the large stalked prostate gland (exceptionally closer to the apical end than to the entrance of the prostatic gland, as in *T. parviductus* and *T. uncinatus* Helgason & Erséus, 1987), and – as a rule – by cap-like ental end of the atrium, histologically differentiated from the rest. The collection studied here confirms that the position of the junction of the vas deferens entering the atrium is apical and not opposite to the prostate, as shown in one of the atria of the holotype specimen (Fig. 4 in Giani & Rodriguez, 1988). This fact is an important objection to the classification of *T. galarzai* within the genus *Tubificoides*.

The systematics of the genus *Tubifex* is complex and still subject to debate. Holmquist (1985) revised the genus and provided a new diagnosis that restricted the classification to a small number of species into it, based on the presence of characters such as the bipartite structure of the vas deferens (ciliated proximal and non-ciliated distal parts), or absence of modified genital setae. In a previous work, Holmquist (1983) discussed the nature of the cuticle that covers the penis in *Tubifex tubifex*, describing it as a common cuticle and not a penial sheath (a term that she recommended be restricted to those specimens in which the penis is left free inside the cuticular structure).

Other species of the *Tubifex* group (*sensu lato*) were classified by Holmquist (1985) into five genera: *Heterochaeta, Lophochaeta, Amerigodrilus, Tasserkidrilus, Christerius*. This revision has met with different levels of acceptance among the specialists and, although it would be inappropriate to discuss the validity of these genera in this paper, it seems reasonable to support our discussion in this revised diagnosis of the group *Tubifex* using a greater number of taxonomic characters.

The species *T. galarzai* clearly differs from *Tubifex* (*sensu stricto*) (i.e., *sensu* Holmquist, 1985) by the presence of moderately long ciliated vasa deferentia, neither coiled nor bipartites, and by the presence of cuticular penial sheaths. The taxon does not fit into the other genera, either by the characteristics of the vas deferens, the relative size of atrium or its shape, or by the presence or the shape of the penial sheath. The entrance of vas deferens into the apical end of atrium is a character shared by several taxa, including *Christerius, Heterochaeta*, and *Tasserkidrilus*. The palmate look of pectinate setae resembles *Tubifex nerthus*, and the distinctly bent distal portion of these pectinate setae resemble *Tubifex ignotus* (*Lophochaeta ignota*, after Holmquist, 1985). The comma-shaped atrium is typical of *Tubifex tubifex* and other *Tubifex* (*sensu lato*) species. The stygobiont species *Tubifex pescei* (Dumnicka, 1981), reported from Italy and Slovenia (Martinez-Ansemil et al., 1997), also resembles *T. galarzai* in the comma-shaped atrium; however, *T. pescei* is clearly a member of *Tubifex* (*sensu stricto*) as derived from the characteristic bipartite structure of the vas deferens joining atrium apically, close to the junction of the prostatic gland. The specimens of *T. galarzai* studied by us do not show this structure, but the vas deferens is homogeneously ciliated along its entire length, and prostate stalk and vas deferens junction are well separated.

Most recently Timm (1998) described the genus *Lamadrilus*, including in it five species from Lake Baikal and Lake Taimyr (Siberia). This genus is close to *Tubificoides* by the position of the prostate entrance to the atrium opposite to the junction of vas deferens, however the blind proximal part of the atrium is not histologically differentiated from the ectal part. The apical end of atrium of *T. galarzai* has not any clear histological separation from the rest of the atrium, and we recognize two types of atrial epithelium similar to those described in *Lamadrilus sorosi* Timm, 1998. The recent description of mature specimens of *Lamadrilus sorosi* from Lake Taimyr, northern Siberia (Timm, 1998), previously ascribed by Timm (1996) to the genus *Isochaetides* Hrabe, 1966, is largely based on serial sections of genitalia, stained after Mallory or according to Hauser (1969). The shape and the mode of repartition of the dark, dense atrial epithelium of variable thickness and the light atrial epithelium of uniform thickness resemble those of *T. galarzai*, except that the light epithelium is totally missing in the apical end of atrium in *Lamadrilus sorosi*. The species of *Lamadrilus* show short penial sheaths with a spade-shaped distal appendage, dissimilar to the penis sheath described in *T. galarzai*, although some of the species included in the genus by Timm (1998) show different and simpler shapes (e.g., *L. crassiseptus* or *L. eximius*, see Figs. 47 & 49 in Timm, 1998). The main morphological divergence of *T. galarzai* from species in the genus *Lamadrilus* is the position of the junction of vas deferens to the atrium, opposite to the prostate in *Lamadrilus*. Geographical distributions of the *Lamadrilus* species known to date are restricted to the central region of Siberia, specifically, lakes Baikal and Taimyr.

Table 2 summarises the affinities and divergences between *T. galarzai* and some species belonging to the genera *Tubificoides*, *Tubifex* (*sensu stricto*), and *Lamadrilus*. Due to the taxonomic dissimilarity of *T. galarzai* from all other tubificine genera described so far, we suggest placement of *T. galarzai* in a new genus, *Troglodrilus* (etymology derived from *troglo*: hole in Greek, and *drilus*: worm in Greek).

*Diagnosis of* Troglodrilus *n. gen.*

Freshwater cave tubificine. No large coelomocytes. Lack of modified genital setae. Body wall with glandular epidermis. Vas deferens ciliated throughout, narrower entally close to the male funnel, longer than atrium but moderately long, neither coiled nor bipartite. Comma-shaped

*Table 2.* Affinities and divergences of *Troglodrilus* with *Tubificoides*, *Lamadrilus*, and *Tubifex* (sensu stricto)

| Tubificoides | Lamadrilus | Tubifex (sensu stricto) |
|---|---|---|
| *Affinities* | | |
| Body wall with more or less ringed glandular epidermis | Atrial epithelium in two parts: one granulated, dense of variable thickness, around the prostate junction, other light of uniform thickness | Comma-shaped atrium |
| Penis sheath slightly swollen ectally | Penis sheath (although of different shape) | Apical junction of vas deferens in some species |
| Bulbous spermathecal duct | Freshwater species | Freshwater species |
| *Divergences* | | |
| Subapical junction of vas deferens more or less opposite to prostate | Junction of vas deferens opposite to prostate | A common cuticle attached to the penis, not a penial sheath |
| Apical part of the atrium histologically differentiated from the rest | Apical part of the atrial epithelium without light non-granulated cells | Vas deferens bipartite, only the ental part ciliated |
| Marine or estuarine species | Body tegument smooth | Body tegument smooth |
| | Non-bulbous spermathecal duct | Non-bulbous spermathecal duct |

atrium with two kinds of epithelial cells: densely granulated in the concave part of the atria, both ental and ectal to the prostate junction; light non granulated cells in the convex part and ectal part of the atria. Large prostate gland joining the atrium subapically, in the concave part, by a stout bundle of canals. No ejaculatory duct; penis in a penial sac with cuticular penial sheath. One pair of spermathecae containing spermatozeugmata with bulbous spermathecal ducts opening through ventral pores.

Type species: *Tubificoides galarzai* Giani & Rodriguez, 1988.

*Remarks*

The morphological analysis of the male genitalia has revealed taxonomic relationships of the new genus *Troglodrilus* with other tubificines of the freshwater genus *Lamadrilus* Timm, 1998 and the estuarine and marine genus *Tubificoides* Lastockin, 1937 as well as some taxa presently ascribed to the genus *Tubifex* Lamarck, 1816 (*sensu lato*). However, the apical junction of the vas deferens to the atrium (opposite to the prostate in *Lamadrilus* and *Tubificoides*), the vas deferens not bipartite (as it is typical in *Tubifex* (*sensu stricto*)), ciliated, of a moderate length, and the presence of a cuticular penis sheath clearly distinguishes this taxon from other tubificine genera. It will be interesting to assess the phylogenetic relationships of these genera in future research using 18S ribosomal DNA sequencing combined with other gene sequences, as Erséus et al. (2002) have performed for other members of the family Tubificidae.

## Acknowledgements

Pilar Rodriguez is indebted to Dr Aitor Galarza and D. Jesus Lopez de Ipiña (Grupo Espeleológico Alavés), for their assistance during collections of specimens from Mairulegorreta. Michel Creuzé des Châtelliers extends his appreciation to Marcel Meyssonnier (federal instructor, SCV – Spéléo Club de Villeurbanne), for serving as a guide during collections from subterranean habitats, for his assistance in sampling under sometimes difficult conditions, and for his assistance in the laboratory during the sorting of samples under a stereomicroscope. The authors extend their gratitude to Mr Henry-Gabriel Dupuy (Ecole Normale Supérieure de Lyon) for his assistance during histological preparation, and to Drs Christer Erséus and Tarmo Timm for sending slides of *Lamadrilus* and for valuable comments on the taxonomic criteria of *T. galarzai* and to Fabiana Castellarini for discussion on statistical analysis. An anonymous reviewer, Mark J. Wetzel and Hong-Zhu Wang provided valuable critique and assistance with the English text of this manuscript.

## References

Brinkhurst, R. O., 1986. Taxonomy of the genus *Tubificoides* Lastockin (Oligochaeta, Tubificida): species with bifid setae. Canadian Journal of Zoology 64: 1270–1279.

Brinkhurst, R. O. & H. R. Baker, 1979. A review of the marine Tubificidae (Oligochaeta) of North America. Canadian Journal of Zoology 57: 1553–1569.

Erséus, C., 1989. Four new west Atlantic species of *Tubificoides* (Oligochaeta, Tubificidae). Proceedings of the Biological Society of Washington 102: 878–886.

Erséus, C., M. Källersjö, M. Ekman & R. Hovmöller, 2002. 18S rDNA phylogeny of the Tubificidae (Clitellata) and its constituent taxa: dismissal of the Naididae. Molecular Phylogenetics and Evolution 22: 414–422.

Giani, N. & P. Rodriguez, 1988. Description de quelques espèces nouvelles de Tubificidae (Oligochaeta) de grottes et de sources karstiques de la péninsule ibérique. Stygologia 4: 121–137.

Hauser, J., 1969. Arbeitsmethoden der Annelidenhistologie. Mikroskopie 25: 136–153.

Helgason, G. V. & C. Erséus, 1987. Three new species of *Tubificoides* (Oligochaeta, Tubificidae) from the North-West Atlantic and notes on geographic variation in the circumpolar *T. kozloffi*. Sarsia 72: 159–169.

Holmquist, C., 1983. What is *Tubifex tubifex* (O. F. Müller) (Oligochaeta, Tubificidae)?. Zoologica Scripta 12: 187–201.

Holmquist, C., 1985. A revision of the genera *Tubifex* Lamarck, *Ilyodrilus* Eisen and *Potamothrix* Vejdovsky & Mrázek (Oligochaeta, Tubificidae) with extensions to some connected genera. Zoologische Jahrbücher, Abteiling für Systematik Oekologie und Geographie der Tiere 112: 311–366.

Juget, J. & M. des Châtelliers, 2001. Taxonomical survey of some stygobiont Oligochaetes from the eastern part of France, including description of a new species. Hydrobiologia 463: 23–28.

Martojat, R. & M. Martojat-Pierson, 1967. Initiation aux techniques de l'histologie animale. Masson (eds), Paris, pp 345.

Martinez-Ansemil, E., B. Sambugar & N. Giani, 1997. Groundwater Oligochaetes from Southern-Europe. I . A new genus and three new species of Rhyacodrilinae (Tubificidae) with a redescription of *Tubifex pescei* (Dumnicka) comb. n. Annales de Limnologie – International Journal of Limnology 33: 33–44.

Milligan, M. R., 1996. Identification manual for the aquatic Oligochaeta of Florida. Vol. II. Estuarine and nearshore marine oligochaetes. Florida Department of Environmental Protection, Bureau of Water Resources Protection, Tallahassee, XVI + 239 pp. + appendix. (free copy available at http://www.dep.state.fl.us/labs/library/keys.htm).

Timm, T., 1996. Oligochaeta of Lake Taimyr: a preliminary survey. Hydrobiologia 334: 89–95.

Timm, T., 1998. Tubificidae (Clitellata, Oligochaeta) from Lake Taimyr, North Siberia, with a description of two new species and *Lamadrilus*, gen. nov. Bulletin de L'Institut Royal des Sciences Naturelles de Belgique, Biologie 68: 23–42.

Hydrobiologia (2006) 564:19–32
P.F.M. Verdonschot, H. Wang, A. Pinder & R. Nijboer (eds), Aquatic Oligochaete Biology IX
DOI 10.1007/s10750-005-1704-1



# Generation of bilateral symmetry in the ectoderm of the *Tubifex* embryo: involvement of cell–cell interactions

Ayaki Nakamoto, Asuna Arai & Takashi Shimizu*
*Division of Biological Sciences, Graduate School of Science, Hokkaido University, 060-0810, Sapporo, Japan*
*(*Author for correspondence: E-mail: stak@sci.hokudai.ac.jp)*

*Key words:* *Tubifex*, bilateral symmetry, ectoderm, teloblastogenesis, cell interactions

## Abstract

In embryos of the oligochaete annelid *Tubifex*, most ectodermal tissues are derived from four bilateral pairs of embryonic stem cells called teloblasts (ectoteloblasts N, O, P and Q). Ectoteloblasts are generated on both left and right sides of the embryo through an invariable sequence of cell divisions of a proteloblast, NOPQ, and they are positioned in a mirror symmetric pattern relative to the embryonic midline. This mirror symmetry of ectoteloblast arrangement gives rise to the generation of bilateral symmetry in the ectoderm. Here we review results of our recent experiments on *Tubifex tubifex* that were designed to gain an insight into the mechanisms underlying the generation of the bilaterally symmetric organization of ectoteloblasts. Cell transplantation experiments have shown that nascent NOPQ cells can be polarized according to positional information residing in the embryo. If a left NOPQ cell is transplanted to the right side of a host embryo, it exhibits polarity comparable to that of right NOPQ cells. It has also been shown that contact between NOPQ cells serves as an external cue for their polarization. Another series of cell transplantation experiments have suggested that the competence of NOPQ cells to respond to external cues becomes undetectable shortly before the production of the first teloblast (N) from the NOPQ cell. Another series of experiments utilizing cell ablation techniques have shown that teloblasts N, P and Q are specified to express the N, P and Q fates, respectively, as early as their birth. In contrast, the O teloblast and its progeny are initially pluripotent and their fate becomes restricted through inductive signals emanating from its sister P lineage. On the basis of these findings, we have proposed a model for polarization of ectodermal teloblastogenesis in the *Tubifex* embryo.

## Introduction

Embryogenesis in clitellate annelids (i.e., oligochaetes and leeches) is characterized by the generation of five bilateral pairs of embryonic stem cells called teloblasts early in development (Shimizu & Nakamoto, 2001). Teloblasts, which are derived from micromeres of the D quadrant, undergo extremely unequal divisions repeatedly to produce a coherent column (bandlet) of smaller daughter cells (referred to as primary blast cells). Four of the five bandlets on each side of the embryo join together to form an ectodermal germ band (GB), while the remaining bandlet becomes a mesodermal GB. Based on results of previous descriptive and cell ablation studies (Whitman, 1878; Penners, 1924, 1926; Devries, 1973a, b), it has been suggested that teloblasts (and their progenies) play a pivotal role in clitellate annelid development. In fact, teloblasts are the only source of ectodermal and mesodermal segmental tissues; none of the non-teloblastic cells can replace missing teloblasts in this respect. Furthermore, morphogenetic events such as body elongation and segmentation depend solely on the presence of teloblasts and their progeny (Blair, 1982; Wedeen & Shankland, 1997; Shain et al., 1998; Goto et al., 1999a; Kitamura & Shimizu, 2000; Nakamoto et al., 2000).

Ectodermal teloblasts (ectoteloblasts N, O, P and Q) on either side of the embryo are produced through an invariable sequence of cell division of a proteloblast, NOPQ, that is derived from the second micromere 2d; a bilateral pair of mesodermal teloblasts (mesoteloblasts M) results directly from equal division of the fourth micromere 4d (Fig. 1a, b). Recent cell lineage analyses of teloblasts have shown that developmental fates of the four ectoteloblasts are not only different from that of the mesoteloblast but also distinct among themselves (Weisblat et al., 1980, 1984; Storey, 1989; Goto et al., 1999b). At present, it is not clear how and when these teloblasts (and their progeny) acquire distinct developmental fates. As has been well documented, teloblasts N–Q emerge at different positions along the embryonic axes and at different times (Fernandez & Olea, 1982; Shimizu, 1982; Sandig & Dohle, 1988). In the oligochaete *Tubifex*, the N teloblast is generated first and located ventralmost, and the Q teloblast, which is generated next, is located dorsalmost; finally the O and P teloblasts are generated by almost equal division of their precursor cell, OP (Fig. 2). There is no doubt that this spatial arrangement of ectoteloblasts gives rise to the generation of bilateral symmetry in the ectoderm. At present, however, nothing is known about the causal relationship between developmental fates of teloblasts and spatiotemporal aspects of their emergence.

In this paper we review results of our recent experiments on embryos of *Tubifex tubifex* that were designed to gain an insight into the mechanisms underlying the generation of the bilaterally symmetric organization of ectoteloblasts. In the first series of experiments, we addressed mechanisms for polarization of NOPQ. For this purpose, we transplaned left NOPQ cells to various ectopic positions of host embryos and examined the polarity of transplanted NOPQ cells (Nakamoto et al., 2004). The second series of experiments was carried out to determine how and when ectoteloblasts are specified. We studied this issue by ablating specific ectoteloblasts from embryos (Arai et al., 2001). Experimental procedures for cell ablation, transplantation, labeling and culture have been described in detail in our previous papers (Kitamura & Shimizu, 2000; Nakamoto et al., 2000).

## Specification of polarity of the ectoteloblast precursor cell NOPQ

Ectoteloblasts N, O, P and Q arise from an invariable sequence of divisions of the cell NOPQ

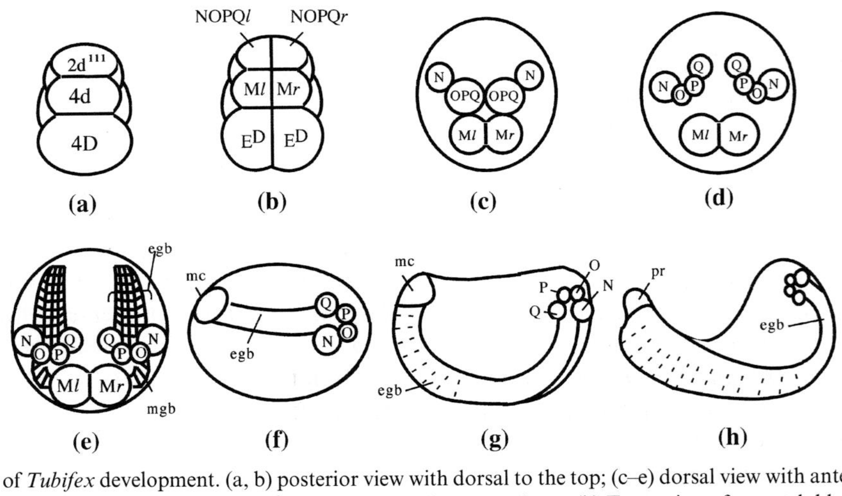


*Figure 1.* Summary of *Tubifex* development. (a, b) posterior view with dorsal to the top; (c–e) dorsal view with anterior to the top; (f–h) side view with anterior to the left and dorsal to the top. (a) A 25-cell stage embryo. (b) Formation of mesoteloblasts (M*l* and M*r*) and ectoteloblast precursors (NOPQ*l* and NOPQ*r*). (c, d) Sequence of the formation of ectoteloblasts N, O, P and Q. For brevity, only teloblast precursors (OPQ and OP) and teloblasts (N–Q) are depicted. (e) A 2-day-old embryo following the bilateral division of 2d$^{111}$. Only teloblasts and associated structures are depicted. egb, ectodermal germ band; mgb, mesodermal germ band. (f–h) Morphogenesis of the ectodermal germ band. Embryos are shown at 2.5 (f), 4 (g) and 6 (h) days following the 2d$^{111}$ division. Ventral and dorsal migration of the germ bands (egb) occurs in an anterior to posterior direction. mc, micromere cap; pr, prostomium.

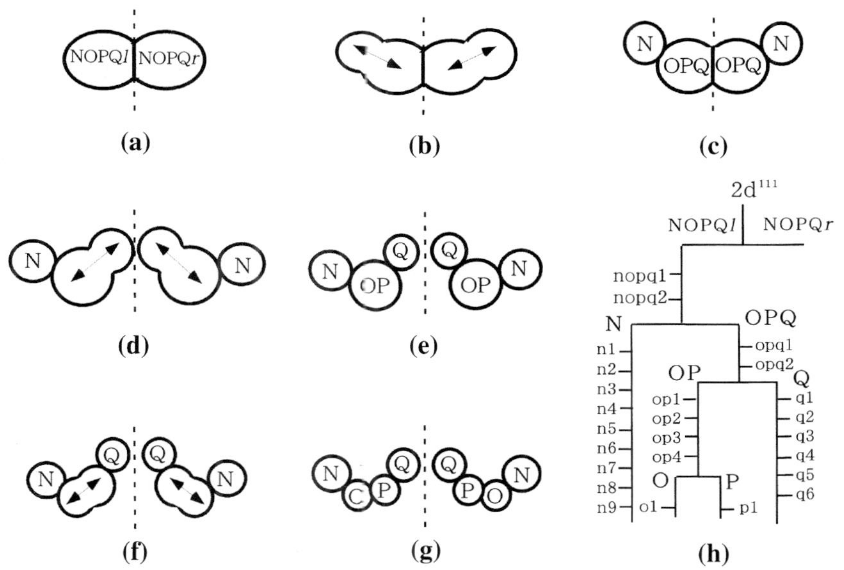


*Figure 2.* Ectodermal teloblastogenesis. (a–g) Spatiotemporal aspects of ectoteloblast formation. Broken lines indicate the dorsal midline. Anterior is to the top. Arrows indicate the direction of cell division. (h) Cell lineage diagram showing the production of ectoteloblasts (N, O, P and Q) on the left side of the embryo. Short horizontal bars added to the vertical thick line indicate the time when small cells (n–q, op, opq, and nopq) are formed. All cell divisions included in this lineage tree occur at 2.5-h intervals (at 22 °C).

on both sides of the embryo (Fig. 1d; Goto et al., 1999b). NOPQ on either side of the embryo undergoes unequal divisions twice after its birth (Fig. 2h) and then divides into a smaller N teloblast and a larger cell, OPQ (Fig. 2b, c). Similarly, after producing small cells twice, OPQ divides into a smaller Q teloblast and a larger cell, OP (Fig. 2d, e). Finally OP undergoes unequal division four times after its birth (Fig. 2h) and then cleaves almost equally, yielding the third-born ectoteloblasts O and P (Fig. 2f, g), at which point teloblastogenesis is complete.

As mentioned before, NOPQ on either side of an intact embryo divides into an N teloblast located distal from and an OPQ located proximal to the dorsal midline (i.e. the contact plane between sister NOPQ cells; see Fig. 2c). We regard this configuration of the two cells (i.e. N and OPQ) as a manifestation of polarity that NOPQ has at the time of its division into these two cells. Similarly, in the following cell transplantation experiments, we assessed polarity of transplanted NOPQ cells according to fates that two large cells (equivalent to N and OPQ in intact embryos) descending from the transplanted NOPQ take on.

We found that when oriented in an appropriate direction in host embryos, transplanted NOPQ underwent cell division in a normal fashion to produce two large cells (which correspond to N and OPQ in intact embryos) at about 8 h after its birth (see Fig. 2h). In this paper, we refer to these two large cells as a Dst-cell and a Pxm-cell (Fig. 3). The Pxm-cell is a cell that inherits the surface of NOPQ originally facing (proximal to) the midline, and the Dst-cell is a cell containing the opposite surface of NOPQ (Fig. 3). It should be noted here that the two large cells resulting from transplanted NOPQ are designated Dst- or Pxm-cells according to their original orientations relative to the midline of a donor embryo (that is, irrespective of their positions relative to the midline of host embryos; Fig. 3).

In intact embryos, the Pxm-cell is always an N teloblast and the Dst-cell is an OPQ proteloblast. Subseqeuntly, the N teloblast generates a single bandlet of n-blast cells (called the N pattern or N fate in this paper). In contrast, the proteloblast OPQ gives rise to the production of three teloblasts and hence three bandlets of blast cells (called the OPQ pattern or OPQ fate in this paper).

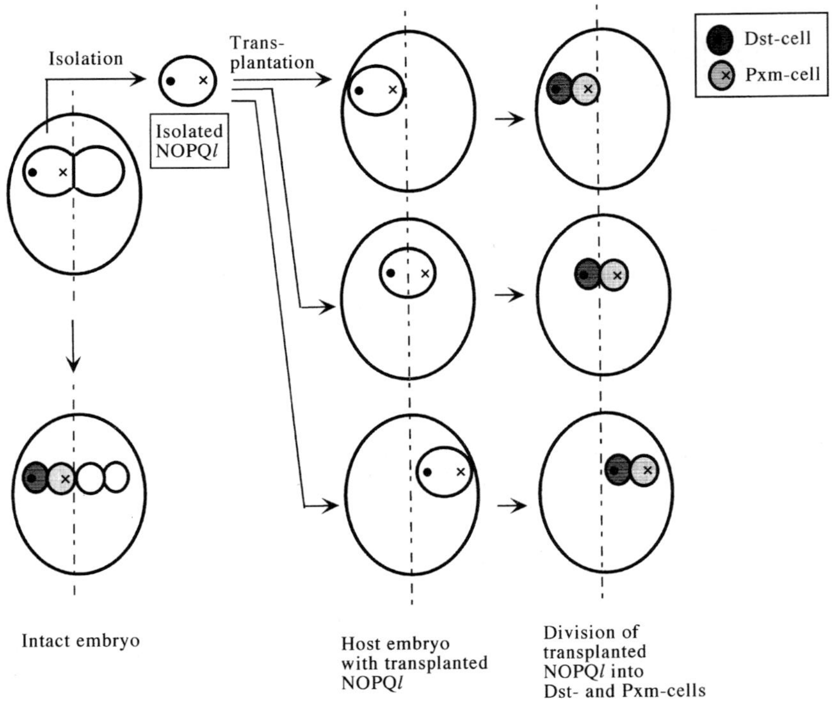


*Figure 3.* Designation of cells Dst and Pxm. The side of an NOPQ cell (NOPQ*l* here) proximal to the donor embryo's midline (broken line) is indicated by a cross, and the opposite side (i.e. distal to the midline) is indicated by a dot. When a transplanted NOPQ cell divides into two large cells, one of them, which has inherited the "distal" surface of NOPQ, is referred to as a Dst-cell, and the other, which has inherited the "proximal" surface, is referred to as a Pxm-cell, irrespective of the position of these cells relative to the midline of host embryos.

### *Development of nascent NOPQ cells transplanted to ectopic positions*

To determine whether NOPQ cells are already polarized at the time of their birth, we isolated a left NOPQ (NOPQ*l*) shortly after its birth and transplanted it to the position of the right NOPQ (NOPQ*r*) of the host embryo (from which NOPQ*r* had been ablated; Fig. 4b). The transplanted cell was oriented in the host embryo so that its anteroposterior polarity remained unchanged. In a preliminary experiment, we found that the progeny of transplanted NOPQ*l* is confined to the right side of a 7-day-old operated embryo and that its distribution pattern is quite similar to that of authentic NOPQ*r* in intact embryos. This indicates that if transplanted to the right side of the embryo, nascent NOPQ*l* adopts the "right side fate".

To verify that NOPQ*l* transplanted to the right side of the embryo acquires the same polarity as that of NOPQ*r*, we followed the fate of Dst- and Pxm-cells resulting from the division of a transplanted NOPQ*l* (see Fig. 3). When either Dst-cells ($n = 8$) or Pxm-cells ($n = 13$) derived from transplanted NOPQ*l* were labeled with DiI and their development was followed, it was found that Dst-cells (7 cases) individually gave rise to three bandlets (see Fig. 5b). (In one case, the number of bandlets was unclear.) On the other hand, Pxm-cells (12 cases) each gave rise to a single bandlet (see Fig. 5a); in no case did Pxm cells produce three bandlets. These findings suggest that Dst- and Pxm-cells descending from NOPQ*l* placed on the right side adopt the OPQ and N fates, respectively. This notion was confirmed by the distribution pattern of the progenies of Dst- and Pxm-cells derived from transplanted NOPQ*l*. In all cases examined ($n = 5$), Dst-cells contributed cells to the dorsal epidermis, dorsal and ventral setal sacs, and ventral ganglia, suggesting that Dst-cells followed the OPQ fate rather than the N fate. Furthermore, progenies of Pxm-cells (7/7 cases) contributed cells mainly to ventral ganglia. In no cases did the progenies differentiate to setal sacs,

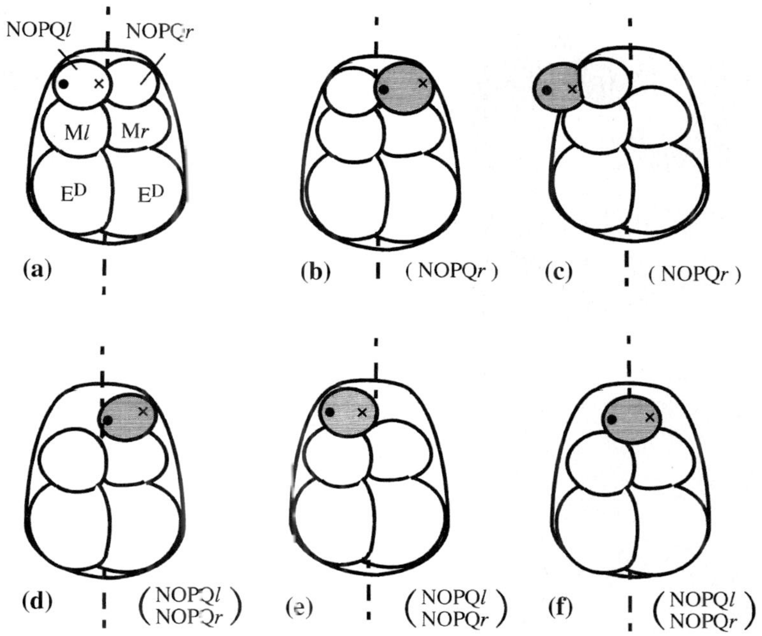


*Figure 4.* Schematic representation of cell transplantation experiments showing positions of transplanted NOPQ*l* (shaded) in host embryos. Cells that have been ablated from host embryos are indicated in parenthesis at the lower right of each panel.

which are known to differentiate from the OPQ cell (Goto et al., 1999b).

In a control experiment, an isolated NOPQ*l* was transplanted to the position of NOPQ*l* of a host embryo (from which NOPQ*l* had been ablated). Dst- and PXm-cells derived from the transplanted NOPQ*l* were found to exhibit N and OPQ patterns of cell division and cell fate, respectively.

On the basis of these observations, it is reasonable to conclude that when transplanted to the right side of the embryo, a nascent NOPQ*l* exhibits features of the authentic NOPQ*r* and that NOPQ*l* is initially plastic in terms of division pattern and fates.

To determine whether a transplanted nascent NOPQ*r* also adopts the "left side fate" of the host embryo, we transplanted a nascent NOPQ*r* to the left side (i.e. the position of NOPQ*l*) of a host embryo (from which NOPQ*l* had been ablated) and followed the division pattern of Pxm-cells derived from the transplanted NOPQ*r*. In most cases (11/15), Pxm-cells generated a single bandlet (that is, in the N pattern); there were no cases in which three bandlets were formed. These results suggest that a nascent NOPQ*r* is not endowed with fixed polarity and that it is able to adopt the "left side fate" of the host embryo if placed to the left side of the embryo.

*The dorso-ventral polarity of NOPQ is gradually determined prior to its division into cells N and OPQ*

The above-described experiments show that NOPQ is initially plastic in terms of division pattern and fates. It is natural to assume that the polarity along the dorso-ventral axis is specified in NOPQ at some time after its birth. We next tried to determine the time when the polarity of NOPQ is fixed. For this purpose, we isolated NOPQ*l* at three different stages [i.e. shortly after its birth (NOPQ *early*), shortly after production of nopq1 (NOPQ *mid*) and shortly after production of nopq2 (NOPQ *late*); see Fig. 2h] and transplanted it to the position of NOPQ*r* of a host embryo at the stage of NOPQ *early* (from which NOPQ*r* had been ablated). Either Dst- or Pxm-cells derived from the transplanted cells were labeled and their fates were followed as described above.

The results are summarized in Figure 6. When NOPQ *early* was transplanted, nearly all of the

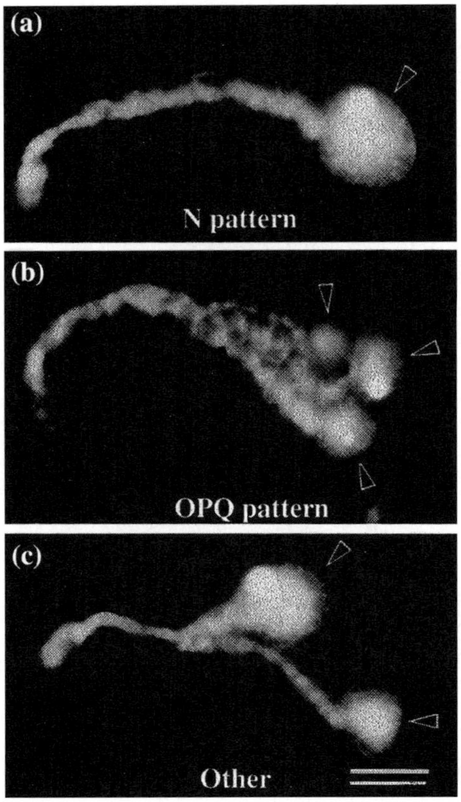


*Figure 5*. Patterns of development of cells descending from transplanted NOPQ cells. Fluorescence images are shown. Two large cells (i.e. Dst- and Pxm-cells) resulting from the division of NOPQ cells were labeled with lipophilic dye DiI and fixed for observation after 2-day culture. (a) N pattern with a single cell bandlet extending from a single teloblast (arrowhead). (b) OPQ pattern with three bandlets associated with three teloblasts (arrowheads). (c) A pattern other than N or OPQ. Two teloblast-like cells (arrowheads) are recognizable, but the number of bandlets is uncertain. Scale bar, 20 $\mu$m (a–c).

resulting Dst-cells (7/8 cases) exhibited the division pattern of an OPQ cell (Fig. 6a). Similarly, Dst-cells (5/7 cases) derived from NOPQ *mid* individually generated three bandlets. It is notable that Dst-cells at these stages never gave rise to a single bandlet. In contrast, most Dst-cells derived from NOPQ *late* individually produced only one bandlet (11 out of 16 cases; see Fig. 5a); there were no cases in which Dst-cells gave rise to three bandlets.

Pxm-cells derived from NOPQ*early* (12/13 cases) generated a single bandlet (i.e. in the N pattern). As development proceeded, however, the frequency of cases in which Pxm-cells gave rise to single bandlets decreased. Conversely, the frequency of cases in which single Pxm-cells gave rise to two or more bandlets significantly increased (Fig. 6b). These results suggest that as NOPQ*l* undergoes cell division twice to produce blast cells nopq (see Fig. 2h), it gradually loses its flexibility to adopt the "right side fate" of the host embryo.

It should be noted, however, that about 25% of the Pxm-cells derived from NOPQ *late* exhibited the N pattern (Fig. 6b), whereas none of the Dst-cells took on the OPQ fate (Fig. 6a). This difference suggests that the distal portion of NOPQ *late* is already specified rigidly, while the remaining (proximal) portion of the cell still retains the flexibility to adopt the "right side fate" to some extent.

### *Contact between NOPQ cells enables them to become polarized*

In the transplantation experiments described above, transplanted NOPQ cells were always placed in contact with NOPQ cells of the host embryos at the midline (Fig. 4b). Under such conditions, the Pxm-cells, which were located distal to the site of contact between the transplanted and host NOPQ cells, were always found to assume the N fate. This suggests that contact between NOPQ cells enables them to become polarized. To test this possibility, we transplanted an isolated NOPQ*l* to the distal side (i.e. on the leftmost surface) of NOPQ*l* of a host embryo from which NOPQ*r* had been ablated (Fig. 4c) and followed the cells descending from the host NOPQ*l*.

At about 8 h after transplantation, the host NOPQ*l* divided, along the dorso-ventral axis, into two large cells (i.e. Dst- and Pxm-cells) in a normal fashion. When Pxm-cells derived from the host NOPQ*l* were followed, nearly all of the Pxm-cells (9/11 cases) that generated cell bandlets successfully were found with single bandlets (see Fig. 5a); in no cases did these cells gave rise to three bandlets. This suggests that these Pxm-cells assume the N fate. Given that the Pxm-cells in intact embryos assume the OPQ fate (see Fig. 3), it is apparent that the fate of NOPQ-derived cells depends on their distance from the site of contact between NOPQ cells rather than their position along the embryonic axis. If located distal to the site of contact between NOPQ cells, cells resulting

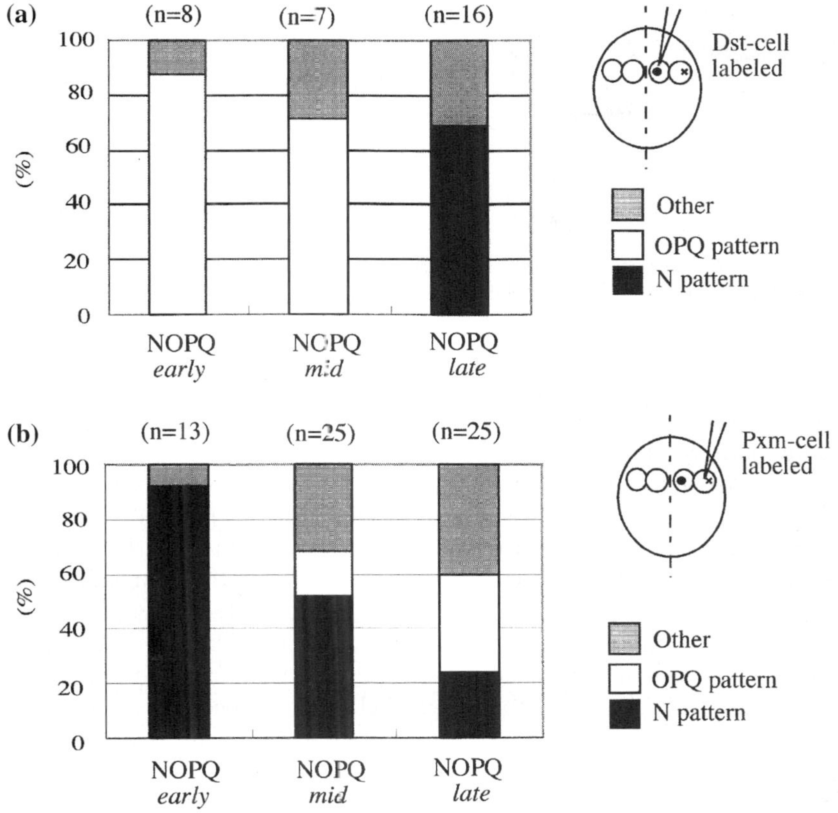


*Figure 6*. Frequency of developmental patterns of Dst- (a) and Pxm-(b) cells descending from NOPQ*l* transplanted to the right side of host embryos (see Fig. 4b). NOPQ cells to be transplanted were isolated from embryos shortly after its birth (NOPQ *early*), shortly after production of nopq1 (NOPQ *mid*) and shortly after production of nopq2 (NOPQ *late*) (see Fig. 2h).

from the division of the NOPQ were always found to assume the N fate.

*Development of "solitary" NOPQ cells placed in various positions*

In previous studies, we found that even if an NOPQ cell is deleted, the remaining sister NOPQ cell, which remains in the position where it has been, undergoes a series of divisions in a normal (or scheduled) fashion and finally generates a full set of ectoteloblasts, which are arranged, along the dorsoventral axis, in the same order as those in intact embryos (Nakamoto et al., 2000; Arai et al., 2001). These observations suggest the existence of some kind of positional cues other than NOPQ cells that are involved in polarization of NOPQ cells. To investigate this issue, we examined the mode of development of NOPQ*l* transplanted to various positions of host embryos from which both NOPQ*l* and NOPQ*r* had been ablated (Fig. 4d–f). In this study, we focused on the fates of Pxm-cells descending from transplanted NOPQ*l*.

The results are summarized in Table 1. When NOPQ*l* was transplanted to the left side of a host embryo (Fig. 4e), the resulting Pxm-cells gave rise to three bandlets (viz. in the OPQ pattern). In contrast, Pxm-cells derived from NOPQ*l* transplanted to the right side (Fig. 4d) individually produced a single bandlet (viz. in the N pattern). This suggests that NOPQ*l* transplanted to the left side and the right side adopt the "left side fate" and "right side fate", respectively, in the absence of host NOPQ cells.

It was interesting to learn how NOPQ*l* placed centrally (i.e. on the dorsal midline of the host embryo) reacts to the position (Fig. 4f). Pxm-cells of some such reconstituted embryos generated one bandlet (2/8 cases) and others generated three

*Table 1.* Division pattern of Pxm-cells derived from transplanted NOPQ*l*

| Position of transplanted NOPQ*l*[a] | No. of embryos[b] | No. Pxm-cells | | |
|---|---|---|---|---|
| | | N pattern | OPQ pattern | Other |
| Right | 13 | 11 | 0 | 2 |
| Left | 6 | 0 | 5 | 1 |
| Center | 8 | 2 | 2 | 4 |

[a]Position of transplanted NOPQ*l* in reconstituted embryos are schematically illustrated in Fig. 4d–f.
[b]Both of NOPQ*l* and NOPQ*r* were ablated from host embryos.

bandlets (2/8 cases). These results reconfirm the previous notion that a nascent NOPQ cell is plastic in terms of division pattern and cell fate. At present, it is not known why Pxm-cells descending from NOPQ*l* placed centrally assume two different fates. Considering the close relationship between the positions where transplanted cells are placed and the fates they follow, however, it is conceivable that NOPQ*l* placed centrally would shift to the left or right during their development to adopt the left or right side fate.

**Specification of ectodermal teloblast lineages**

To determine the extent to which specification of ectoteloblast lineages depends on external cues, we followed the development of ectoteloblasts that had been forced to be "solitary". In this study, we assessed the fates of operated ectoteloblast lineages according to compositions and spatial distribution of terminally differentiated cells descending from these ectoteloblasts. In intact embryos, each ectoteloblast makes a topographically characteristic contribution to the ectodermal tissues, which exhibit a segmentally repeated distribution pattern (Fig. 7; Goto et al., 1999b).

*Fates of "solitary" ectodermal bandlets*

Our previous study showed that a bandlet derived from a "solitary" O teloblast (resulting from removal of all of its ipsilateral sister teloblasts) exhibited early morphogenetic features (e.g. shapes of bandlets) characteristic to the P lineage rather than the O lineage, while bandlets derived form "solitary" N, P and Q teloblasts were very similar to the respective bandlets in intact embryos (Nakamoto et al., 2000). To extend this observation to more advanced developmental stages when cells are terminally differentiated, we labeled one of the four ectoteloblasts with DiI and ablated the other three ipsilateral ectoteloblasts (or their precursors) simultaneously, and after 5-day culture, we examined the composition and distribution of labeled cells descending from "solitary" teloblasts.

Figure 8a, b show the organization of labeled cells derived from an intact and a "solitary" N teloblast, respectively. These two cases are indistinguishable from each other in that nearly all of the labeled cells were located in the ventral region of the embryo and occupied each hemiganglion. Similarly, labeled cells derived from "solitary" P and Q teloblasts are organized in a pattern comparable with that in intact P and Q lineages, respectively (Fig. 8e, f; not shown for the Q lineage).

In contrast, organization and composition of cells derived from "solitary" O teloblasts are distinct from those in normal o bandlets. As Fig. 8d shows, "solitary" o bandlets apparently exhibited a P pattern rather than an O pattern of progeny cells (also see Fig. 8c). This result suggests that "solitary" o bandlets adopt the P fate rather than the O fate.

In another series of experiments, we followed the fate of "solitary" left bandlets in embryos that had been subjected to bilateral ablations of ectoteloblasts. To do this, we ablated the right NOPQ (i.e. exclusive source of the right ectodermal GB; see Fig. 1e) and three of the four left ectoteloblasts, leaving a single (DiI-labeled) ectoteloblast in each embryo. After 5-day culture, the operated embryos (5–7 for each lineage) were examined for the composition and distribution of labeled cells. We found that even after bilateral ablations of

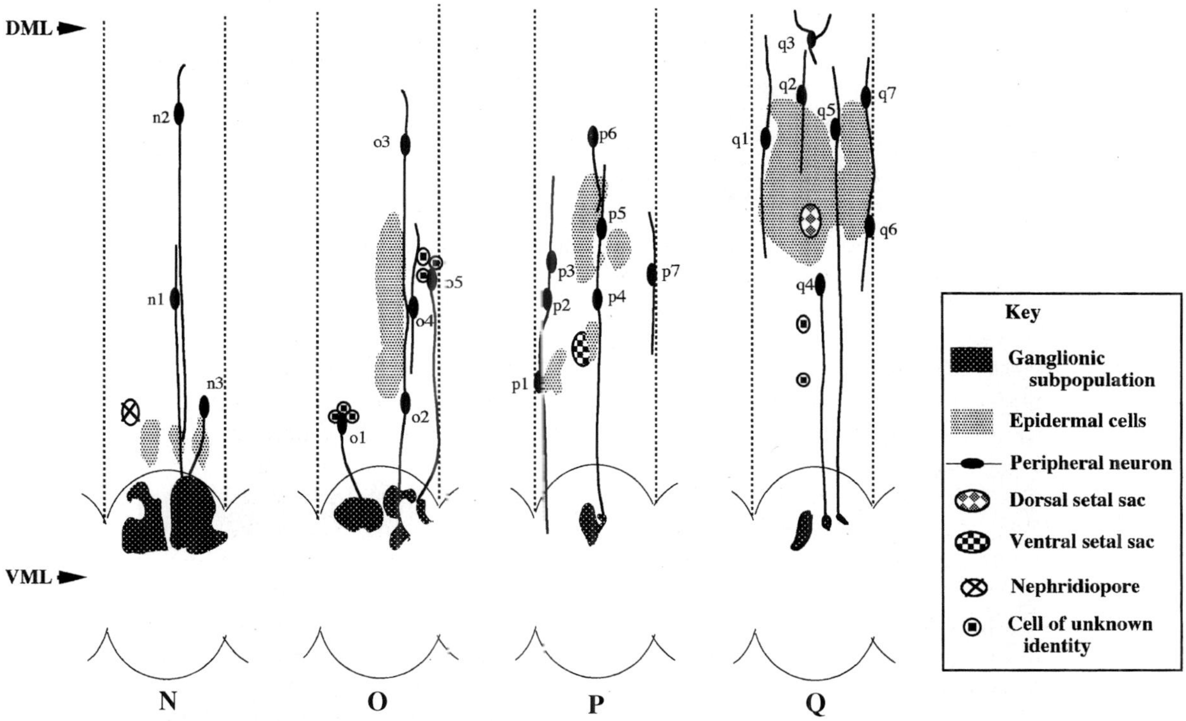


*Figure 7.* Diagrammatic summary of cellular contributions of the teloblasts N, O, P and Q to a mid body segment of *Tubifex*. For each pattern, the left half of one segment is shown, with the ventral midline (VML) and ganglion (shown in outline) to the bottom, dorsal midline (DML) to the top, and anterior to the left Dashed lines indicate segmental boundaries.

ectoteloblasts, "solitary" n, p and q bandlets exhibited N, P and Q patterns of distribution of progeny cells, respectively, and that o bandlets followed the P fate rather than the O fate. These results are the same as those obtained in the unilateral ablation experiments. The results therefore suggest that the presence of contralateral GBs does not influence the fate decision of "solitary" bandlets. In the following experiments, we used embryos in which right GBs were intact.

### *O fate in o bandlets is induced by interaction with p bandlets*

The above-described results indicate the possibility that in intact GB, o bandlets are induced to assume O fate by interactions with other bandlets. To test this possibility and to determine which bandlet acts as such an inducer, we ablated teloblasts in various combinations, leaving an O teloblast plus one or two other teloblasts in each embryo and followed the fates of the progenies of O teloblasts.

The results are summarized in Table 2. It was only when p bandlets survived that o bandlets assumed the O fate. Neither n nor q bandlets were effective at all in this respect; even when both n and q bandlets coexisted with o bandlets, they failed to induce o bandlets to assume the O fate. It is unlikely that this failure resulted from separation of o bandlets from n and/or q bandlets in operated embryos, since bandlets in a GB from which one bandlet had been deleted were found to be aligned tightly with each other. These results suggest that p bandlets exclusively serve as an inducer of O fate in o bandlets.

### *P teloblast may be specified to assume the P fate at the time of its birth*

Unlike O teloblasts, fates of N, P and Q teloblasts do not appear to be affected by neighboring bandlets. This suggests that these three teloblasts are specified as early as their birth. Considering the fact that the O teloblast is the sister of the P teloblast (see Fig. 2f, g), however,

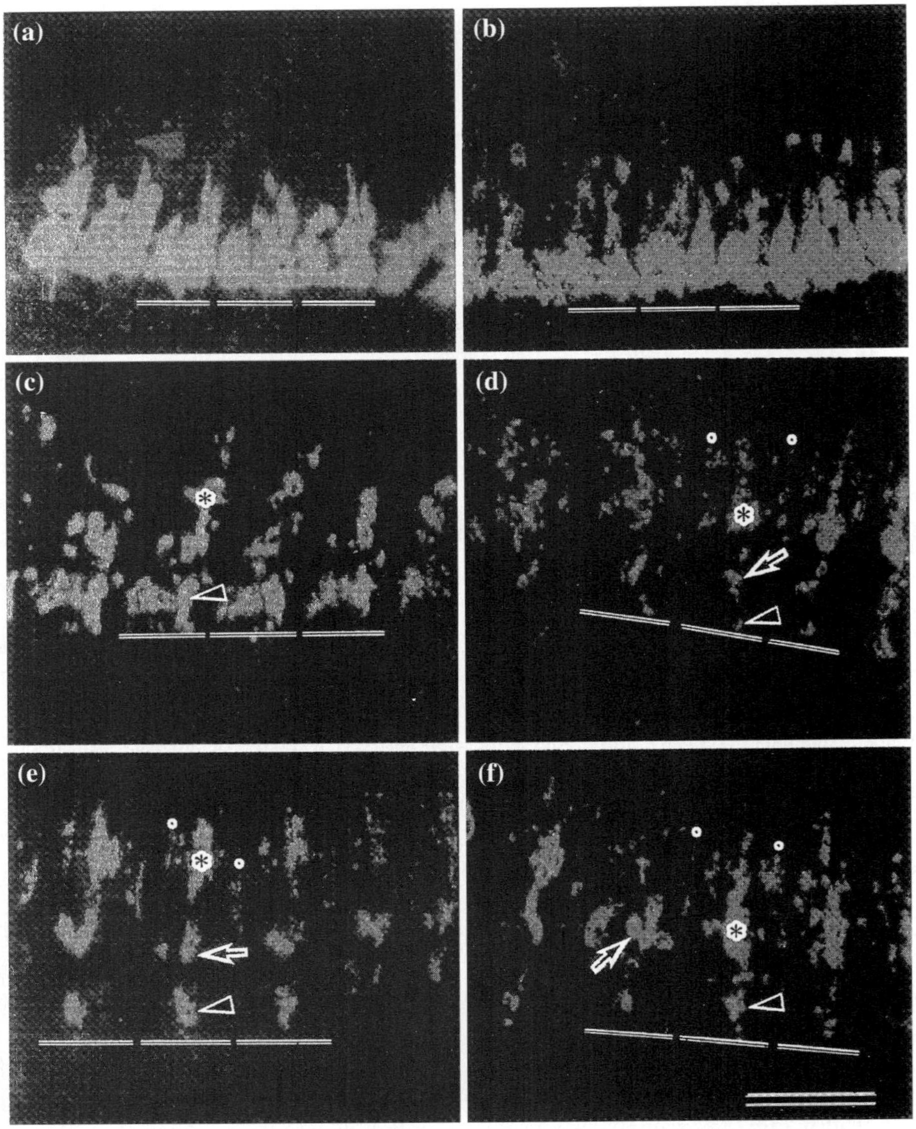


*Figure 8.* Fate of ectodermal bandlets in the absence of neighboring bandlets. One of the left ectoteloblasts was injected with DiI shortly after its birth. Control embryos (a, c, e) were allowed to develop without any further treatment after DiI injection. In experimental embryos (b, d, f), ipsilateral ectoteloblasts (or precursors) other than the DiI-injected one were all ablated, so that DiI-labeled ectoteloblasts were forced to be "solitary". Control and experimental embryos were allowed to develop for 5 days before fixation and observed in whole-mounts. In all panels, the mid region of the embryo is viewed from the side. Anterior is to the left and dorsal is to the top. In each panel, three horizontal lines indicate the position of three consecutive segments and the approximate level of the ventral midline. (a, b) n bandlets. (c, d) o bandlets. (e, f) p bandlets. Arrows, deep cells; arrowheads, central neurons; asterisks, epidermis; dots, peripheral neurons. Scale bar, 100 $\mu$m (a–f).

it is also possible that, like the O teloblast, the P teloblast is pluripotent and can express fates other than the P fate in an appropriate environment. To distinguish these possibilities, we labeled P teloblasts with DiI shortly after birth and simultaneously ablated other teloblasts in various combinations, leaving a labeled P teloblast plus one or two other teloblasts in each embryo, and after 5-day culture, we examined the distribution of labeled cells.

The results are summarized in Table 3. Irrespective of the presence of any other teloblasts, p

*Table 2.* Effects of neighboring bandlets on the fate of O-derived bandlets

| Combination of Teloblasts[a] | Teloblast(s) ablated | No. of embryos | No. O-derived bandlets | | |
|---|---|---|---|---|---|
| | | | O fate | P fate | N or Q fate |
| O | N, P, Q | 15 | 0 | 15 | 0 |
| O+N | P, Q | 8 | 0 | 8 | 0 |
| O+P | N, Q | 8 | 8 | 0 | 0 |
| O+Q | N, P | 6 | 0 | 6 | 0 |
| O+N+Q | P | 5 | 0 | 5 | 0 |

[a]Left O teloblasts were injected with DiI and simultaneously other ipsilateral teloblasts were ablated in combinations as indicated.

bandlets assumed the P fate. The p bandlets did not show any sign of N, O or Q pattern of progeny cells in any of the combinations with other teloblasts. These results suggest that p bandlets are unlikely to receive inductive or inhibitory signals from neighboring bandlets. Thus, it is more likely that P teloblasts are specified to assume the P fate at the time of their birth.

Taken together, the results suggest that the N, P and Q teloblasts of the *Tubifex* embryo are committed to their respective fates at the time of their birth. In contrast, it is reasonable to assume that O teloblasts are pluripotent. As found in the above-described experiments, o bandlets assume the O fate in the presence of p bandlets; otherwise they express the P fate. Apparently, for the O lineage, the P fate is the primary fate and the O fate is the secondary fate.

The O and P teloblasts are sister blastomeres resulting from the equal division of an OP proteloblast. As stated above, the O teloblast is pluripotent and its progeny cells are able to respond to inductive signals from the neighboring p bandlet. In contrast, P teloblasts do not appear to be affected by the O teloblast. Furthermore, P teloblasts assumed only the P fate under any of the experimental conditions that included ablation of teloblasts in all possible combinations. Based on these observations, we suggest that sister teloblasts O and P in the *Tubifex* embryo are not equivalent but are distinct from each other in their developmental potency. It appears that the OP proteloblast undergoes an asymmetric division giving rise to two equal-sized teloblasts.

## Cellular basis for bilaterally symmetric organization of the ectoderm

The bilaterally symmetric organization of the ectoderm in *Tubifex* is traced back to mirror symmetry of ectodermal germ bands (GBs) and ectoteloblasts. The right and left GBs, which coalesce with each other along the ventral midline, contribute ectodermal cells, respectively, to the right and left halves of the embryo. During

*Table 3.* Effects of neighboring bandlets on the fate of P-derived bandlets

| Combination of Teloblasts[a] | Teloblast(s) ablated | No. of embryos | No. O-derived bandlets | |
|---|---|---|---|---|
| | | | P fate | N, O or Q fate |
| P | N, O, Q | 20 | 20 | 0 |
| P+N | O, Q | 7 | 7 | 0 |
| P+O | N, Q | 8 | 8 | 0 |
| P+Q | N, O | 5 | 5 | 0 |
| P+N+O | Q | 10 | 10 | 0 |
| P+N+Q | O | 6 | 6 | 0 |

[a]Left P teloblasts were injected with DiI and simultaneously other ipsilateral teloblasts were ablated in combinations as indicated.

gastrulation, ectodermal cells originating from the ventrally-located GB migrate toward the dorsal midline; during this process, these cells do not cross over the midline (Goto et al., 1999b). Organization of bandlets comprising the left and right GBs exhibits mirror symmetry. Apparently this mirror symmetry gives rise to mirror symmetry of the ectoderm, and it is a direct consequence of the bilateral arrangement of ectoteloblasts, which are organized in the order of N, O, P and Q along the ventrodorsal axis on either side of the embryo.

In the present study, we examined the possibility of the involvement of cell–cell interactions in the generation of mirror symmetry in the ectodermal teloblastogenesis. The results obtained suggest that cell–cell interactions are involved at least in two stages during teloblastogenesis, viz., prior to the division of NOPQ into N and OPQ and shortly after the division of OP into O and P. NOPQ cells, which do not initially have a rigidly fixed polarity, become polarized through external cues, including signals emanating from their sister NOPQ cells (Fig. 9b). It is conceivable that during this polarization process, some kind of a gradient is generated in NOPQ cells (Fig. 9c); a Dst-cell descending from an NOPQ cell might be determined as an N teloblast, and a Pxm-cell might be determined as an OPQ cell (Fig. 9d). The OPQ cell subsequently undergoes "polarized" cell divisions so that it produces a dorsal-most cell that is determined as a Q teloblast. Apparently the OPQ cell acquires polarized properties. In view of the fact that OPQ cells remain in contact with each other shortly after their birth (Fig. 9d), it seems possible that these cells undergo the second round of the polarization process mediated by the contact between themselves. However, this point remains to be studied experimentally.

At present, nothing is known about the ground (default) state of nascent NOPQ cells. Upon its birth, the NOPQ cell may be determined as a precursor of ectoteloblasts. However, it is unclear whether nascent NOPQ cells already possess polarized properties that foreshadow the polarity of teloblastogenesis. Therefore, it remains to be determined whether the external cues endow NOPQ cells with *de novo* polarity or strengthen (or stabilize) the preexisting polarity of NOPQ cells.

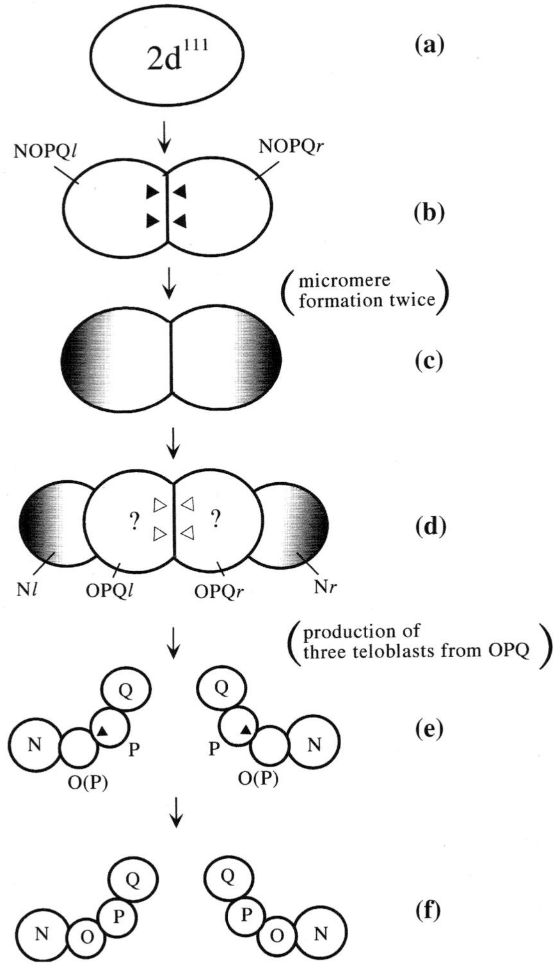


*Figure 9.* Model for patterning of ectodermal teloblastogenesis in the *Tubifex* embryo. Filled arrowheads (b) indicate the interactions between NOPQ cells that polarize themselves. Open arrowheads (d) indicate tentative interactions between OPQ cells that are responsible for further polarization of OPQ cells. An inductive signal from the P teloblast is indicated by a filled arrowhead (e).

## Comparison with other clitellate annelids

As mentioned before, in *Tubifex*, a pair of ectoteloblast precursor NOPQ cells are derived from the 2d cell and make contact with each other along the midline after their birth. Each NOPQ, which is associated with an anterior arc of micromeres, gives rise to four ectoteloblasts. Similar cellular configurations of NOPQ and sequence of ectodermal teloblastogenesis have been reported in a wide variety of clitellate annelids, including enchytraeid *Lumbricillus lineatus* (Penners, 1930), tubificid *Peloscolex benedeni* (Penners, 1929), lumbricid *Eisenia foetida* (Devries, 1973a), branchiobdellid *Bdellodrilus philadephicus* (Tannreuther, 1915), and leeches *Glossiphonia complanata*, *Theromyzon rude* and *Helobdella stagnalis* (Whitman, 1878; Fernandez & Olea, 1982). Given that leeches and branchiobdellidans form a monophyletic group of oligochaetes (Siddal et al., 2001), such a mode of ectodermal teloblastogenesis as seen in *Tubifex* is an ancestral form of oligochaete embryogenesis and has been preserved widely in the Clitellata. At present, however, it is not known whether these cell–cell interactions in annelids such as those demonstrated in *Tubifex* are involved in the generation of bilateral symmetry in the organization of the ectoderm. As discussed later, it is equally possible that evolutionary changes in cell interaction networks produce embryonic structures with similar morphological pattern.

It is interesting to note that the lumbriculid *Rhynchelmis* and naids *Stylaria* and *Chaetogaster* exhibit developmental patterns distinct from those in other clitellates. *Rhynchelmis* embryos have four bilateral pairs of ectoteloblasts, of which one originates from 2d and the remaining three from 3d (Svetlov, 1923). *Stylaria* and *Chaetogaster* form the segmental ectoderm without production of ectoteloblasts; these two oligochaetes also differ from each other in that the source of the segmental ectoderm is 2d for *Stylaria* and 1d for *Chaetogaster* (Svetlov 1926; Davydov, 1942). Given that these oligochaetes have a bilaterally symmetric body plan as other clitellates do, these exceptional cases suggest it unlikely that the bilaterally symmetric organization of 2d-derived ectoteloblasts is solely responsible for the generation of bilateral symmetry in the ectoderm. It will be of interest to determine, in future studies, the extent to which mechanisms for generation of bilateral symmetry in the ectoderm are diverse in the Clitellata.

As stated before, the N, P and Q teloblasts in the *Tubifex* embryo are committed to their respective fates at the time of their birth. In contrast, the O teloblast is pluripotent at the time of its birth and is induced to assume the O fate by the P teloblast lineage; otherwise it expresses the P fate. As to the mode of specification of ectodermal teloblast lineages, the only thing that is known in other clitellate annelids is that in the leech *Helobdella*, bandlets derived from a pair of O/P teloblasts (equivalent to the O and P teloblasts in oligochaetes) are initially equipotent and differentiate from each other according to their position within the GB (Weisblat & Blair, 1984; Zackson, 1984). Recently, Huang & Weisblat (1996) have suggested that blast cells derived from either O/P teloblast assume the P fate if they interact with a bandlet derived from the Q teloblast and that otherwise o/p blast cells express the O fate. Thus, it is apparent that distinct cell interaction networks are involved in patterning the ectodermal GB in *Tubifex* and *Helobdella*.

In spite of these differences, however, the resulting (final) pattern of the ectodermal GB is strikingly similar in *Tubifex* and *Helobdella* (Weisblat & Shankland, 1985; Goto et al., 1999b). Thus, we suggest that during their evolutionary isolation, oligochaetes and leeches have preserved an ancestral pattern of the ectodermal GB despite the divergence of cell interaction networks through which this pattern is brought about.

## References


Arai, A., A. Nakamoto & T. Shimizu, 2001. Specification of ectoermal telolablast lineages in embryos of the oligochaete annelid *Tubifex*: involvement of novel cell–cell interactions. Development 128: 1211–1219.

Blair, S. S., 1982. Interactions between mesoderm and ectoderm in segment formation in the embryo of a glossiphodiid leech. Developmental Biology 89: 389–396.

Davydov, C., 1942. Etude sur l'embryologie des Naididae indochinos. Archies de Zoologie experimentale et generale 81: 173–194.

Devries, J., 1973a. La formation et la destinée des feuillets embryonnaires chez le lombricien *Eisenia foetida* (Annélide Oligochete). Archives de l'anatomie microscopique 62: 15–38.

Devries, J., 1973b. Détermination précoce du développement embryonnaire chez le lombricien *Eisenia foetida*. Bulletin de la Societe Zoologique de France 98: 405–417.

Fernandez, J. & N. Olea, 1982. Embryonic development glossiphoniid leeches. In Harrison, F. W. & R. R. Cowden, (eds), Developmental Biology of Freshwater Invertebrates. Alan R. Liss, New York, 317–361.

Goto, A., K. Kitamura & T. Shimizu, 1999a. Cell lineage analysis of pattern formation in the *Tubifex* embryo. I. Segmentation in the mesoderm. International Journal of Developmental Biology 43: 317–327.

Goto, A., K. Kitamura, A. Arai & T. Shimizu, 1999b. Cell fate analysis of teloblasts in the *Tubifex* embryo by intracellular injection of HRP. Development Growth & Differentiation 41: 703–713.

Huang, F. Z. & D. A. Weisblat, 1996. Cell fate determination in an annelid equivalence group. Development 122: 1839–1847.

Kitamura, K. & T. Shimizu, 2000. Analyses of segment-specific expression of alkaline phosphatase activity in the mesoderm of the oligochaete annelid *Tubifex*: implications for specification of segmental identity. Developmental Biology 219: 214–223.

Nakamoto, A., A. Arai & T. Shimizu, 2000. Cell lineage analysis of pattern formation in the *Tubifex* embryo. II. Segmentation in the ectoderm. International Journal of Developmental Biology 44: 797–805.

Nakamoto, A., A. Arai & T. Shimizu, 2004. Specification of polarity of teloblatogenesis in the *Tubifex* embryo: cellular basis for bilateral symmetry in the ectoderm. Developmental Biology 272: 248–261.

Penners, A., 1924. Über die Entwicklung teilweise abgetöteter Eier von *Tubifex rivulorum*. Verhandlungen der Deutschen Zoologischen Gesellschaft 29: 69–73.

Penners, A., 1926. Experimentelle Untersuchungen zum Determinationsproblem am Keim von *Tubifex rivulorum* Lam. II. Die Entwicklung teilweise abgetöteter Keime. Zeitschrift fur Wissenschaftliche Zoologie 127: 1–140.

Penners, A., 1929. Entwicklungsgeschichtliche Untersuchungen an marinen Oligochaten. I. Furchung, Keimstreif, Vordendarm und Urkeimzellen von *Pelescolex benedeni* Udekem. Zeitschrift fur Wissenschaftliche Zoologie 134: 307–344.

Penners, A., 1930. Entwicklungsgeschichtliche Untersuchungen an marinen Oligochaten. II. Furchung, Keimstreif und Keimbahn von *Pachydrilus* (*Lumbricillus*) *lineatus* Mull. Zeitschrift fur Wissenschaftliche Zoologie 137: 55–119.

Sandig, M. & W. Dohle, 1988. The cleavage pattern in the leech *Theromyzon tessulatum* (Hirudinea, Glossiphoniidae). Journal of Morphology 196: 217–252.

Shain, D. H., F.-A. Ramirez-Weber, J. Hsu & D. A. Weisblat, 1998. Gangliogenesis in leech: morphogenetic processes leading to segmentation in the central nervous system. Development Genes and Evolution 208: 28–36.

Shimizu, T., 1982. Development in the freshwater oligochaete *Tubifex*. In Harrison, F. W. & R. R. Cowden, (eds), Developmental Biology of Freshwater Invertebrates. Alan R. Liss, New York, 286–316.

Shimizu, T. & A. Nakamoto, 2001. Segmentation in annelids: cellular and molecular basis for metameric body plan. Zoological Science 18: 286–298.

Siddal, M. E., K. Apakupakul, E. M. Burreson, K. Coates, C. Erseus, S. R. Gelder, M. Kallersjo & H. Trapido-Rosenthal, 2001. Validating Livanow: molecular data agree that leeches, branchiobdellians, and *Acanthobdella peledina* form a monophyletic group of oligochaetes. Molecular Phylogenetics & Evolution 21: 346–351.

Storey, K. G., 1989. Cell lineage and pattern formation in the earthworm embryo. Development 107: 519–531.

Svetlov, P., 1923. Sur la segmentation de l'oeuf de *Rhynchelmis limosella* Hoffmstr. Izvestia biologicheskogo nauchno-issledovatelskogo Insttuta biologicheskoi stantsii pri Permskom gosudarstvennom Universitete 1: 141–152.

Svetlov, P., 1926. Über die Embryonalentwichlung bei den Naididen. Izvestia biologicheskogo nauchno-issledovatelskogo Insttuta biologicheskoi stantsii pri Permskom gosudarstvennom Universitete 4: 359–372.

Tannreuther, G. W., 1915. The embryology of *Bdellodrilus philadelphicus*. Journal of Morphology 26: 143–216.

Wedeen, C. J. & M. Shankland, 1997. Mesoderm is required for the formation of a segmented endodermal cell layer in the leech *Helobdella*. Developmental Biology 191: 202–214.

Weisblat, D. A. & S. S. Blair, 1984. Developmental interdeterminacy in embryos of the leech *Helobdella triserialis*. Developmental Biology 101: 326–335.

Weisblat, D. A., G. Harper, G. S. Stent & R. T. Sawyer, 1980. Embryonic cell lineages in the nervous system of the glossiphoniid leech *Helobdella triserialis*. Developmental Biology 76: 58–78.

Weisblat, D. A., S. Y. Kim & G. S. Stent, 1984. Embryonic origin of cells in the leech *Helobdella striserialis*. Developmental Biology 104: 65–85.

Weisblat, D. A. & M. Shankland, 1985. Cell lineage and segmentation in the leech. Philosophical Transactions of the Royal Society (London) B312: 39–56.

Whitman, C. O., 1878. The embryology of *Clepsine*. Quarterly Journal of Microscopical Science 18: 215–315.

Zackson, S. L., 1984. Cell lineage, cell–cell interaction, and segment formation in the ectoderm of a glossiphoniid leech embryo. Developmental Biology 104: 143–160.

Hydrobiologia (2006) 564:33–43
P.F.M. Verdonschot, H. Wang, A. Pinder & R. Nijboer (eds), Aquatic Oligochaete Biology IX
DOI 10.1007/s10750-005-1705-0


# Distribution pattern of γ-amino butyric acid immunoreactive neural structures in the central and peripheral nervous system of the tubificid worm, *Limnodrilus hoffmeisteri*

László Molnár[1,2,*], Gábor Kiszler[1], Edit Pollák[1,2] & László Deres[1]
[1]*Research Group of Comparative Anatomy, Department of General Zoology and Neurobiology, University of Pécs, Ifjúság u. 6, H-7624 Pécs, Hungary*
[2]*Adaptation Biology Research Group of the Hungarian Academy of Sciences and University of Pécs, Ifjúság u. 6, H-7624 Pécs, Hungary*
*(*Author for correspondence: E-mail: molnar@ttk.pte.hu)*

*Key words:* immunohistochemistry, GABA, neurotransmitters, neuronal patterns, interneurons, reflex circuits

**Abstract**

By means of whole-mount immunohistochemistry, putative inhibitory (GABAergic) neural structures were identified in the central and peripheral nervous system of the tubificid worm, *Limnodrilus hoffmeisteri.* In the supraoesophageal ganglion (brain) only few strongly labelled cells were observed. However, in its commissural part a high number of stained nerve fibres, arising mainly from the ventral nerve cord and prostomium, occurred. Except for the suboesophageal ganglion the arrangement of γ-amino butyric acid-immunoreactive (GABA-IR) structures proved to be identical in each VNC ganglion. Behind the first segmental nerves three pairs of heavily stained neurones were located. Their processes (both ipsi- and contralateral) form four bundles of fine-fibred polysegmental interneuronal tracts that run close to the dorsal giant axons from the terminal ganglion to the suboesophageal one without interruption. A few small motoneurons and a pair of large ones with contralateral processes were also identified. A bipolar (presumably sensory) neuron was located at the root of each second segmental nerve. GABA-IR neurons were also found in the stomatogastric ganglia and pharyngeal wall; however, the latter structure had a well-developed fibre network, as well. Present results suggest that GABA acts as a common neurotransmitter in sensory, interneuronal and motor system of *L. hoffmeisteri.* The possible functional role of the identified GABA-IR neural structures in locomotion, escape and withdrawal reflexes in tubificid worms is discussed.

## Introduction

The anatomical and neurochemical organization of both central (CNS) and peripheral nervous system (PNS) of oligochaetes has long been investigated. The gross anatomy and histology of the CNS, especially that of the ventral nerve cord (VNC) ganglia, are well known (for review see Bullock & Horridge, 1965; Dorsett, 1978; Jamieson, 1981), but the function of most of the neurons and axon bundles is yet unknown. The function of certain neural structures is better understood. We know, for example, that the sensory longitudinal axon bundles conduct sensory inputs from extero- and interoreceptors (Mill, 1982) and that the giant motoneurons, innervating the longitudinal muscles of the body wall (Günther, 1972), together with the dorsal giant axons (often called giant fibres) mediate locomotion, escape and withdrawal reflexes (Dorsett, 1978). These latter structures have already been studied in some tubificid worms, using neuroanatomical, ultrastructural and electrophysiological methods. These studies suggested that they are basically similar, in their organization and function, to those found in earthworms (Zoran & Drewes, 1987, 1988; Drewes & Zoran, 1989), except for some species-specific characteristics (Zoran et al., 1988). Nevertheless, not much is known about the transmitter-specific neurons of freshwater oligochaetes.

There is growing evidence that the synchronization of sensory inputs and motor outputs of the CNS is mediated by transmitter-specific neural circuits that produce simple reflexes or more complex motor activity like locomotion, eating and mating behaviour (Dorsett, 1978; Mill, 1982). In contrast to the more investigated crustaceans, insects and molluscs (Sepherd, 1994), there are little data in the literature concerning the occurrence and pattern of transmitter-specific neural structures in annelids, considered to be the phylogenetic ancestors of some of the groups above (Wilmer, 1990). Therefore, the identification of these neural networks could contribute not only to the understanding of the structure and the functioning of annelid nervous system but also to increase our knowledge on the phylogeny of the group.

At present, the amino acid glutamate (Molnár et al., unpublished), the neuropeptides (Gardner & Cashin, 1975; Gardner, 1981; Curry et al., 1989; Fuji et al., 1989; Al-Yousuf, 1992; Lubics et al., 1997; Reglődi et al., 1997; Molnár et al., 2001), the gaseous neurotransmitter nitric oxide (Zsombok et al., 1999; Kitamura et al., 2001; Molnár et al., 2001; Zsombok & Molnár, 2001), the monoamines (Rude, 1966; Ehinger & Myhrberg, 1971; Bieger & Hornykiewicz, 1972; Gardner & Cashin, 1975; Gardner, 1981; Spörhase-Eichmann et al., 1987a, b, 1998) and the inhibitory neurotransmitter $\gamma$-amino butyric acid (GABA) (Telkes et al., 1996; Spörhase-Eichmann et al., 1997; Molnár et al., 2000, 2001) have been identified in earthworms. GABA, which is thought to be the major inhibitory neurotransmitter in both invertebrates and vertebrates (Sepherd, 1994), acts as neurotransmitter in earthworms, having been identified in a population of primary sensory cells (Spörhase-Eichmann et al., 1997; Solt et al., 2001) and in some motoneurons (Telkes et al., 1996). It is also present in certain polysegmental interneuronal tracts located at the dorsal neuropile of the VNC ganglia (Molnár et al., 2000).

This study focuses on the observation of GABA-immunoreactive (IR) neural structures in both the CNS and in part, the PNS (e.g. stomatogastric system) of the tubificid worm, *Limnodrilus hoffmeisteri*. Further, it proposes some possible physiological roles to the labelled neurons and axon bundles. This study also compares GABA-IR structures in soil and freshwater oligochaetes, showing that these neural structures are partly conservative and that they are characteristic of oligochaetes, especially those that regulate the activity of dorsal giant axons.

## Materials and methods

The experiments were performed on sexually mature specimens of *L. hoffmeisteri* Claparède 1868, individually selected from samples of tubificid worms on the basis of their species-specific anatomical characteristics (Stephenson, 1930). Ten to fifteen worms were sacrificed in each experiment. Each experiment has been repeated three times.

All experimental animals were anaesthetized by cooling for 5–10 min using ice and then fixed for 3 h at 4 °C in the dark in a freshly prepared mixture of saturated picric acid (3 ml), 25% glutaraldehyde (1 ml) and concentrated acetic acid (40 $\mu$l). Most of the specimens were used as whole-mounts but the CNS and various regions of both the body wall and the alimentary canal were dissected out from a few specimens. Both whole-mounted and dissected specimens have been handled according to the following protocol. After several rinses in 70% ethanol, fixed samples were washed in several changes of phosphate-buffered saline (PBS) then treated with 2% Triton X-100 diluted in PBS for two nights at 4 °C. To decrease background staining, the samples were incubated for 2 h in 10% normal goat serum in 2% Triton X-100. Then they have been incubated in 1:500 dilution of anti-GABA primary antibody (Sigma Chemical Company, Budapest, Hungary) in 2% Triton X-100 buffer for 4 days at 4–6 °C.

The specificity of the antibody to GABA was confirmed by the preincubation of the antibody together with GABA–glutaraldehyde–bovine serum albumin complex (1 mM of GABA). After preincubation the immunohistochemical protocol resulted in no staining. Labelled structures were visualized using diaminobenzidine. (Avidin–biotin–horseradish peroxidase method, Extravidin kit, Sigma Chemical Company, Budapest, Hungary) with the aid of a stereomicroscopy.

Control samples were processed utilizing the above immunocytochemical staining protocol except for the omission of the primary antibody.

Stained samples were processed in two different ways. Whole mounts were cleared with buffered glycerol (50% PBS and 50% glycerol) for 2–3 days then mounted between cover-slips in pure glycerol for microscopy. Some specimens were dehydrated in graded ethanol series and embedded in epoxy-resin (Durcupan ACM). 1.5-$\mu$m thick serial sections were made with a Reichert ultramicrotome, attached to gelatine-chrome-alum covered slides and stained with toluidine blue.

Immunostained samples were observed and photographed by a Nikon Optiphot-2 microscope. Camera lucida drawings were taken by attached Olympus equipment.

## Results

Similarly to earthworms, the CNS of *L. hoffmeisteri* is localized in the body cavity. The supra-esophageal ganglion (brain) is located in the first segment behind the prostomium and it is connected to the suboesophageal ganglion by long circumpharyngeal connectives. The stomatogastric ganglia are located at the anterior edge of the circumpharyngeal connectives (Fig. 1). The ventral nerve cord (VNC) is a long chain of repeated ganglia, located in each segment and separated by connectives, resulting in a spatial segregation of neuronal cell bodies (perikarya) and nerve fibres.

As shown in our microphotographs, the applied immunohistochemical protocol resulted in a clear-cut specific staining of GABA-IR neurons in whole-mount preparations. This allowed us to determine both the exact anatomical position and geometry of all labelled structures in some body parts.

GABA-IR perikarya and nerve processes occurred in both the CNS and PNS. The latter one however has not been investigated in detail: the alimentary canal was partly covered by chloragogenous tissue, and also in the body wall only weak staining has been observed thus the unequivocal identification of labelled structures was not always possible. As a consequence, we only focused on the segmental nerves of the CNS and, with respect to the stomatogastric nervous system, on stomatogastric ganglia and stomodaeum neural structures.

### *GABA-IR structures of the CNS*

The brain has only four pairs of GABA-IR neurons located at the posterolateral part of both hemiganglia. These neurons have a thick axon that sprouts into several fine ramifications: their main processes, however, run directly to the anterior neuropile. In the posterior neuropile (identified in the commissural part of the brain) a high number of incoming nerve fibres of variable thickness was

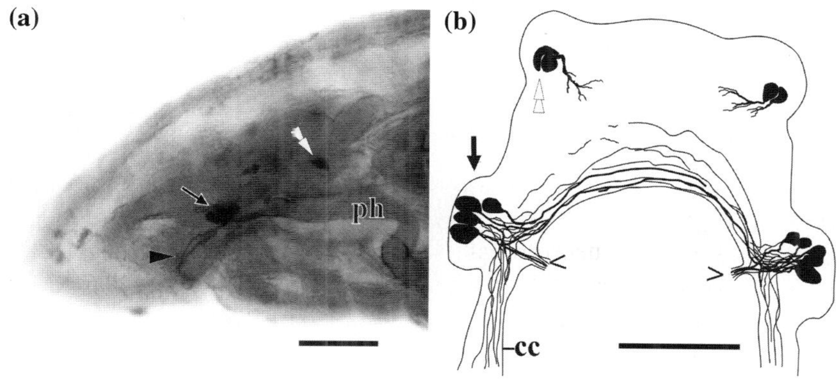


*Figure 1.* Anatomy and pattern of GABA-IR structures in the anteriormost segments of *Limnodrilus hoffmeisteri*. (a) Lateral view of labelled structures in situ. (b) Camera lucida drawing of dissected parts of the central nervous system. The anterior is on the top. Note the brain has only few labelled cells but its commissural part is rich in nerve fibres originated from stomatogastric ganglia and ventral nerve cord. *cc*: circumpharyngeal connective; *ph*: pharynx; *arrows*: GABA-IR cells of stomatogastric ganglia; *double arrowheads*: labelled cells in the brain; *arrowhead*: labelled fibres in the roof of the buccal cavity; *open arrowheads*: pharyngeal connectives of stomatogastric ganglia. Scale bars: 100 $\mu$m.

observed. A larger part of the incoming fibres could be followed coming from both the ventral nerve cord and the stomatogastric ganglia via circumpharyngeal connectives; others derive from the prostomium and from the roof of the buccal cavity (Fig. 1).

Significantly more labelled neurons, showing a repeated pattern in each ganglion, occurs in the VNC (Figs. 2a–d, 3a). The number of labelled cells, except for the suboesophageal ganglion, seems to be identical in all ganglia, including those innervating genital organs or situated in clitellar segments. The most characteristic structures are a medial and a lateral pair of fine polysegmental fibre bundles. These bundles are located at the dorsal part of the VNC and they run, without interruption and parallel to the dorsal giant axons, from the terminal ganglion to the suboesophageal one (Fig. 2a–d). They then enter the circumpharyngeal connectives to end in the commissural part of the brain (Fig. 1b). The neurons that form the polysegmental interneuronal fibre bundles are located behind the first segmental nerves, in the lateral and dorsolateral parts of each ganglion, (Figs. 2a–d, 3a, b). These neurons have a well-developed arborizing axon which consists of several thin ascending ramifications and a thick descending one. Most of the ascending axons remain within the same side of the ganglia and they enter the ipsilateral polysegmental fibre tracts; however, a few fine axons cross the midline of the ganglia and they attach to the contralateral medial fibre tracts (Fig. 2c, d). Thick descending axons always cross the contralateral hemiganglion. They pass through the neuropile of the ganglion before entering the medial contralateral polysegmental fibre bundles. Nevertheless, some processes enter the lateral polysegmental fibre bundles of the other side, as well. In the best preparations a well-developed neural network, located between the perikarya of interneurons and their crossing fibres, has been observed (Fig. 2d). This basic pattern is

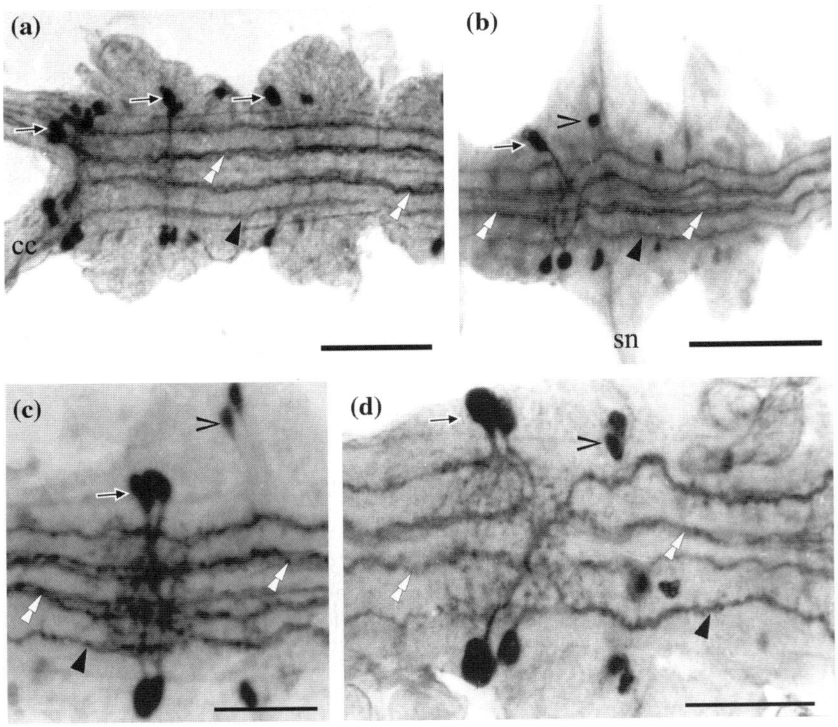


*Figure 2.* GABA-IR structures of the ventral nerve cord ganglia. Dorsal view. The anterior part is on the left. (a) Suboesophageal ganglion and the first segmental ganglion. (b) A segmental ganglion from a postclitellar segment. (c, d) High magnification micrographs of interneurons located behind the first segmental nerves. Note GABA-IR polysegmental fibre bundles consist of ipsi- and contralateral processes of the above interneurons. *cc*: circumpharyngeal connective; *sn*: 2nd segmental nerve; *arrows*: interneurons; *arrowheads*: lateral polysegmental fibre bundles; *double arrowheads*: medial polysegmental fibre bundles; *open arrowheads*: bipolar neurons on the root of the 2nd segmental nerves. Scale bars: 100 $\mu$m.

modified at the anterior end of the suboesophageal ganglion, where six pairs of interneurons are connected with the polysegmental fibre bundles (Fig. 2a).

Observing these fibre bundles in the VNC, from the terminal to the suboesophageal ganglion or vice versa, it is possible to see the axons of interneurons join successively into the polysegmental fibre bundles replacing those that terminate in fine ramifications. As a result of this the diameter of these bundles is constant in each segment.

In agreement with Zoran & Drewes (1987), no myelin-like sheath was detectable around dorsal giant fibres and the diameter of lateral ones was about five times greater than that of the medial ones in midbody segments (Fig. 3b–f).

Apart from the above described interneurons ten pairs of labelled neurons characterized by small soma size were identified in the anterior and posterior parts of each ganglion (Fig. 3a). Some of them have contralateral axon processes, but the axonal pathways cannot be followed in all cases. At the level of the second segmental nerves a pair of large neurons is located in the ventromedial part of the ganglion and these have pear shape perikaryon and axons running contralaterally (Fig. 3c). Two GABA-IR neurons are located at the root of each second segmental nerve (Figs. 2b–d, 3a). One of them has only peripheral processes, while the second is a bipolar cell with peripheral and central processes. The latter processes enter the neuropile and take connection with the lateral polysegmental fibre bundles on the same side (Fig. 3d).

Inside the neuropile not only the polysegmental interneuronal fibre bundles but also a further pairs of longitudinally directed structures show GABA-immunoreactivity. Their identification is problematic because of their moderate labelling and their thin diameter but in some cases we could follow the organization of these polysegmental fibre bundles. A thin fibre bundle enters each ganglion via the 2nd segmental nerves (Fig. 3a) that form a polysegmental fibre bundle located at the same side of the ganglion. One of these polysegmental structures is located at the ventrolateral part of the ganglion close to the cellular layer, while the second is positioned more medially. Their position strongly resembles that of the ventromedial and ventrolateral sensory longitudinal axon bundles found in earthworm CNS (Günther, 1971; Molnár et al., 2001; Solt et al., 2001).

Both the dorsal and ventral parts of segmental nerves of the VNC ganglia contain GABA-IR fibres, indicating that the labelled fibres belong to functionally different (motor and sensory) neural structures in *L. hoffmeisteri*.

### *GABA-IR structures of the stomatogastric nervous system*

Inside the stomatogastric ganglia that innervate the anterior part of the alimentary canal (Millot, 1943) five pairs of neurons are labelled for GABA (Fig. 1). They have a large perikaryon and a richly arborized axon. Some of the axon processes enter the commissural part of the brain and merges with the neural processes originating from the VNC and the prostomium; the other ones supply the pharynx via pharyngeal connectives and form a dense neural network in its wall (Figs. 1b and 4a). Few GABA-IR neurons are located in the pharyngeal wall, mainly in its lateral parts, and they send their axons into the neural network (Fig. 4a).

A high number of GABA-IR cells characterized by elongated somata and seemingly with a random distribution was found in the oesophagus. These cylindrical cells are mainly located in the oesophageal epithelium. In the best preparations we observed one or sometimes a few thin outgoing processes of the labelled cells; their pathway, however, could not be followed in whole-mount preparations (Fig. 4b).

## Discussion

We have applied a controlled GABA immunohistochemical protocol to identify a distinct set of inhibitory (GABAergic) neurons in the nervous system of *L. hoffmeisteri*. Since no staining was observed in control samples, which were either not treated with primary antibodies or were treated with preabsorbed antibodies, we propose that all labelled neural structures contain GABA as transmitter.

GABA synthesizing cells were clearly labelled in the whole CNS and partly in the PNS. Further

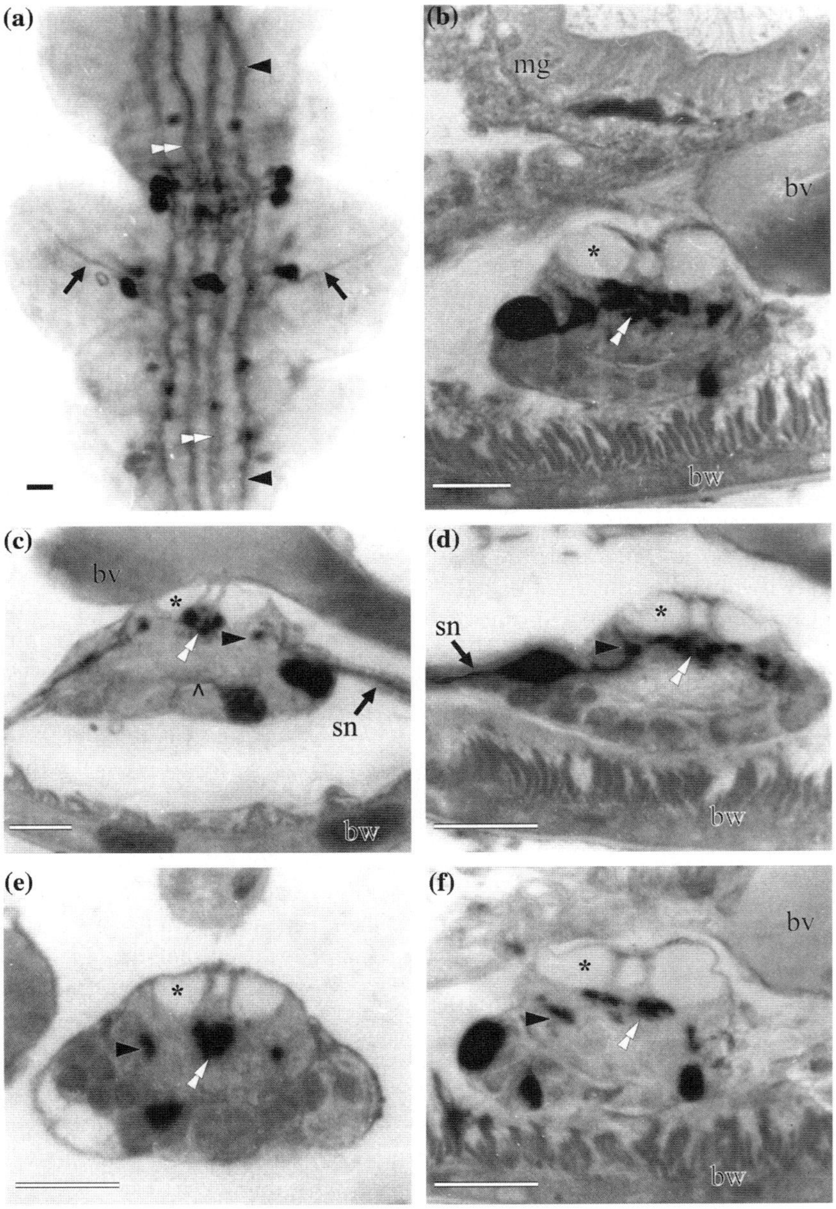


*Figure 3.* Anatomical position of both GABA-IR cells and fibre bundles in the ventral nerve cord ganglion from a posteriormost segment in whole-mount (the anterior is on the top) ventral view (a) and in its sections (b–f). (b) An interneuron located at the dorsolateral part of the ganglion has direct connections with both lateral and medial polysegmental fibre bundles. (c) A giant motoneuron with contralateral axon, and two intensely labelled cells at the second nerve root. (d) A bipolar neuron at the second nerve root. Note the centrally directed neural process connected with the lateral polysegmental fibre bundle. (e, f) GABA-IR neurons with various soma size and shape from the posterior part of the ganglion. *bv*: blood vessel; *bw*: body wall; *mg*: midgut; *sn*: 2nd segmental nerve; *arrowheads*: lateral polysegmental fibre bundles; *double arrowheads*: medial polysegmental fibre bundles; *open arrowhead*: contralateral axon of a giant motoneuron; *arrows*: incoming sensory fibres in the 2nd segmental nerves, *asterisks*: lateral giant axon. Scale bars: 25 $\mu$m.

GABA staining was always seen not only in the cytoplasm of cell bodies but also in their processes, showing three-dimensional arrangement of labelled structures. This strong staining in whole-mount preparations allowed us to follow the organization of certain GABAergic systems in the

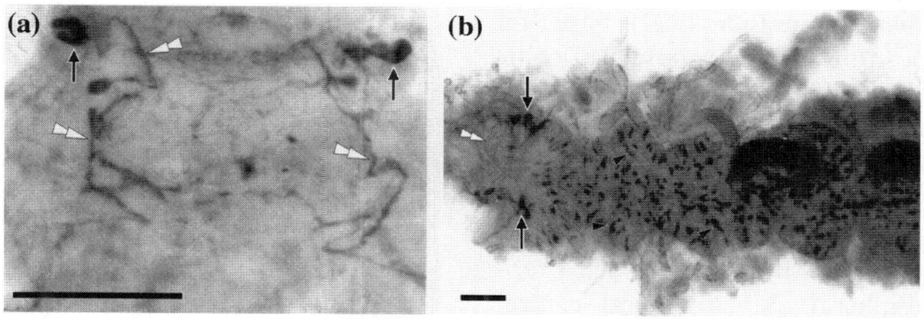


*Figure 4.* GABA-IR structures of the pharynx (a) and the oesophagus (b). Ventral view. *Arrows*: stomatogastric ganglia; *arrowheads*: labelled cells in the oesophagus; *double arrowheads*: labelled fibre-network in the pharyngeal wall. Note the extremely high number of labelled cells with cylindrical soma shape in the oesophagus. Scale bars: 100 $\mu$m.

smallest details, in contrast to the results of earlier work (Telkes et al., 1996).

Our results show that GABA-IR cells are unequally distributed in the CNS of *L. hoffmeisteri.* The limited number of labelled cells in the brain suggests that local GABAergic inhibition has little importance in the brain. Similar results and conclusions have been obtained from the investigation of the terrestrial species *Lumbricus terrestris*, but in that species a higher number of labelled cells was counted (Telkes et al., 1996). At this stage of our experiments it is not clear why the number of GABA-IR cells differs in these species, since in tubificid worms the brain is said to be larger in proportion to the body than in the much larger terrestrial worms (Bullock and Horridge, 1965).

In contrast to earlier work (Telkes et al., 1996) this study clearly demonstrates that the neuropile of the brain receives afferent labelled fibres from stomatogastric and VNC ganglia. These fibres are heterogeneous both in their origin and function: (i) labelled cells of stomatogastric ganglia with both central and peripheral processes could be vegetative neurons and the presence of their fibres in brain neuropile suggests that the brain has a direct role in the organization of alimentary canal motility and (ii) the projection of polysegmental interneuronal fibre bundles of the VNC to the brain neuropile indicates that interneuronal inhibitory impulses also act in the brain. Distinct representations of GABA-IR sensory fibres (Molnár et al., unpublished) coming from the prostomium and the ventral nerve cord have been established in *Lumbricus terrestris.* A similar organization may occur in *L. hoffmeisteri* since GABA-IR fibre bundles were observed in its prostomial nerves. As there are no labelled projection neurons in the brain, labelled fibres may only be central processes of primary sensory cells located in the body wall epithelium. Based on the small number of serotoninergic neurons in the brain of freshwater oligochaetes, Hessling et al. (1999) concluded that the oligochaete brain is mainly sensory in function. In contrast to this, we propose that the oligochaete brain, thought to be relatively simple in its histological organization (Ogawa, 1939; Bullock & Horridge, 1965), may be a complex transmitting centre of the oligochaete CNS in which the vegetative, interneuronal and probably sensory impulses regulate and modulate the activity of the neuronal structures located in, or run across, the brain.

As shown in the Results section (Figs. 2 and 3) GABA is present in anatomically distinct sets of neurons inside the VNC ganglia of *L. hoffmeisteri.* However, at this stage of our experiments there are no data to characterize in detail all the labelled structures. According to the widely accepted classification of neurons (for review see Bullock & Horridge; 1965, Dorsett, 1978; Sepherd, 1994) those GABA-IR cells of VNC ganglia having peripherally directed axons should only be motoneurons which supply peripheral organs and/or tissues. Those motoneurons situated at the medial part of the VNC ganglion, with large pear-shaped soma and thick contralateral axon, may be a special type of motoneuron. In the VNC ganglia of earthworm four pairs of giant motoneurons that supply the longitudinal muscles of the body wall have been identified. A pair of these neurons is located at the level of the 3rd segmental nerves and they send their crossing axons to the periphery via these segmental nerves (Günther, 1972). Based on the anatomical position and morphological characteristics of the above-mentioned large neuron of *L. hoffmeisteri*, we propose that this may be homologous with the above giant motoneurons identified in the earthworm VNC ganglion.

The bipolar cells located at the root of the second segmental nerves show sensory cell properties having peripheral and ipsilateral central processes. Though sensory cell somata in invertebrates are generally located at the periphery (Bullock & Horridge, 1965), centrally located sensory cells, as a specific characteristic to annelid nervous system, were identified in both leeches and earthworms (Günther, 1971; Dorsett, 1978) as touch and pressure receptors that are situated between the 1st and 2nd segmental nerves.

The most characteristic GABA-IR cells are the interneurons located behind the first segmental nerves of each ganglion. Their processes form polysegmental fibre bundles running from the suboesophageal to the terminal ganglion without interruption. Certain fibres of these bundles have also representation in the commissural part of the brain. The anatomical position and organization of these interneuronal systems show the same characteristics to those already identified in terrestrial oligochaetes, like *Lumbricus terrestris* and *Eisenia fetida* (Molnár et al., 2000; Solt et al., 2001). However, in the CNS of earthworms, many more neurons contribute to the formation of the GABAergic polysegmental tracts since 6–8 pairs of interneurons, located behind the first segmental nerves, and 10–12 pairs of smaller interneurons, at the level of second and third segmental nerves, send their processes to the medial and lateral tracts. Although the number of neurons involved may vary from species to species their homology in oligochaetes seems to be self-evident. The number of interneurons should be determined by the complexity of locomotion, characteristic for each species. In earthworms, characterized by a more intensive locomotion than freshwater oligochaetes, the reflex centre in each segment, namely the ventral cord ganglion, needs more effective neural activity. Their interneurons and motoneurons are affected by a higher number of inhibitory neurons.

The facts that GABAergic polysegmental interneuronal fibre bundles (i) came from the suboesophageal ganglion, thought to be the highest motor centre of oligochaete CNS (Dorsett, 1978) and (ii) they run in close to the dorsal giant axons, possibly making numerous synaptic contacts with their collaterals in the neuropile, convey the suggestion that these polysegmental interneuronal tracts have an inhibitory influence on the activity of the dorsal giant axons. These latter structures make synaptic contacts with the giant motoneurons of earthworms (Günther & Walther, 1971; Günther, 1972; Günther & Schürmann, 1973), mediating both escape and withdrawal reflexes besides locomotion of oligochaetes (Dorsett, 1978). If the GABA-IR bipolar cell located at each 2nd segmental nerve is a real sensory neuron there is a possibility to investigate the function of a distinct neural circuit consisting of a bipolar sensory cell, the GABA-IR polysegmental interneuronal fibre bundles, and the giant axons besides the giant motoneurons as effector cells.

In the oligochaete nervous system the motoneurons of each segmental ganglion form the final output path. The VNC contains not only the neural circuits required for the basic alternating contraction of segments but also the neural structures that produce intersegmental coordination under changing conditions by anterior-to-posterior propagation of the contraction along the body axis. To produce this pattern, the reflex centre in each segment, namely the VNC ganglion, must be activated at fixed time. The activity of reflex circuits of a ganglion is controlled by commands from other segmental ganglia and from the supra and suboesophageal ganglion. The identified GABAergic polysegmental interneuronal fibre bundles could inhibit the activity of the reflex circuits located in segmental ganglia. Considering the labelled fibres in segmental nerves, the GABAergic sensory inputs from peripheral organs and tissues could further modify the activity of GABA-IR interneuronal structures. To clarify the exact connections and function of GABA-IR neurons and fibre bundles ultrastructural and electrophysiological investigations are needed.

The relatively high number of GABA-IR neurons in stomatogastric ganglia and the well-developed GABAergic fibre network in pharynx indicate the significance of inhibitory regulation in muscle contraction of the alimentary canal. The enteric nervous system of mammals (Jessen et al.,

1986; Fourness et al., 1989) and of some invertebrate species like lobster (Cournil et al., 1990), the isopod *Porcellio scaber* (Pollák et al., 2003), and the earthworm *Lumbricus terrestris* (Telkes et al., 1996) contains GABA-IR cells and their processes. The occurrence of GABA in enteric nervous system of animals shows that it should be a major inhibitory transmitter in the alimentary canal and that it is conserved during the phylogeny. Though GABA does not always have a direct inhibitory effect on muscles it may modify the activity of other neurons that make direct synaptic contacts on muscles (Atwood & Kwan, 1979). There is growing evidence that the enteric nervous system has distinct neuron sets working with various transmitters. For example, in the isopod *Porcellio scaber* the neuropeptide proctolin, the gaseous transmitter nitric oxide, and the monoamine serotonin (Pollák et al., 2003); in the earthworm *Eisenia fetida* serotonin, octopamin, FRMFamid and GABA (Barna et al., 2001), while in some naidid species only serotonin (Hessling et al., 1999) have been identified.

Because of technical difficulties, no GABA-IR fibre network has been identified inside the alimentary canal, except for in the pharynx. Nevertheless, neural networks with GABA-IR fibres could be distributed in the whole alimentary canal. Ultrastructural observation of the midgut showed neural profiles of variable in diameter under the epithelium. The origin of these structures, however, remains unknown (Lindner, 1965).

The cylindrical cells found in high number in the oesophagus could be heterogeneous in function. Those cells having centrally directed processes could be primary sensory cells, sensitive to mechanical and/or chemical irritations and modulating the muscle contraction in certain parts of the oesophagus. A putative role of GABA in sensory information processing of environmental stimuli has been suggested in earthworms (Spörhase-Eichmann et al., 1997; Solt et al. 2001) and echinoderms (Newman & Thorndyke, 1994). In part, the labelled cells can be entero-endocrine cells that secrete GABA as hormone and modulate the activity of some gland cells in gut epithelium. GABA-IR non-neuronal cells have been observed in the epithelium of rat alimentary canal, suggesting they are entero-endocrine cells (Jessen et al., 1988). There is growing evidence that GABA produced by non-neural cells is involved in hormonal regulation as well (Erdö & Wolff, 1990).

In conclusion, this study demonstrates not only the presence of GABA as an inhibitory transmitter in certain neural structures of *L. hoffmeisteri*, but also it presented detailed three-dimensional organization of all labelled neurons and neural processes in the whole CNS and in some parts of PNS. Our results show that GABA acts as a common neurotransmitter in certain sensory and motor neurons and interneurons. Furthermore, the basic pattern of labelled structures is the same in the tubificid worm, *L. hoffmeisteri* and in the earthworms *Lumbricus terrestris* and *Eisenia fetida*. Since the organization of the nervous system is phylogenetically conservative (Lacalli, 1984), studies such as the one reported here could contribute not only to knowledge of the functional organization of the nervous system of oligochaetes, but also to an understanding of phylogenetic relationships within the Oligochaeta.

## References

Al-Yousuf, S. A., 1992. Neuropeptides and immunocytochemistry in annelids. International Review of Cytology – A Survey of Cell Biology 133: 231–308.

Atwood, H. L. & I. Kwan, 1979. Reciprocal axo-axonal synapses between excitatory and inhibitory neurons in crustaceans. Brain Research 174: 324–328.

Barna, J., M. Csoknya, Lázár Zs., L. Barthó, J. Hámori & K. Elekes, 2001. Distribution of some putative neurotransmitters in the stomatogastric nervous system of the earthworm, *Eisenia fetida* (Oligochaeta, Annelida). Journal of Neurocytology 30: 313–325.

Bieger, D. & O. Hornykiewicz, 1972. Dopamine in the earthworm, *Lumbricus terrestris*, enhancement of rhythmic contractile activity. Neuropharmacology 11: 745–748.

Bullock, T. H. & G. A. Horridge, 1965. Structure and Function in the Nervous System of Invertebrates, Vol. 1. W. H. Freeman and Company, San Francisco, 798 pp.

Cournil, I. R. C., P. Meyrand & M. Moulins, 1990. Identification of all GABA-immunoreactive neurons projecting to the lobster stomatogastric ganglion. Journal of Neurocytology 19: 478–493.

Curry, W. J., I. Fairweather, C. F. Johnston, D. W. Halton & K. D. Buchanan, 1989. Immunocytochemical demonstration of vertebrate neuropeptides in the earthworm *Lumbricus terrestris* (Annelida, Oligochaeta). Cell and Tissue Research 257: 577–586.

Dorsett, D. A., 1978. Organization of the nerve cord. In Mill, P. J. (ed.), Physiology of Annelids. Academic Press, London: 115–159.

Drewes, C. D. & M. J. Zoran, 1989. Neurobehavioral specializations for respiratory movements and rapid escape from predators in posterior segments of the tubificid *Branchiura sowerbyi*. Hydrobiologia 180: 65–71.

Ehinger, B. & H. E. Myhrberg, 1971. Neuronal localization of dopamine, noradrenaline and 5-hydroxytryptamine in the central and peripheral nervous system of *Lumbricus terrestris* (L). Histochemie 28: 265–748.

Erdö, S. L. & J. R. Wolff, 1990. $\gamma$-Aminobutyric acid outside the mammalian brain. Short review. Journal of Neurochemistry 54: 363–372.

Fourness, J. B., D. C. Trussel, S. Pompolo, J. C. Bornstein, B. E. Maley & J. Storm-Mathisen, 1989. Shapes and projections of neurons with immunoreactivity for gamma-aminobutyric acid in the guinea-pig small intestine. Cell and Tissue Research 256: 293–301.

Fuji, K., N. Ohta, T. Sasaki, Y. Sekizawa, C. Yamada & H. Kobayashi, 1989. Immunoreactive FMRFamide in the nervous system of the earthworm, *Eisenia foetida*. Zoological Science 65: 951–961.

Gardner, C. R., 1981. Effects of active amino acids and monoamines on the neuromuscular transmission of *Lumbricus terrestris*. Comparative Biochemistry and Physiology 68C: 85–90.

Gardner, C. R. & C. H. Cashin, 1975. Some aspects of monoamine function in the earthworm (*Lumbricus terrestris*). Neuropharmacology 14: 493–500.

Günther, J., 1971. Mikroanatomie des Bauchmarks von *Lumbricus terrestris* L. Zeitschrift für Morphologie (und Ökologie) der Tiere 70: 141–182.

Günther, J., 1972. Giant motor neurons in the earthworm. Comparative Biochemistry and Physiology 42A: 967–973.

Günther, J. & F. W. Schürmann, 1973. Zur Feinstruktur des dorsalen Riesenfasersystems im Bauchmark des Regenwurms. II. Synaptische Beziehungen der proximalen Riesenfaserkollateralen. Zeitschrift für Zellforschung und mikroskopische Anatomie 139: 369–396.

Günther, J. & J. B. Walther, 1971. Funktionelle Anatomie der dorsalen Riesenfaser-Systeme von *Lumbricus terrestris* L. (Annelida, Oligochaeta). Zeitschrift für Morphologie (und Ökologie) der Tiere 70: 253–280.

Hessling, R., M. C. Müller & W. Westheide, 1999. CLSM analysis of serotonin-immunoreactive neurons in the central nervous system of *Nais variabilis*, *Slavina appendiculata* and *Stylaria lacustris* (Oligochaeta: Naididae). Hydrobiologia 406: 223–233.

Jessen, K. R., J. M. Hills & M. J. Saffrey, 1986. Immunohistochemical demonstration of GABAergic neurons in the enteric nervous system. Journal of Neuroscience 6: 1628–1634.

Jessen, K. R., J. M. Hills & A. R. Limbrick, 1988. GABA imunoreactivity and $^{3}$H-GABA uptake in mucosal epithelial cells of the rat stomach. Gut 29: 1549–1556.

Jamieson, B. G. M., 1981. The Ultrastructure of the Oligochaeta. Chapter 9. The Nervous System: General Ultrastructure. Academic Press, London, pp 211–244.

Kitamura, Y., Y. Naganoma, H. Horita, N. Tsuji, R. Shimizu, H. Ogawa & K. Oka, 2001. Visualization of nitric oxide production in the earthworm ventral nerve cord. Neuroscience Research 40: 175–181.

Lacalli, T. C., 1984. Structure and organization of the nervous system in trochophore larva of *Spirobranchus*. Philosophical Transactions of the Royal Society of London 306B: 79–135.

Lindner, E., 1965. Ferritin und Hämoglobin im Chloragog von Lumbriciden Oligochaeta. Zeitschrift für Zellforschung und mikroskopische Anatomie 66: 891–913.

Lubics, A., D. Reglődi, S. Slezák, M. Szelier & I. Lengvári, 1997. Co-localization of serotonin and FMRFamide-like immunoreactivities in the central nervous system of the earthworm *Eisenia fetida*. Acta Histochemica 99: 459–467.

Mill, P. J., 1982. Recent development in earthworm neurobiology. Comparative Biochemistry and Physiology 73A: 641–661.

Millot, N., 1943. The visceral nervous system of the earthworm: I. Nerves controlling the tone of the alimentary canal. Proceeding of the Royal Society of London 131B: 271–295.

Molnár, L., A. Zsombok & E. Pollák, 2000. Fine-fibred polysegmental GABA-immunoreactive interneuronal tracts in the central nervous system of *Lumbricus terrestris*. Neurobiology 8: 367–368.

Molnár, L., A. Zsombok, Zs. Solt & E. Pollák, 2001. Functional anatomy of the NADPH-diaphorase positive neural structures in the central nervous system of lumbricid earthworms. Proceedings of 28th Göttingen Neurobiology Conference and the 4th Meeting of the German Neuroscience Society, George Thieme Verlag, Stuttgart, Vol. II: 477.

Newman, S. J. & M. C. Thorndyke, 1994. Localization of gamma aminobutiric acid (GABA)-like immunoreactivity in the echinoderm *Asterias rubens*. Cell and Tissue Research 278: 177–185.

Ogawa, F., 1939. The nervous system of earthworm (*Pheretima communissima*) in different ages. Science Report of the Tohoku Imperial University 13: 395–488.

Pollák, E., L. Molnár, Zs. Hunyadi & Zs. Herbert, 2003. Histochemically identified extrinsic and intrinsic neural structures in the hindgut of the isopod, *Porcellio scaber*. In Sfenthourakis S., et al. (eds), The Biology of Terrestrial Isopods, V, Crustaceana Monographs 2. Brill, Leiden-Boston: 347–367.

Reglődi, D., S. Slezák, A. Lubics, M. Szelier, K. Elekes & I. Lengvári, 1997. Distribution of FMRFamide-like immunoreactivity in the nervous of *Lumbricus terrestris*. Cell and Tissue Research 288: 575–582.

Rude, S., 1966. Monoamine-containing neurons in the nerve cord and the body wall of *Lumbricus terrestris* L. Journal of Comparative Neurology 128: 397–412.

Sepherd, G. M., 1994. Neurobiology, 3rd edn. Oxford University Press, New York-Oxford.

Solt, Zs., A. Zsombok, E. Pollák & L. Molnár, 2001. Transmitter specific sensory pathways of lumbricid earthworms: a histochemical study with functional correlations. Proceedings of 28th Göttingen Neurobiology Conference and the 4th Meeting of the German Neuroscience Society, George Thieme Verlag, Stuttgart, Vol. II: 478.

Spörhase-Eichmann, U., H. Grass & F. -W. Schürmann, 1987a. Patterns of serotonin-immunoreactive neurons in the central nervous system of the earthworm *Lumbricus terrestris* L. I.

Ganglia of the ventral nerve cord. Cell and Tissue Research 249: 601–614.

Spörhase-Eichmann, U., H. Grass & F. -W. Schürmann, 1987b. Patterns of serotonin-immunoreactive neurons in the central nervous system of the earthworm *Lumbricus terrestris* L. II. Rostral and caudal ganglia. Cell and Tissue Research 249: 625–632.

Spörhase-Eichmann, U., M. Winkler & F. -W. Schürmann, 1997. GABAergic, dopaminergic and peptidergic sensory cells in the epidermis of earthworms. Society for Neuroscience Abstracts 23: 38813.

Spörhase-Eichmann, U., M. Winkler & F. -W. Schürmann, 1998. Dopaminergic sensory cells in the epidermis of the earthworm. Naturwissenschaften 85: 547–550.

Stephenson, J., 1930. The Oligochaeta. Clarendon Press, Oxford.

Telkes, I., M. Csoknya, P. Buzás, R. Gábriel, J. Hámori & K. Elekes, 1996. GABA-immunoreactive neurons in the central and peripheral nervous system of the earthworm, *Lumbricus terrestris* (Oligochaeta, Annelida). Cell and Tissue Research 285: 463–475.

Wilmer, P., 1990. Invertebrate Relationships. Patterns in Animal Evolution. Cambridge University Press, Cambridge, 400 pp.

Zoran, M. J. & C. D. Drewes, 1987. Rapid escape reflexes in aquatic oligochaetes: variations in design and function of evolutionarily conserved fiber system. Journal of Comparative Physiology A 161: 729–738.

Zoran, M. J. & C. D. Drewes, 1988. The rapid tail withdrawal reflex of the tubificid worm, *Branchiura sowerbyi*. Journal of Experimental Biology 137: 487–500.

Zoran, M. J., C. D. Drewes, C. R. Fourtner & A. J. Siegel, 1988. The lateral giant fibers of the tubificid worm, *Branchiura sowerbyi*: structural and functional asymmetry in a paired interneuronal system. Journal of Comparative Neurology 275: 76–86.

Zsombok, A. & L. Molnár, 2001. Pattern of NADPH-diaphorase containing structures in the subintestinal ganglia of *Lumbricus terrestris*. Neurobiology 9: 67–69.

Zsombok, A., Zs. Solt & L. Molnár, 1999. Putative NOS expressing neurons in lumbricid oligochaetes: NADPH-diaphorase histochemistry. Proceedings of the 27th Göttingen Neurobiology Conference and the 3rd Meeting of the German Neuroscience Society, George Thieme Verlag, Stuttgart, Vol. II: 613.

Hydrobiologia (2006) 564:45–50
P.F.M. Verdonschot, H. Wang, A. Pinder & R. Nijboer (eds), Aquatic Oligochaete Biology IX
DOI 10.1007/s10750-005-1706-z



# Identification and pattern of primary sensory cells in the body wall epithelium of the tubificid worm, *Limnodrilus hoffmeisteri*

László Molnár[1,2,*], Gábor Kiszler[1] & Edit Pollák[1,2]
[1]*Department of General Zoology and Neurobiology, Research Group of Comparative Anatomy, University of Pécs, Ifjúság u. 6, H-7624 Pécs, Hungary*
[2]*Adaptation Biology Research Group of the Hungarian Academy of Sciences and University of Pécs, Ifjúság u. 6, H-7624 Pécs, Hungary*
*(*Author for correspondence: E-mail: molnar@ttk.pte.hu)*

*Key words:* primary sensory cells, sensory pathways, scanning electron microscopy, NADH-diaphorase histochemistry, histochemical markers

## Abstract

This study compares the effectiveness of two examination methods suitable for morphological observations, scanning electron microscopy (SEM) and light microscopic NADH-diaphorase (NADP-d) histochemistry, to identify location and distribution pattern of primary sensory cells situated in the body wall epithelium of *Limnodrilus hofmeisteri*. SEM observations revealed that grouped sensory cells forming two distinct types of sensillas (or sense organs) are scattered over the body surface. The easily identifiable first type (ciliated sensilla) contains numerous penetrative ciliate sensory cells while no ciliate sensory cells were seen in the second type (sensory bud). NADH-d staining proved to be a suitable method to identify both structures further staining solitary sensory cells, and certain sensory fibres that could not be visualised by SEM preparations. Our results show that NADH-d staining, as a histochemical marker of primary sensory cells, is a suitable and effective method to reveal their morphology and distribution pattern in whole mount preparations in the smallest detail, suggesting that the application of this relatively simple and cheap method may contribute to understand the basic organisation and evolution of the peripheral sensory structures of oligochaetes.

## Introduction

The body wall of oligochaetes is rich in primary sensory cells, and free nerve endings of centrally located sensory cells that mediate external stimuli to the central nervous system (CNS) and modify the activity of the fixed motor pattern of a reflex centre, namely a segmental ganglion, resulting an appropriate reaction to the changing environment (Mill, 1978, 1982). Applying histological and electron microscopic observations five distinct types of primary sensory cells have been identified in earthworms (for review see Mill, 1978, 1982; Jamieson, 1981). These cells are phaosomal photoreceptors (Hess, 1925; Röhlich et al., 1970), penetrative uni- and multiciliate sensory cells (Aros et al., 1971b; Knapp & Mill, 1971), non-penetrative multiciliate sensory cells (Aros et al., 1971a) and basal ciliated sensory cells (Myhrberg, 1979).

Photoreceptors and non-penetrative multiciliate sensory cells are solitary cells and are scattered over the epidermis. The prostomium and a few anteriormost and posteriormost segments are especially provided with in the former structures (Hess, 1925; Mill, 1978; Jamieson, 1981) while the latter are more numerous in the prostomium and oral cavity (Aros et al., 1971a). Penetrative uni- and multiciliate sensory cells and basal ciliated sensory cells accompanied by some supporting cells are concentrated in sensillas/sense organs (Aros et al., 1971b; Knapp & Mill, 1971).

By the aid of methylene blue staining Langdon (1895) determined and described first the distribution pattern of sensillas on various body regions of *Lumbricus agricola*: On the prostomium large sensillas with seemingly random distribution were found, while in the body wall of each segments most sensillas formed three belt-like structures encircled the animal, and only a few small sensillas were evenly distributed on the segment surface. Later on, SEM observations supported her data (Aros et al., 1978; Moment & Johnson, 1979), suggesting that the distribution pattern of sensillas is identical in lumbricid worms.

In contrast to earthworms, less is known about primary sensory cells of aquatic oligochaetes although some reflexes, like those involved in escape and withdrawal, in which primary sensory cells form the afferent or sensory pathway, have been investigated in detail (Zoran & Drewes, 1987, 1988; Drewes & Zoran, 1989). Since neither the anatomy nor the distribution pattern of primary sensory cells is known in tubificid worms this study focuses on possibilities of their identification and characterization in whole mount preparations with SEM and a relatively simple and cheap enzyme histochemical (NADH-diaphorase staining) method.

## Materials and methods

Sexually mature specimens of *Limnodrilus hoffmeisteri* Claparede 1868, selected from field collected samples of tubificid worms based on the species-specific anatomical characteristics (Stephenson, 1930) were used for microscopical observations. Before fixation all animals were anaesthetised by cooling on ice for 5–10 min and the body cut into 7–10 mm long pieces using a razor blade.

### *Scanning electron microscopy*

For scanning electron microscopy (SEM) the body pieces were fixed in freshly prepared 2.5% glutaraldehyde in 0.1 M phosphate buffer (PB) (pH 7.4) at room temperature for 2 h. After a thorough washing in PB, samples were postfixed in 1% osmium tetroxide at 4 °C for 2 h then washed in 50% ethanol and dehydrated in a graded ethanol series. Specimens were then exposed to amyl-acetate overnight. The amyl-acetate was removed by the critical-point drying method, using carbon dioxide and the specimens mounted on copper stubs with silver conducting tape. They were then coated with a thin layer of gold used an evaporative-coating method. SEM examinations were performed using a JEOL electron microscope.

### *NADH-diaphorase (NADH-d) histochemistry*

To identify sensory cells of the body wall, dissected body pieces were fixed with freshly prepared 4% paraformaldehyde (PFA) in PB (0.1 M, pH 7.4) for 3 h at 4 °C. After fixation, a thorough washing (6×10 min with continuous stirring in PB) was applied and the samples were treated with 1% Triton-X 100 in 0.1 M PB for 2 days at 4 °C. Thereafter they were incubated in a freshly prepared mixture of 0.25 mg/ml nitro blue tetrazolium (NBT) and 1 mg/ml $\beta$-NADH (dissolved in 0.1 M PB, pH 7.4) in the dark at 37 °C. Staining was controlled under a dissecting microscope and interrupted by washing in PB. In control experiments either NADH or NBT was omitted from incubating solution, but no specific staining was seen in the samples. The NBT-formazan stained samples were post-fixed in PFA overnight, and then washed in PB. Stained samples were processed in two different ways. Whole-mounts were cleared with buffered glycerol (50% PBS and 50% glycerol) for 2–3 days then mounted between cover-slips in pure glycerol for microscopy. Some preparations were dehydrated in graded ethanol series and embedded in paraffin as usual. Ten $\mu$m thick serial sections were made with a Reichert microtome, attached to gelatine-chrome-alum covered slides and observed without counter staining. In order to determine the exact anatomical positions of NADH-d positive structures in the body wall both whole-mount preparations and paraffin sections were investigated by a Nikon microscope equipped with UFX camera.

## Results

*Scanning electron microscopy*

Sensillas bearing apical projections (cilia) are large and prominent on the prostomium showing seemingly random distribution (Fig. 1A, B). On the anteriormost and posteriormost segments high number of smaller ciliated sensillas can be observed. A part of them are located in the rows of chaetae encircling the segments while the others are scattered over the surface (Fig. 1F).

In midbody segments two distinct types of sensillas have been observed: ciliated and non-ciliated sensillas (sensory bud). The easily identifiable small ciliated sensillas are located in the rows of chaetae and around chaetae (Fig. 1C–D) while the more numerous sensory buds can be seen in the posterior part of each segment (Fig. 1E), although their identification is rather complicated.

*NADH-diaphorase histochemistry*

Staining with $\beta$-NADH gives Golgi-like staining in primary sensory cells thus anatomical position of the stained cells can be observed in whole-mounts showing their pattern in finest detail (Fig. 2). Someone unexplained inconsistencies were revealed with this technique. For example, intensity of staining was heterogeneous and variable signal-to-noise ratio occurred in samples even when they were derived from the same incubation. However, consistent, distinct groups of sensory cells could always be identified with this method.

A high number of labelled structures was identified in the prostomium and in the anteriormost and posteriormost segments. Their distribution pattern (not shown) was identical with that was found with SEM observations. In contrast to SEM results enzyme histochemical staining revealed many more sensillas as well as solitary sensory cells and fibre-like structures in other body segments. As shown in Figure 2A a few labelled sensillas can be observed in the rows of chaetae and numerous sensillas are situated in the posterior part of each segment. Location and number of labelled cells are different from sensilla to sensilla: between 25 and 60 labelled cells were counted per sensilla (Fig. 2B, C). Solitary sensory cells are randomly distributed in segments, though some of them show constant location especially those close to chaetae (Fig. 2D).

## Discussion

In this work, applying SEM and enzyme histochemical method, we described the pattern of solitary and grouped sensory cells (sensillas) in the body wall of *L. hoffmeisteri*. SEM is widely used to identify sensillas in oligochaete species (Aros et al., 1978; Moment & Johnson, 1979), but this is mainly suitable for observation of ciliated primary sensory cells. Those structures having no penetrating cilia across the cuticle (phaosomal photoreceptors, non-penetrative multiciliate sensory cells and basal ciliated sensory cells) could not be, or hardly be, observed. Therefore, a more suitable method is needed for observation of those structures (either solitary or grouped sensory cells or fibres) situated deeper in the body wall.

We have established that the two different methods revealed the same pattern of sensillas. Therefore, we propose that both the expensive SEM observation and the relatively cheap NADH-d histochemistry are suitable for identifying epidermal sensory structures of oligochaetes. Moreover, the latter method reveals the anatomical location of some non-superficial structures like non-ciliated solitary sensory cells as well as certain fibre-like structures. The identification of these fibres is uncertain at this stage of our experiments. They could be central processes of NADH-d stained primary sensory cells that enter the subepidermal plexus and run to the central nervous system but the subepidermal plexus contains efferent nerve fibres as well (Mill, 1978, 1982).

NADH-d staining is a widely used method to identify peripheral neural structures (Gabella, 1969) since they are rich in a NADH-requiring enzyme believed to be lypoamid-dehydrogenase concentrated in cell bodies, mainly in mitochondria (Caballero-Bleda et al., 1991). It is likely that primary sensory cells of the body wall are also rich in NADH-requiring enzyme and can therefore also be labelled in a Golgi-like manner.

There are a few methods in neurobiology that are used to empirically stain certain sensory

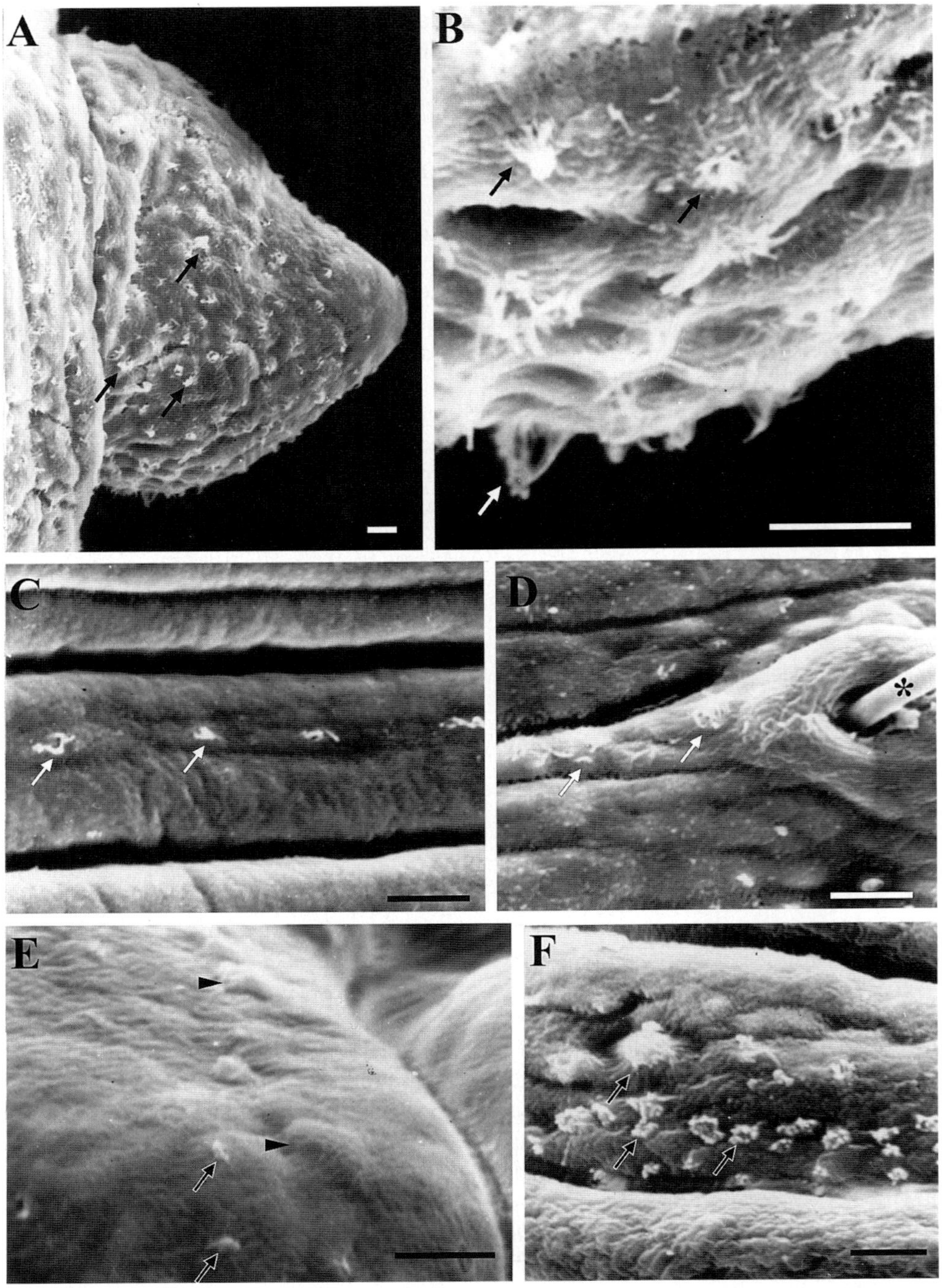


*Figure 1.* Sensillas of the prostomium (A, B), midbody segments (C–E) and a posteriormost segment (F). Arrows: ciliated sensillas, arrowheads: sensory buds. Scale bars 10 $\mu$m.

structures. For example S-100 protein is a selective marker for sensory hair cells of the lateral line system in teleosts (Abbate et al., 2002) and fluoride-resistant acid phosphatase histochemistry is suitable to label a distinct set of primary sensory neurons in the dorsal root ganglia of rats and mice (Vincent, 1992). Although, at this stage of our experiments, the histochemical processes behind

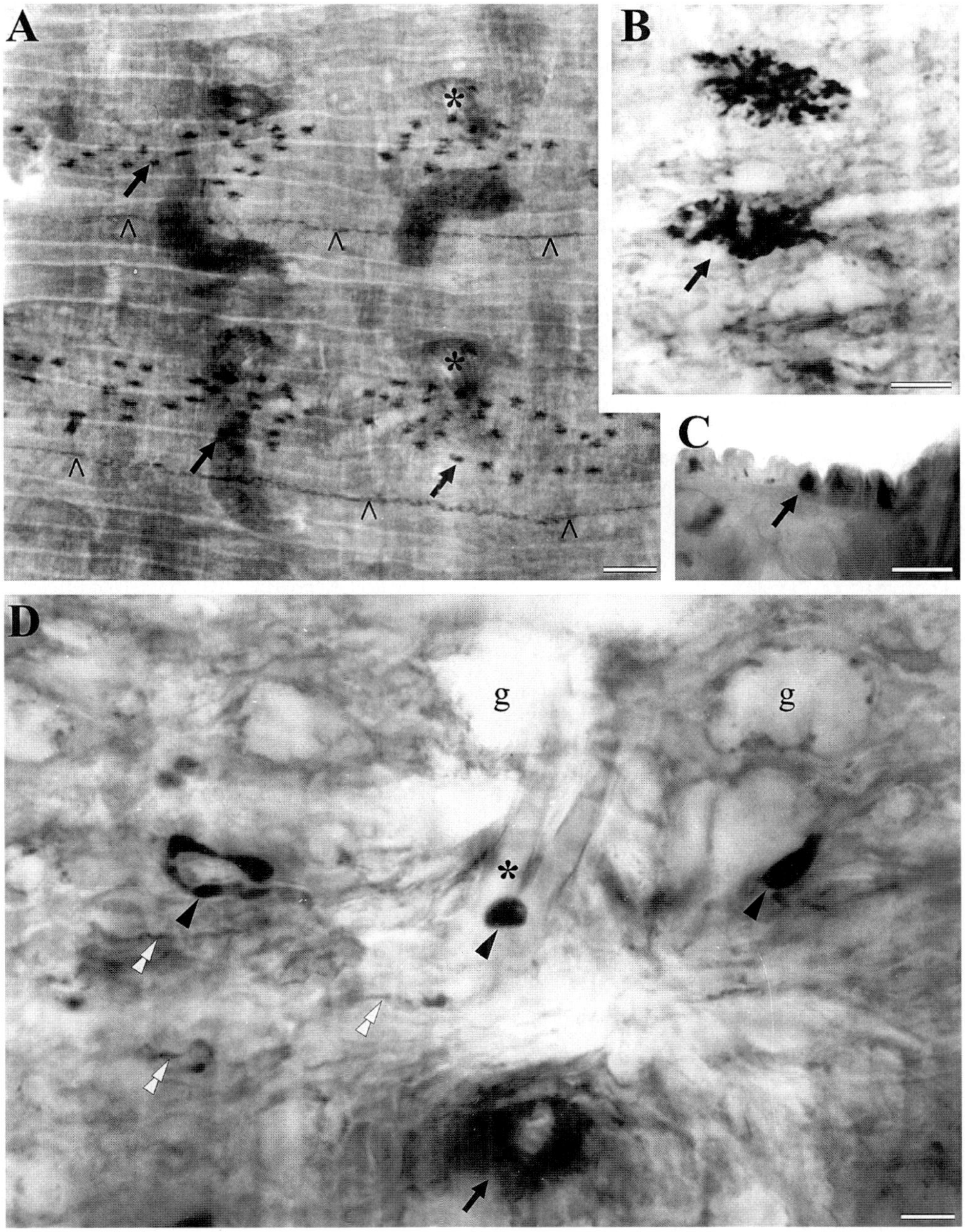


*Figure 2.* NADH-diaphorase stained sensillas (A–C) and solitary sensory cells (D) in midbody segments. Arrows: sensillas, arrowheads: solitary sensory cells, double arrowheads: stained fibres of the subepidermal plexus, open arrowheads: segmental groves, asterisk: chaeta. Scale bars: A: 100 $\mu$m, B: 10 $\mu$m, C: 5 $\mu$m, D: 10 $\mu$m.

NADH-staining of primary sensory cells is not clear, its use in controlled experiments can be a relatively cheap and quick method to identify primary sensory cells in tubificid worms and in other aquatic oligochaetes.

## References

Abbate, F., S. Catania, A. Germana, T. Gonzáles, B. Diaz-Esnal, G. Germana & J. A. Vega, 2002. S-100 protein is a selective marker for sensory hair cells of the lateral line system in teleosts. Neuroscience Letters 329: 133–136.

Aros, B., P. Röchlich & B. Vigh, 1971a. Fine structure of the peripheral sensory cells in the earthworm *Dendrobaena octaedra*. I. The solitary sensory cells. Acta Biologica Hungarica 22: 141–153.

Aros, B., P. Röhlich & B. Vigh, 1971b. Fine structure of the peripheral sensory cells in the earthworm *Dendrobaena octaedra*. II. The epidermal sensory organs. Acta Biologica Hungarica 22: 443–456.

Aros, B., I. Vigh-Teichmann, B. Vigh & J. Kovács, 1978. Scanning electron microscopy of the prostomium and anterion segments of the earthworm (*Lumbricus terrestris* L., *Eisenia foetida* Sav.). Zeitschrift für Mikroskopisch-Anatomische Forschung 92: 753–769.

Caballero-Bleda, M., B. Fernandez & L. Puelles, 1991. Comparative mapping of acetylcholinesterase and reduced nicotinamide adenine dinucleotide diaphorase in the rabbit dorsal thalamus. Acta Anatomica 140: 224–234.

Drewes, C. D. & M. J. Zoran, 1989. Neurobehavioral specializations for respiratory movements and rapid escape from predators in posterior segments of the tubificid *Branchiura sowerbyi*. Hydrobiologia 180: 65–71.

Gabella, G., 1969. Detection of nerve cells by a histochemical technique. Experientia 25: 218–225.

Hess, W. N., 1925. Photoreceptors of *Lumbricus terrestris* with special reference to their distribution, structure, and function. Journal of Morphology 41: 63–95.

Jamieson, B. G. M., 1981. The ultrastructure of the oligochaeta. Chapter 10. Ultrastructure of sensory cells. Academic Press, London, 245–267.

Knapp, M. F. & P. J. Mill, 1971. The fine structure of ciliated sensory cells in the epidermis of the earthworm *Lumbricus terrestris*. Tissue & Cell 3: 623–636.

Langdon, F. E., 1895. The sense-organs of *Lumbricus agricola* Hoffm. Journal of Morphology 11: 193–234.

Mill, P. J., 1978. Sense organs and sensory pathways. In Mill, P. J. (ed.), Physiology of Annelids. Academic Press, New York: 63–114.

Mill, P. J., 1982. Recent developments in earthworm neurobiology. Comparative Biochemistry and Physiology 73A: 641–661.

Moment, G. B. & J. E. Johnson, 1979. The structure and distribution of external sense organs in newly hatched and mature earthworms. Journal of Morphology 159: 1–6.

Myhrberg, H. E., 1979. Fine structural analysis of the basal epidermal receptor cells in the earthworm (*Lumbricus terrestris*). Cell and Tissue Research 203: 257–266.

Röhlich, P., B. Aros & S. Virágh, 1970. Fine structure of photoreceptor cells in the earthworm, *Lumbricus terrestris*. Zeitschrift für Zellforschung und mikroskopische Anatomie 104: 345–357.

Stephenson, J., 1930. The Oligochaeta. Clarendon Press, Oxford.

Vincent, S. R., 1992. Histochemistry of endogenous enzymes. In Bolam, J. P. (ed.), Experimental Neuroanatomy. A Practical Approach. IRL Press, Oxford-New York-Tokyo, 153–171.

Zoran, M. J. & C. D. Drewes, 1987. Rapid escape reflexes in aquatic oligochaetes: variations in design and function of evolutionarily conserved fiber system. Journal of Comparative Physiology A 161: 729–738.

Zoran, M. J. & C. D. Drewes, 1988. The rapid tail withdrawal reflex of the tubificid worm, *Branchiura sowerbyi*. Journal of Experimental Biology 137: 487–500.

Hydrobiologia (2006) 564:51–58
P.F.M. Verdonschot, H. Wang, A. Pinder & R. Nijboer (eds), Aquatic Oligochaete Biology IX
DOI 10.1007/s10750-005-1707-y



# Divergence of the $F_1$-ATP synthase complex in the ice worm, *Mesenchytraeus solifugus* (Annelida, Clitellata, Enchytraeidae)

Kristi A. Hohenstein & Daniel H. Shain*
*Biology Department, Rutgers The State University of New Jersey, 315 Penn Street, Camden, NJ 08102, USA*
*(*Author for correspondence: Tel.: +1-856-225-6144; Fax: +1-856-225-6312; E-mail: dshain@camden.rutgers.edu)*

*Key words:* psychrophile, ATP, oligochaete, adaptation, evolution, amino acid substitution

**Abstract**

The $F_1$-ATP synthase complex constitutes the catalytic component of $F_1F_0$-ATP synthase, the primary ATP synthetic enzyme in the cell. Previous studies indicate that the glacier ice worm, *Mesenchytraeus solifugus*, maintains unusually high ATP levels that continue to rise as temperatures decline, suggesting that molecular changes within ice worm $F_1$-ATP synthase subunits may contribute to this energetic anomaly. In this report, we compared ice worm $F_1$-ATP synthase subunits ($\alpha$, $\beta$, $\gamma$) with homologues across metazoan phyla (arthropod, chordate, nematode) and among a group of clitellate annelids (*Enchytraeus albidus, Enchytraeus buchholzi, Lumbriculus variegatus, Theromyzon tessulatum*). Amino acid alignments indicated that ice worm $F_1$-ATP $\alpha$ and $F_1$-ATP $\beta$ subunits share strong sequence homology with their mesophilic counterparts, respectively, but that ATP $\gamma$ has diverged more rapidly. Moreover, $F_1$-ATP $\alpha$ and $F_1$-ATP $\beta$ displayed amino acid compositional changes consistent with trends observed in other cold adapted proteins, while $F_1$-ATP $\gamma$ diverged in unexpected directions (e.g., gains in size, charged residues). Several ice worm-specific amino acid substitutions map to positions near the $F_1$-ATP $\beta$ catalytic site while others occur near subunit contact sites.

## Introduction

The mitochondrial $F_1F_0$-ATP synthase constitutes the primary ATP (adenosine 5′ triphosphate) synthetic machinery in eukaryotic cells under aerobic conditions. The $F_1$-ATP synthase complex comprises three major subunits ($\alpha$, $\beta$, $\gamma$); the catalytic $\beta$ and regulatory $\alpha$ subunits are arranged in alternation (three subunits each per complex) forming a hydrophobic sleeve around which the $\gamma$ subunit rotates (Abrahams et al., 1994; Menz et al., 2001). The membrane-bound $F_0$ component channels $H^+$ down an electrochemical gradient which in turn propels the rotation of the $\gamma$ subunit within the catalytic $F_1$ complex, leading to ATP formation (Nichols & Ferguson, 2002; Leyva, 2003).

The fundamental requirement for ATP in all living cells requires that the $F_1F_0$-ATP synthase functions at levels sufficient to meet basic energy demands in all cell types. In organisms inhabiting extreme environments (e.g., psychrophiles, thermophiles), energy levels must be adjusted accordingly to maintain biological processes at viable levels. For instance, the rate of biochemical reactions double every ~10 °C rise in temperature (i.e., the $Q_{10}$ relationship); consequently, most organisms increase ATP production as temperatures rise in order to keep up with energy demands (e.g., Fedorow et al., 1998; English & Storey, 2000). An exception to this paradigm is observed in the ice worm, *Mesenchytraeus solifugus* (Emery, 1898), an oligochaete that survives exclusively in glacier ice/snow (Wright, 1887; Welch, 1916). Ice worms paradoxically increase ATP levels as temperatures fall in an apparent mechanism that off-sets reductions in molecular diffusion and enzyme kinetics at low temperature (Napolitano et al., 2004). Simply, elevated ATP levels appear to drive biochemical reactions forward at low physiological temperatures, possibly enhancing the ability of ice worms to maintain biological processes (e.g., growth, reproduction, metabolism) at ~0 °C.

One aspect of ice worm adaptation is likely to involve changes in membrane fluidity, a well-documented component of cold temperature adaptation (Nichols et al., 1997; Russell, 1997). Strikingly, however, ice worm ATP levels rise sharply (i.e., within minutes) when worms are transferred to sub-zero temperatures, an observation that is difficult to explain by changes in membrane composition, but rather suggests a role for components of the ice worm $F_1$-ATP synthase complex (Napolitano et al., 2004, Napolitano & Shain, 2004). An independent line of evidence also supports this notion; namely, two subunits within ice worm $F_1$-ATP synthase ($\beta$ and $\gamma$) appear to have evolved at accelerated rates from their mesophilic counterparts, as compared with a broad sampling of other ice worm genes (Farrell et al., 2004). In this study, we report the full-length sequence of the third major subunit within the ice worm $F_1$ complex, ATP synthase $\alpha$. We have also cloned cDNA fragments of $F_1$-ATP synthase subunits from three mesophilic oligochaete species (i.e., *Enchytraeus albidus, E. buchholzi, Lumbriculus variegatus*) and the hirudinean, *Theromyzon tessulatum*, in an effort to identify ice worm-specific amino acid substitutions that may contribute to their energetic adaptation to low temperature.

## Materials and methods

### *Specimens*

*Mesenchytraeus solifugus* specimens were collected from Byron Glacier, Alaska and maintained as described in Shain et al. (2000). *Enchytraeus albidus, Enchytraeus buchholzi* and *Lumbriculus variegatus* were purchased from the Bug Farm (San Rafael, CA), and although their species designation was not confirmed, their external features (e.g., size, coloration, setal morphology) were consistent with previous descriptions (e.g., Brinkhurst & Jamieson, 1971). Preserved specimens from each stock have been stored by the authors. Stocks of *E. albidus* and *E. buchholzi* were maintained at 12 °C and 22 °C, respectively, in soil provided by the supplier. *L. variegatus* was maintained at 22 °C in 0.3% Instant Ocean Salt (PetSmart). *Theromyzon tessulatum* was collected in the ponds of Golden Gate Park (San Francisco, CA) and maintained at 12 °C in 0.3% Instant Ocean Salt (PetSmart).

### *cDNA construction and PCR (polymerase chain reaction)*

Total RNA was isolated from oligochaete species (~100 specimens per species) as described (Chomczynski & Sacchi, 1987). cDNA was synthesized with a SMART cDNA Construction Kit as specified by the manufacturer (Clontech). Degenerate primer sets for each respective subunit are listed below along with annealing temperature range and expected product size: $F_1$-ATP synthase $\alpha$ – ATHGGNDAYGGNATHGCNMG, CCNGCN ACYTGYTTCATNGC (56 → 49 °C; 1062 bp); $F_1$-ATP synthase $\beta$ – ATHGCNATG GARGGNACNGA, CCNGTRAANACRTCNGCNAC (56 → 49 °C; 1065 bp); $F_1$-ATP synthase $\gamma$ – ACNMARWSNATGAARATGGT, GCNGCNS WDATDATYTC (50 → 44 °C; 765 bp). Touchdown PCR (i.e., drop of 0.2 °C annealing temperature per cycle) was conducted with NovaTaq DNA polymerase (Novagen) using the following parameters: 94 °C (15 s); annealing temperature as specified for each primer set (1 min); 72 °C (1 min) for 30–35 cycles. Ice worm cDNA ends were amplified by RACE-PCR (rapid amplification of cDNA ends by PCR); briefly, nested gene-specific primers designed from the amplified fragments above were used in conjunction with Clontech anchor primers CDSIII (3′ end) and 5′ PCR primer (5′ end) in standard PCR reactions. Gene-specific primers are listed below with annealing temperature and fragment size: $F_1$-ATP synthase $\alpha$5′ end – CAATGTCACCCTCTTTAATG (outside), GTT GTCAGGCTCCAAGTTC (inside) (56 °C; 500 bp); $F_1$-ATP synthase $\alpha$3′ end – ATCATTGA CTGCTCTGCCA (outside), TGTCTGCATACATCCCAAC (inside) (56 °C; 800 bp); $F_1$ ATP synthase $\beta$5′ end – ACGCTCATCTCGAT GAATG (outside), GTTCTCCAATGACGTTAATG (inside) (56 °C; 500 bp); $F_1$-ATP synthase $\beta$3′ end – GGNCARATGAAYGAYCCNCC

(56 → 49 °C; 700 bp); ice worm $F_1$-ATP synthase $\gamma$ was isolated from an ice worm cDNA library (Farrell, 2002). All oligonucleotides were synthesized commercially (Sigma-Genosys).

*DNA sequencing and analysis*

Amplified cDNA fragments were sequenced directly at Northwoods DNA, Inc (Becida, MN). DNA sequence chromatograms were analyzed by Chromas 2.21 (Technelysium Pty. Ltd.) and GenBank BLAST (Basic Local Alignment Search Tool; Altschul et al., 1990). Amino acid alignments were constructed with Clustal X software (Chenna et al., 2003).

*Protein structural analysis*

The Protein Database (Berman et al., 2000) was employed to locate crystalized protein structure templates for the construction of predicted candidate protein models using Swiss Model in the ExPASy server (Gasteiger et al., 2003). Ice worm-specific amino acid changes and predicted tertiary structures were manipulated in Swiss-Pdb Viewer (Kaplan & Littlejohn, 2001) and RasMol (Sayle & Milner-White, 1995) to highlight amino acid changes within alignments. Ligand binding sites were determined using Ligand Protein Contacts software (LPC; Sobolev et al., 1999).

*Amino acid composition analysis*

Positions that contained an identical amino acid in both ice worm and a mesophilic counterpart were excluded from the comparison. In a variable position (i.e., different amino acids present among mesophilic counterparts, all of which differed from the ice worm sequence), the most abundant amino acid was chosen; in the absence of an abundant residue, the *D. melanogaster* sequence served as default in 'animal' comparisons and *E. albidus* was default in 'annelid' comparisons.

## Results

Full length $F_1$-ATP synthase subunits were cloned from *Mesenchytraeus solifugus*, and cDNA fragments were isolated from two enchytraeid species (*Enchytraeus albidus, E. buchholzi*), one lumbiculid (*Lumbriculus variegatus*) and the glossiphoniid leech, *Theromyzon tessulatum*, by degenerate PCR. Lack of a mesophilic *Mesenchytraeus* source, and current uncertainty about the phylogenetic position of *M. solifugus* among enchytraeidae precluded the possibility of obtaining sequence data from a terrestrial, sister taxa most closely related to *M. solifugus*. Nonetheless, two commercially available enchytraeids (likely *E. albidus* and *E. buchholzi*) provided representative mesophilic species in the same family. Primer sets for degenerate PCR were designed from amino acid alignments between chordate (e.g., *Homo sapiens*), arthropod (*Drosophila melanogaster*) and nematode (*Ceanorhabditis elegans*) homologues for each respective subunit. Not all templates amplified with some primer sets, presumably the result of subtle nucleotide differences in target sequences; consequently, full length cDNAs were not isolated for all species.

Representative amino acid alignments of $F_1$-ATP synthase homologues are presented in Figure 1. ATP-$\alpha$ and ATP-$\beta$ homologues displayed striking sequence conservation across all species compared (e.g., the most distant sequences were ~80% identical). In contrast, ATP-$\gamma$ homologues were clearly more divergent, displaying between ~50–75% sequence identity (Fig. 1). The vertical alignment of amino acid residues among all homologues permitted the identification of positions that were selectively modified in ice worms, thereby identifying potential targets of structural/functional change. Based on the criteria that all mesophilic residues within a given position contained the identical amino acid or a conservative change with respect to charge or polarity (e.g., R → K, T → S), ice worm-specific substitutions were most abundant in ATP-$\beta$ (8), followed by ATP-$\gamma$ (6) and ATP-$\alpha$ (3).

The linear amino acid sequences of ice worm $F_1$-ATP synthase homologues were modeled in an effort to predict potential structure/function effects of ice worm-specific amino acid substitutions (Fig. 2). Contact sites between subunits appeared unchanged between ice worms and their mesophilic counterparts. However, two ice worm-specific substitutions at the base of the ATP-$\beta$ DELSEED loop (R → K461, V → T423;

$F_1$-ATP synthase β

```
                                     150
H.sapiens      A..K...........M............IKT.Q.........E.M.........V.......L...
D.melanog      Y..R....A...................IDTDKT.A......E.VQ........V.......L...
C.elegans      D..K...........M............IAS.N..A......E.V.........V.......L...
M.solifugus    SPITIPVGPETLGRIINVIGEPIDERGPLNAKAFAPIHAEAPAFIEMSVEQEILETGIKVVDMLAP
E.albidus      ............................V....T........E...................L...
E.buchholzi    ............................V...HT........E...................L...
L.variegatus   T..R........................V...H..E...D..E.VD....................
T.tessulatum   A..R.....G..................V.S.N..E...D..E.V.....................
```

$F_1$-ATP synthase γ

```
                                         150
B.taurus       KEV...GV...I..I.H.THSDQF.VT.KEV.R...T.G..SVIA..L.NS.....E.S.IF.R.R
D.melanog      A...VFCV.....AI.S...G.NI..VANEV.RL..T.L..SKIAN.V.QT..DYTE...VY.R..
C.elegans      KEIRVV.......AG.Q...ANSI.LSGNE..RA..S....S.AAKA..DS..D.ET.T.LF.R..
M.solifugus    TNTKIIAIGDKSRSMLARLYPKDMLMHFVDIGKRPPQFADAAIVTLEILKCGYEFDVGKIYHNIFK
E.albidus      .SF.............G..FKN...........K..L.E..SVIAQ....S..QY...QL.....R
E.buchholzi    ..L.F...........G..FKN...........K..L.E..STIA.....S..QY...QL.....R
T.tessulatum   .SV.F......A.G..G.FFGQ.LI...M....K..V.D..SKIA..L..L...Y.Y.QL.F.V.R
```

*Figure 1.* Amino acid alignment of $F_1$-ATP synthase homologues. Representative blocks are shown for $F_1$-ATP synthase $\beta$ and $\gamma$ ($\alpha$ not shown). The ice worm homologue (*Mesenchytraeus solifugus*) is shown in its entirety while differences in other homologues are indicated. Shaded residues indicate ice worm-specific amino acid substitutions (E → A171 in $F_1$-ATP $\beta$; S → A163, A → T169 in $F_1$-ATP $\gamma$). Ice worm-specific amino acid substitutions not shown are: N → I156, D → N265, A → E521 in $F_1$-ATP $\alpha$; A → K55, A → S58, I → V62, V → I63, V → T423, R → K461, H → S476 in $F_1$-ATP $\beta$; and S → A37, N → S40, L → T93, L → I233 in $F_1$-ATP $\gamma$. GenBank accession numbers for annelid $F_1$-ATP synthase $\alpha$ subunits are: *E. albidus* (AY838445); *E. buchholzi* (AY838448); *L. variegatus* (AY83849); *M. solifugus* (AY838446); *T. tessulatum* (AY838447). For annelid $F_1$-ATP synthase $\beta$ subunits: *E. albidus* (AY838450); *E. buchholzi* (AY838451); *L. variegatus* (AY83852); *M. solifugus* (AY838454); *T. tessulatum* (AY838453). For annelid $F_1$-ATP synthase $\gamma$ subunits: *E. albidus* (AY838455); *E. buchholzi* (AY838456); *M. solifugus* (AY613864); *T. tessulatum* (AY838457). *D. melanog* (*D. melanogaster*).

Fig. 2) are in positions that could influence enzymatic activity. In a dynamic series of conformational changes, the DELSEED loop interacts with the rotating $\gamma$ subunit causing the transient appearance of three distinct binding conformations ($\beta_E$, $\beta_{DP}$, $\beta_{TP}$; Abrahams et al., 1994; Menz et al., 2001). Although R → K461 is a conservative change with respect to charge, it results in the loss of two bulky amide groups and thus the potential to increase local flexibility. The V → T423 substitution introduces a hydroxyl group near the surface which is predicted to increase surface hydrophilicity, a trend that has been observed in other cold adapted proteins (Smalas et al., 2000). Also, an ice worm-specific substitution near the ADP binding site (H → S476; Fig. 2), results in a large size reduction and loss of a positive charge. Loss of charged residues near the catalytic binding pocket could, in principle, change the ATP off-rate and therefore influence the enzyme's catalytic efficiency. Although some non-conservative substitutions were observed in the ATP-$\alpha$ subunit (e.g., N → I156; D → N265) and ATP-$\gamma$ subunit (e.g., S → A37, N → S40), they do not appear to be in positions critical to enzyme function. Nonetheless, it is noteworthy that a single S → A conversion caused an increase in enzymatic activity of the cold adapted protein, triosephosphate isomerase, from *Vibrio marinus* (Alvarez et al., 1998).

Overall amino acid compositional changes in ice worm $F_1$-ATP synthase subunits are presented in Table 1. Ice worm ATP-$\alpha$ and ATP-$\beta$ subunits showed reductions in the volume of side chains (i.e., Mwt), consistent with trends observed in other cold adapted proteins (Smalas et al., 2000). Surprisingly, ice worm ATP-$\gamma$ appeared larger and more charged than its mesophilic counterparts, a profile more similar to a thermophilic-adapted protein (Haney et al., 1999). Otherwise, ice worm $F_1$-ATP synthase subunits displayed compositional changes that were generally consistent with increases in molecular flexibility (e.g., gains in Ala), with some exceptions (e.g., gains in Thr).

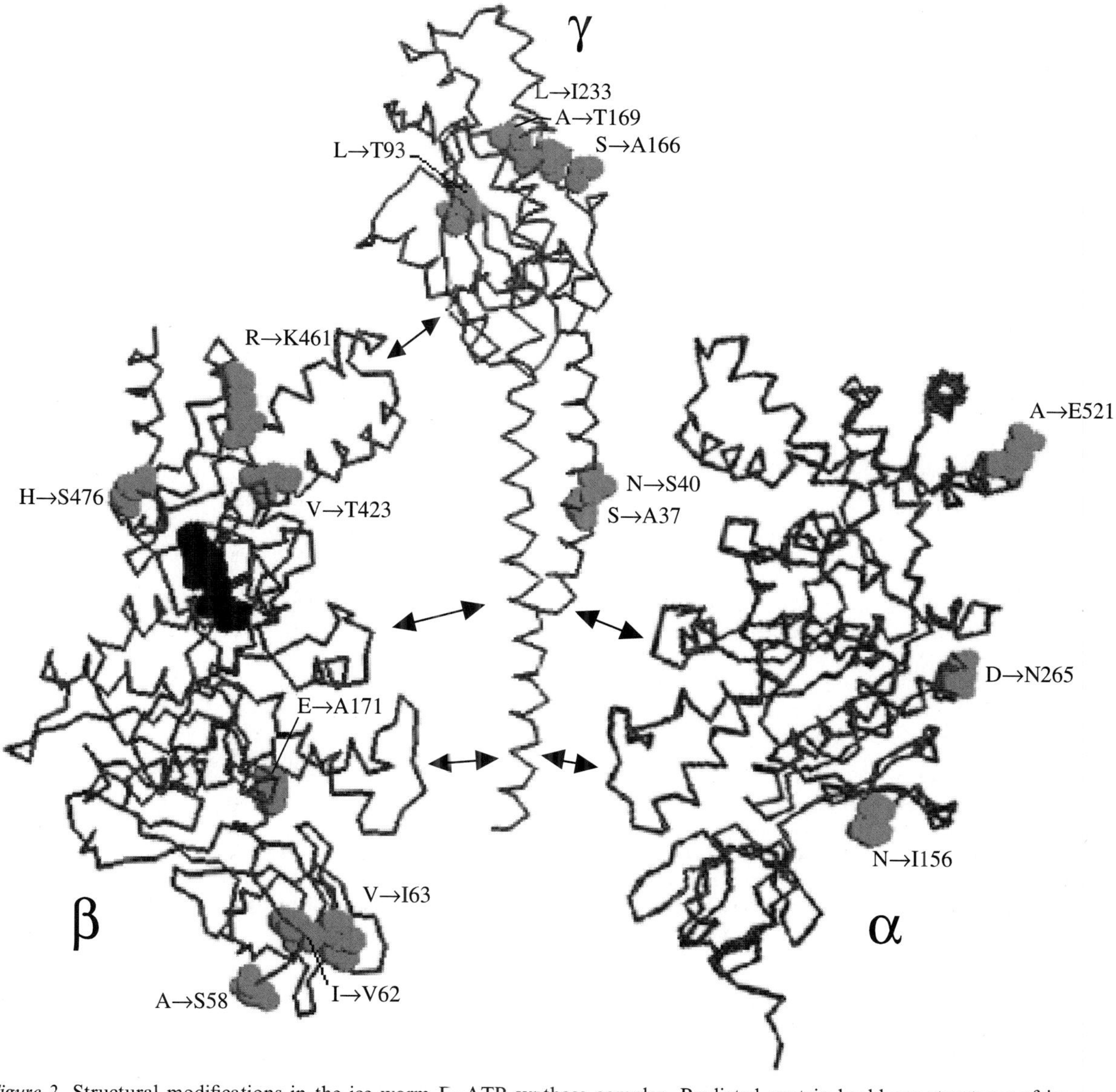


*Figure 2.* Structural modifications in the ice worm $F_1$-ATP synthase complex. Predicted protein backbone structures of ice worm $F_1$-ATP $\alpha$, $\beta$ and $\gamma$ subunits based on bovine templates, respectively (Menz et al., 2001). Rotation of the $\gamma$ subunit provides energy for the synthesis of ATP at the $\beta$ subunit catalytic domain (the dark gray, space-filling molecule depicts ADP within the catalytic site). Ice worm-specific amino acid substitutions appear as light gray, space-filling molecules with amino acid transitions indicated (i.e., mesophilic → ice worm). Numbers correspond to positions within the linear amino acid sequence, respectively (see Fig. 1). Note the proximity of several ice worm-specific substitutions to the site of ATP synthesis (i.e, $\beta$ subunit catalytic region). Arrows indicate subunit contact sites; upper arrow – DELSEED loop (a highly conserved stretch of seven amino acids on the $\beta$ subunit) undergoes a conformational change upon each revolution of the $\gamma$ subunit, middle arrows – salt bridge catch region forms transient ionic interactions between the $\beta$ and $\gamma$ subunits during rotation, lower arrows – hydrophobic sleeve facilitates rotation of the $\gamma$ subunit.

## Discussion

The physiological mechanism(s) by which ice worms survive and thrive in glacier ice/snow remains unknown. Clearly, a variety of physiological and biochemical adaptations are likely to be crucial for ice worm survival (e.g., membrane viscosity, mitochondrial density, metabolic rate), some of which may be shared with their enchytraeid relatives who endure annual permafrost in the

*Table 1.* Amino acid composition changes in ice worm $F_1$-ATP synthase subunits

| Amino acid | ATP-$\alpha$ | | ATP-$\beta$ | | ATP-$\gamma$ | | Total | |
|---|---|---|---|---|---|---|---|---|
| | Animal | Annelid | Animal | Annelid | Animal | Annelid | Animal | Annelid |
| Ala | +2 | +1 | 0 | +2 | +2 | +3 | +4 | +6 |
| Arg | +1 | 0 | −1 | −1 | −2 | +1 | −2 | 0 |
| Asn | +1 | −1 | 0 | +2 | −1 | −1 | 0 | 0 |
| Asp | −4 | −3 | +1 | 0 | +1 | 0 | −2 | −3 |
| Cys | 0 | 0 | +1 | 0 | +1 | 0 | +2 | 0 |
| Gln | 0 | +2 | −2 | 0 | +2 | 0 | 0 | +2 |
| Glu | +1 | +1 | −2 | −1 | −1 | −1 | −2 | −1 |
| Gly | 0 | −1 | 0 | 0 | 0 | −1 | 0 | −2 |
| His | 0 | 0 | 0 | 0 | 0 | 0 | 0 | 0 |
| Ile | −1 | 0 | −1 | +1 | −1 | −1 | −3 | 0 |
| Leu | +1 | +1 | +1 | 0 | −1 | −3 | +1 | −2 |
| Lys | −2 | −2 | +2 | +1 | +1 | +2 | +1 | +1 |
| Met | +1 | −1 | 0 | 0 | +2 | 0 | +3 | −1 |
| Phe | +1 | 0 | 0 | 0 | 0 | 0 | +1 | 0 |
| Pro | 0 | −1 | 0 | +1 | 0 | −1 | 0 | −1 |
| Ser | −1 | 0 | +2 | −2 | −3 | −2 | −2 | −4 |
| Thr | −1 | +4 | +1 | +1 | +1 | +4 | +1 | +9 |
| Trp | 0 | 0 | 0 | 0 | 0 | 0 | 0 | 0 |
| Tyr | −1 | 0 | 0 | 0 | +1 | 0 | 0 | 0 |
| Val | +2 | 0 | −2 | −4 | −2 | 0 | −2 | −4 |
| Polar | −2 | +5 | +1 | 0 | 0 | +1 | −1 | +6 |
| Charged | −4 | −4 | 0 | 0 | −1 | +2 | −5 | −2 |
| Mwt. (Da) | −50 | −27 | −119 | −46 | +44 | +6 | −125 | −67 |

Numbers indicate gains (+) or losses (−) in ice worm subunits in comparison with sequences across animal phyla (i.e., chordate, arthropod, nematode) and among clitellate annelids (*E. albidus*, *E. buchholzi*, *L. variegatus*, *T. tessulatum*). Polar amino acids refer to Asn, Gln, Ser, Thr, Tyr; charged amino acids refer to Arg, Asp, Glu, His, Lys. Predicted molecular weight change (Mwt.) is in Daltons (Da).

Arctic tundra (e.g., Sømme & Birkemoe, 1997; Bauer, 2002; Pederson & Holmstrup, 2003). But recent biochemical evidence suggests that enhanced ATP production via the $F_1F_0$-ATP synthase complex may play an important role in ice worm adaptation (Napolitano et al., 2004; Farrell et al., 2004).

The current study identifies a few candidate amino acid substitutions in the ice worm $F_1$-ATP synthase (while eliminating most others) that could alter the structure/function of the enzyme complex. Among these include modifications near subunit contact sites (e.g., DELSEED loop) and near the catalytic binding pocket in ATP-$\beta$. While any amino acid change could, in principle, influence $F_1$-ATP synthetic activity at reduced temperature, no definitive amino acid substitutions in the subunits examined seem obvious. Of course, our analysis is limited by the absence of a closely related, temperate mesenchytraeid, preferably collected from within the ice worm's geographic range (current fieldwork is aimed at addressing this deficiency). In other studies of a similar nature, amino acid substitutions between two closely related species have been interpreted as evidence of adaptive change (e.g., Rentier-Delrue et al., 1993). In our analyses, however, many sequences appear to contain stochastic changes that could be misleading in direct comparisons between only two species. For example, *M. solifugus* vs. *E. albidus* amino acid alignments identified several more potential ice worm-specific changes than if additional oligochaete sequences were included, even the distantly related hirudinean, *T. tessulatum*. Thus, candidate amino acids in this case are probably best identified by the inclusion of two or more mesophilic, *Mesenchytraeus* species (unfortunately an under-represented genus in modern oligochaete studies). That is not to say that amino acid comparisons across metazoan phyla (e.g., annelid, arthropod, chrodate, nematode) are not useful; indeed, these alignments identified similar amino acid composition trends in comparison to clitellate-specific alignments (see Table 1).

Studies on the modifications of cold adapted proteins are limited in number but a few general trends have emerged. Most importantly, protein activity at low temperature is accompanied by increases in molecular flexibility at the expense of thermal stability (Marshall, 1997; Smalas et al., 2000; D'Amico et al., 2002). The strategies for arriving at this end-point are varied and include reductions in the volume of side chains, reductions in the number of interactions between structural domains or subunits, increased surface hydrophilicity, fewer salt bridges, reductions in arginine and increased methionine content. In general, the ice worm $F_1$-ATP $\alpha$ and $\beta$ subunits conformed with these trends, although changes were subtle and sometimes absent (e.g., arginine, methione content). $F_1$-ATP $\gamma$ subunits, on the other hand, diverged more rapidly from their mesophilic counterparts, and the ice worm $F_1$-ATP $\gamma$ subunit was larger and more charged than expected. The significance of these observations remain unclear, but it underscores the need to examine $F_1$-ATP $\gamma$ subunits from additional species.

Previous studies targeted the $F_1F_0$-ATP synthase complex as a likely component of ice worm energetic adaptation (Napolitano et al., 2004; Farrell et al., 2004), and therefore an examination of the ATP synthetic domain of this assembly ($F_1$) represented a logical starting point. The remainder of the $F_1F_0$-ATP synthase complex should also be considered, particularly in light of the unexpected divergence of the ice worm $F_1$-ATP $\gamma$ subunit, which provides a critical structural/functional link between the $F_1$ and $F_0$ complexes.

## References


Abrahams, J. P., A. G. W. Leslie & R. J. E. Lutter Walker, 1994. Structure at 2.8 Å resolution of $F_1$-ATPase from bovine heart mitochondria. Nature 370: 621–628.

Altschul, S. F., W. Gish, W. Miller, E. W. Myers & D. J. Lipman, 1990. Basic Local Alignment Search Tool. Journal of Molecular Biology 3: 403–410.

Alvarez, M., J. P. Zeelen, V. Mainfroid, F. Rentier-Delrue, J. A. Martial, L. Wyns, R. K. Wierenga & D. Maes, 1998. Triose-phosphate isomerase (TIM) of the psychrophilic bacterium *Vibrio marinus*. Kinetic and structural properties. Journal of Biological Chemistry 273: 2199–2206.

Bauer, R., 2002. Survival of frost and drought conditions in the soil by enchytraeids (Annelida; Oligochaeta) in Artci, subalpine and temperate areas. European Journal of Soil Biology 38: 251–254.

Berman, H. B., J. Westbrook, Z. Feng, G. Gilliland, T. N. Bhat, H. Weissig, I. N. Shindyalov & P. E. Bourne, 2000. The Protein Data Bank. *Nucleic Acids Research* 28: 235–242.

Brinkhurst, R. O. & B. G. M. Jamieson, 1971. Aquatic Oligochaeta of the World . University of Toronto Press, Toronto.

Chenna, R., H. Sugawara, T. Koike, R. Lopez, T. J. Gibson, D. G. Higgins & J. D. Thompson, 2003. Multiple sequence alignment with the Clustal series of programs. Nucleic Acids Research 31: 3497–3500.

Chomczynski, P. & N. Sacchi, 1987. Single-step method of RNA isolation by acid guanidinium thiocyanate-phenol-chloroform extraction. Analytical Biochemistry 162: 156–159.

D'Amico, S., P. Claverie, T. Collins, D. Georlette, E. Gratia, A. Hoyoux, M. Meuwis, G. Feller & C. Gerday, 2002. Molecular basis of cold adaptation. Philosophical Transactions of the Royal Society, London B 357: 917–925.

Emery, C., 1898. Diagnosi di un nuovi genere e vuova specie di annelidi della famiglia degli Enchytraeidae. Atti R. Accademei Lincei 5: 100–111.

English, T. E. & K. B. Storey, 2000. Enzymes of adenylate metabolism and their role in hibernation of the white-tailed prairie dog, *Cynomys leucurus*. Archives of Biochemistry and Biophysics 376: 91–100.

Farrell A. K., 2002. Molecular evolution in the ice worm, *Mesenchytraeus solifugus*. Masters Thesis, Rutgers, The State University of New Jersey.

Farrell, A. H., K. A. Hohenstein & D. H. Shain, 2004. Molecular adaptation in the ice worm, *Mesenchytraeus solifugus*: divergence of energetic-associated genes. Journal of Molecular Evolution 59: 1–8.

Fedorow, C. A., T. A. Churchill & N. M. Kneteman, 1998. Effects of hypothermic hypoxia on anaerobic energy metabolism in isolated anuran livers. Journal of Comparative Physiology B 168: 555–561.

Gasteiger, E., A. Gattiker, C. Hoogland, I. Ivanyi, R. D. Appel & A. Bairoch, 2003. ExPASy: the proteomics server for in-depth protein knowledge and analysis. Nucleic Acids Research 31: 3784–3788.

Haney, P. L., J. H. Badger, G. L. Buldak, C. I. Reich, C. R. Woese & G. J. Olsen, 1999. Thermal adaptation analyzed by comparison of protein sequences from mesophilic and extemely thermophilic *Methanococcus* species. Proceedings of the National Academy of Sciences of the USA 96: 3578–3583.

Kaplan, W. & T. G. Littlejohn, 2001. Swiss-PDB Viewer (Deep View): biomolecular graphics. Briefings in Bioinformatics 2: 195–197.

Leyva, J. A., 2003. Understanding ATP synthesis: structure and mechanism of the $F_1$-ATPase. Molecular Membrane Biology 20: 27–33.

Marshall, C. J., 1997. Cold-adapted enzymes. Trends in Biotechnology 15: 359–364.

Menz, R. I., J. E. Walker & A. G. W. Leslie, 2001. Structure of bovine mitochondrial $F_1$-ATPase with nucleotide bound to all three catalytic sites: implications for the mechanism of rotary catalysis. Cell 106: 331–341.

Napolitano, M. J., R. O. Nagele & D. H. Shain, 2004. Ice worms elevate adenylate levels at low physiological temperature. Comparative Biochemistry and Physiology Part A 137: 227–235.

Napolitano, M. J. & D. H. Shain, 2004. Four kingdoms on glacier ice: convergent energetic processes boost energy levels as temperatures fall. Proceedings of the Royal Society London B (Suppl.) 271(suppl.): S273–S276.

Nicholls, D. G. & S. J. Ferguson, 2002. Bioenergetics3. Academic Press, London.

Nichols, D. S., P. D. Nichols, N. J. Russell, N. W. Davies & T. A. McMeekin, 1997. Polyunsaturated fatty acids in the psychrophilic bacterium *Shewanella gelidimarina* ACAM 456T: molecular species analysis of major phospholipids and biosynthesis of eicosapentaenoic acid. Biochim et Biophys Acta 1347: 164–176.

Pedersen, P. G. & M. Holmstrup, 2003. Freeze or dehydrate: only two options for the survival of subzero temperatures in the arctic enchytraeid *Fridericia ratzeli*. Journal Comparative Physiology B 173: 601–609.

Rentier-Delrue, F., S. C. Mande, S. Moyens, P. Terpstra, V. Mainfroid, K. Goraj, M. Lion, W. G. Hol & J. A. Martial, 1993. Cloning and overexpression of the triosephosphate isomerase genes from psychrophilic and thermophilic bacteria. Structural comparison of the predicted protein sequences. Journal of Molecular Biology 229: 85–93.

Russell, N. J., 1997. Psychrophilic bacteria – molecular adaptations of membrane lipids. Comparative Biochemistry and Physiology Part A. 118: 489–493.

Sayle, R. A. & E. J. Milner-White, 1995. RasMol: Biomolecular graphics for all. Trends in Biochemical Sciences 20: 374–376.

Smalas, A. O., H. S. Leiros, V. Os & N. P. Willassen, 2000. Cold adapted enzymes. Biotechnology Annual Reviews 6: 1–57.

Shain, D. H., M. R. Carter, K. P. Murray, K. A. Maleski, N. R. Smith, T. R. McBride, L. A. Michalewicz & W. M. Saidel, 2000. Morphologic characterization of the ice worm, *Mesenchytraeus solifugus*. Journal of Morphology 246: 192–197.

Sobolev, V., A. Sorokine, J. Prilusky, E. E. Abola & M. Edelman, 1999. Automated analysis of interatomic contacts in proteins. Ligand – Protein Contacts. Bioinformatics 15: 327–332.

Sømme, L. & T. Birkemoe, 1997. Cold tolerance and dehydration in Enchytraeidae from Svalbard. Journal Comparative Physiology B 167: 264–269.

Welch, P. S., 1916. Glacier oligochaeta from Mt. Rainier. Science 43: 143.

Wright, G. F., 1887. The Muir glacier. American Journal of Science 33: 5.

# Part Two
# Life History

Hydrobiologia (2006) 564:61–71
P.F.M. Verdonschot, H. Wang, A. Pinder & R. Nijboer (eds), Aquatic Oligochaete Biology IX
DOI 10.1007/s10750-005-1708-x



# From demographic strategies to mathematical models: trends in population dynamics studies of aquatic Oligochaeta

Giuliano Bonomi[1] & Andrea Pasteris[2,*]
[1]*Dipartimento di Biologia Evoluzionistica Sperimentale, Università di Bologna, via Selmi 3, 40126 Bologna, Italy*
[2]*Centro interdipartimentale di Ricerca per le Scienze Ambientali, Università di Bologna, via S. Alberto 163, 48100 Ravenna, Italy*
*(*Author for correspondence: Tel.: +39-0544-937393; Fax: +39-0544-937411; E-mail: andrea.pasteris@unibo.it)*

*Key words:* aquatic Oligochaeta, demography, life tables, mathematical models, population dynamics

## Abstract

The study of the population dynamics of aquatic oligochaetes is still largely a descriptive subject. However, going beyond description is essential in any field of science. This paper analyses the methodological approaches that can be used in this process, together with some examples from each of these approaches as they have been applied to aquatic oligochaetes. Observation and experiments in the field, laboratory experiments, and mathematical models are reviewed. Increased use of tools that are regarded as standard in the field of population dynamics at large and interaction with applied disciplines are advocated as ways to improve the quality of research in this field.

## Introduction

Population dynamics may be defined as the study of temporal changes within populations and the factors that cause or influence those changes.

Oligochaetes are an important component of the benthos, especially in freshwaters, and are often dominant in the deep parts of lakes. Virtually any quantitative study on freshwater soft bottom macrofauna provides data on the number or biomass of oligochaete populations. However studies designed to analyse in detail the dynamics of single species populations of oligochaetes are much less common. Most of them are accounts of how population numbers and structure have varied over time in a specific ecosystem. However quantitative and detailed these studies may be, they are basically descriptive: the actual mechanisms that determined the observed patterns are investigated in a limited number of cases. The word 'descriptive' is not intended as a criticism. Accurate description of the dynamics of an aquatic oligochaete population is in itself a difficult task, and description must obviously come before explanation.

However, going beyond description is essential in any field of science. This paper analyses the methodological approaches that can be used in this process, together with some examples from each of these approaches as they have been applied to aquatic oligochaetes. We hope to encourage other researchers to move in this direction.

## Field observations

Field observations make up the bulk of studies on what is or could be labelled as population dynamics (Stites, 1987; Erséus & Diaz, 1989; Lindegaard, 1992; Diaconu et al., 1993; Jenderedjian, 1994; Lindegaard et al., 1994; Marchese, 1994; Löhlein, 1996; Armendariz, 2000; Risnoveanu & Vadineanu, 2002). These studies often describe population structure as well as total numbers. However, since age is not measurable in the field, collected individuals are classified and enumerated on the basis of their size or the development of the reproductive organs. This information is often used to infer life cycle traits such as life span, generation time, or number of broods per year. In our opinion, these estimates should be viewed with some caution: observations of laboratory cultures showed that oligochaetes can survive several years (Timm, 1984), that individuals of the same age may vary greatly in size (Bonomi & Di Cola, 1980), and that sexually mature individuals may undergo regression of the reproductive system and look like juveniles. Moreover, many field studies do not take account of cocoons and eggs.

One major obstacle to a detailed analysis of field populations is taxonomic: juveniles are difficult to determine to species, if not to genus, and the same holds for cocoons. For this reason some authors have selected one-species sites (e.g., Lake Orta, N Italy) or young, disturbed water bodies hosting only a very limited number of species. This is the case of some freshwater reservoirs in which the deep zone is colonised by *Tubifex tubifex* and *Limnodrilus hoffmeisteri*, whose cocoons are clearly different and whose free living individuals are easy to distinguish using the characteristics of their chaetae (Archipova, 1976; Bonomi, 1979; Bonomi & Di Cola, 1980; Bonacina et al., 1996).

Few authors have tried to estimate growth, death and birth rates from field populations. In our opinion this is a major shortcoming, since estimation of rates is crucial in population dynamics. It can be misleading to infer the factors that control a population by correlating numbers of individuals and environmental variables, as is usual. As Edmondson (1960) pointed out, the conditions observed in concomitance with the highest population numbers are usually described as optimal. However, a population maximum occurs at a time when conditions have become so unfavourable that the population is about to decrease. Actually, optimal conditions correspond to high birth and growth rate values and low death rate values.

The population growth rate can be estimated from the time series of numbers of a field population. However, to estimate birth and death rate in a species with continuous reproduction, in which the two processes occur simultaneously, laboratory cultures or field enclosures are required.

**Laboratory experiments**

A major contribution towards understanding the distribution and ecological succession as well as the demographic strategies of some important species may be made through laboratory observations and laboratory cultures kept under controlled conditions (Kosiorek, 1974; Timm, 1974; Thorhauge, 1975; Jonasson & Thorhauge, 1976; Poddubnaya, 1980; Timm, 1980; Adreani et al., 1984; Dumnicka, 1984; Monti, 1986; Bonacina et al., 1987). The different kinds of laboratory cultures can provide a spectrum of results as exemplified in Figure 1 and supply basic data for setting up realistic models of population dynamics.

Simple cultures can yield information on, for example, mode of reproduction, fecundity and generation time, as well as identification of cocoons (if any). An example of this last point is provided by the laboratory-produced cocoons of *Spirosperma ferox*. The consequent description of their typical shape and the unusual persistence of their wall in the sediments resulted in what is probably the only contribution of aquatic Oligochaeta investigations to paleolimnology (Bonacina et al., 1986).

The embryonic developmental time ($T_E$, time spent inside the cocoon) was measured for several species at different temperatures, the final results showing that 'eutrophic' species (e.g., *Tubifex tubifex*) display a shorter $T_E$ than 'oligotrophic' species (e.g., *Spirosperma ferox*; Bonacina et al., 1987).

However, it is only by studying cohort cultures that we can make real progress in understanding the traits of population biology. The study of cohorts, groups of individuals born at the same time, is the most reliable procedure for the construction of life tables, one of the standard methods used by population ecologists to report age specific information on survival and fecundity (Wilson & Bossert, 1971; Begon and Mortimer, 1986). Life tables provide an analytical representation of the variation in demographic traits as a function of the age of the individual, and are the starting point for estimating a number of integrative and ecologically meaningful population

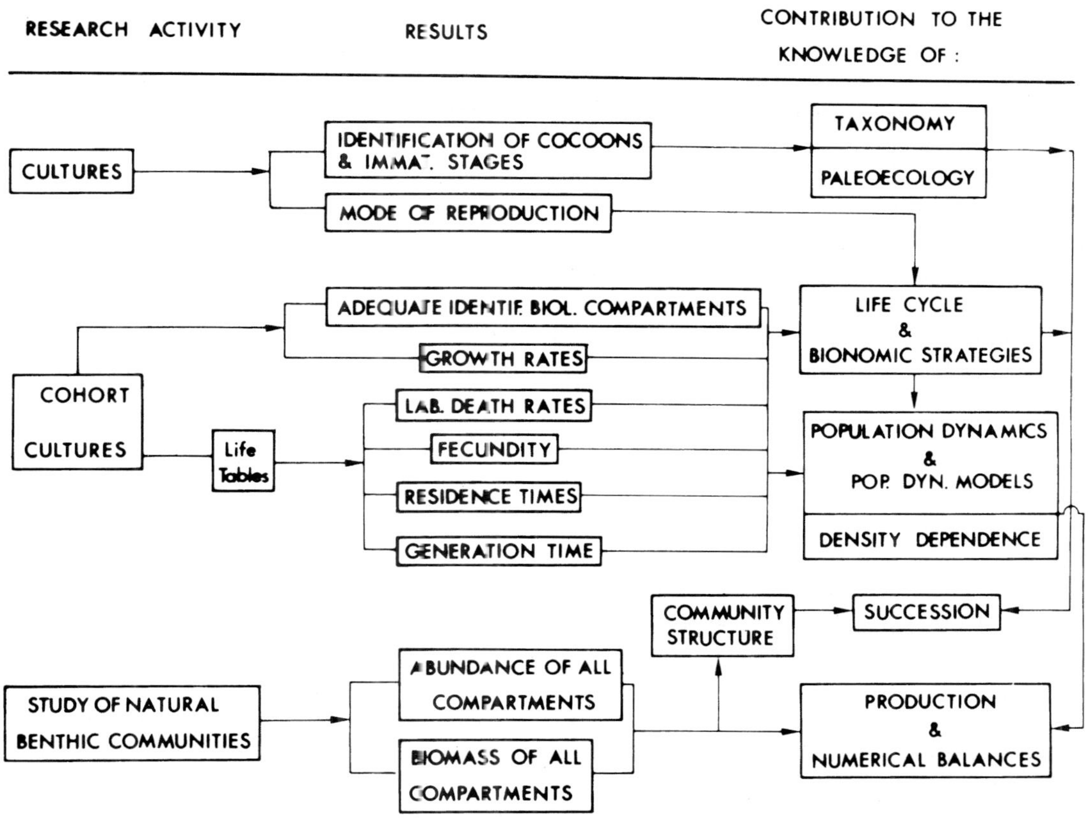


*Figure 1.* Schema showing how culturing aquatic Oligochaeta can contribute to furthering knowledge of their population and community ecology (from Bonacina et al., 1987).

parameters. Ideally, the study of a cohort begins when the individuals forming the cohort are born and ends when the last survivor dies. Over this period and at an appropriate frequency, the numbers of surviving individuals and new-borns are recorded.

Laboratory experiments on cohort cultures are among the best methods for assessing the influence of environmental factors at the population level, allowing direct measurement of the effect of experimental treatments on life table statistics. There are, however, major practical problems in the application of the life table approach to aquatic oligochaetes. Oligochaetes are long-lived organisms with a potential life span of several years (Timm, 1984). Furthermore, if environmental conditions are favourable, tubificids breed continuously and their embryonic development is rapid in comparison to their life span (250 degree-days for *T. tubifex*). This requires the number of eggs laid by the cohort to be recorded frequently, e.g., weekly for *T. tubifex* at 23 °C. A longer period between observations would result in embryos dying or hatching as young worms in the time between observations, and fecundity would be grossly underestimated. This combination of long duration and high frequency of observation makes it logistically difficult to conduct experiments on cohorts with tubificids, which is possibly one of the reasons why this technique is not widely used for oligochaetes. Nevertheless, while cohort experiments should ideally end when the last survivor dies (which could mean a duration of several years), even a more limited experiment lasting for only a few months can provide relevant information.

Working with cohort cultures, Bonacina et al. (1987) measured $T_{GM}$, the age required to lay the first cocoon, of several species. They were able to demonstrate a strong linear correlation between $T_E$ and the $T_{GM}$ of each species. Species regarded as 'eutrophic' had shorter $T_E$ and $T_{GM}$ than 'oligotrophic' species and consequently a higher potential population growth rate. They were also able to learn how to identify 'regressed' individuals, i.e., worms which, after laying cocoons, undergo a regression of the reproductive system and look like juveniles.

These results enabled the authors to make a more adequate identification of biological "stages" (compartments) and led to the first application of population dynamics models for estimating field population statistics, including death and birth rates (Bonomi & Di Cola, 1980, see below).

Cohort cultures and life tables were also used to assess the effect of a toxicant (copper) on the population dynamics of *Tubifex tubifex* (Pasteris et al., 2003).

However, cohort cultures and life tables have one major limitation: age-based demographic statistics cannot be applied to field populations of oligochaetes. Thus some different experimental designs have been proposed to measure size- and stage-classified demographic traits. These are basically modifications of the standard life table experiment in which the individuals of the cohort were periodically sorted into weight and reproductive stage classes. Then growth, survivorship and fecundity of these homogeneous groups are measured separately. Data from such experiments were used to parameterise matrix models (Pasteris et al., 1996, 1999, see below).

## Field experiments

Field manipulative experiments are a powerful tool, allowing us to estimate population parameters, assess the effects of specific factors and test hypotheses in a more realistic situation than laboratory experiments. This potential has not yet been fully explored in population dynamics of aquatic oligochaetes.

Some authors have used small field enclosures to estimate the population growth rate of naidids (Finogenova, 1984; Juget et al., 1989).

In our opinion the paper by Hentschel & Jumars (1994) is a good example of the care and creativity required in planning good field experiments. The goal of this study was to determine whether the meiofaunal oligochaete *Amphichaeta leydigii* and juveniles of the polychaete *Hobsonia florida* share benthic diatoms as a limiting resource during recruitment.

The authors developed a method for inhibiting benthic diatom growth by perfusing pore waters with herbicide (DCMU) diffusing from a gel contained in bottom-sealed chambers buried in the sediments. The abundances of both oligochaetes and larger juveniles of polychaetes were significantly reduced in the herbicide treated patches. No other dominant meiofauna showed significant responses to the treatment. Calculated concentrations of DCMU in superficial sediments were one or two orders of magnitude below those that cause lethal or sublethal effects on animals, implying that abundances of *A. leydigii* and *H. florida* varied because of the effect of DCMU on diatoms rather than as a direct response to the chemical itself.

## Mathematical models

### *Compartmental models*

In compartmental models the population is divided into compartments (such as eggs, juveniles, adults) and the dynamic of each compartment is described with a differential equation (Argentesi et al., 1974, 1981).

After some attempts at using simple (but inadequate) models (Bonomi, 1979), a complex bi-dimensional model was set up in collaboration with a mathematician (Bonomi & Di Cola, 1980). The model graph is given in Figure 2.

For each monthly set of samples (20 replicates) from Pertusillo Reservoir, the cocoons were split into two developmental stages; immature individuals were divided by weight, the $y$ compartment (individual weight less than 1 mg) including individuals regarded as 'true' juveniles (not regressed). Mature and ovigerous individuals were equally positioned by weight. As the model was highly complex and some of the transfer rates (between compartments) difficult to measure, a simplified model was devised (Fig. 3) and used; in this model, however, the transfer rates (horizontal arrows) are the weighted means of the rates of the different component weight classes. This model was then used to estimate secondary production and birth and death rates of the *Tubifex tubifex* and *Limnodrilus hoffmeisteri* populations in the field.

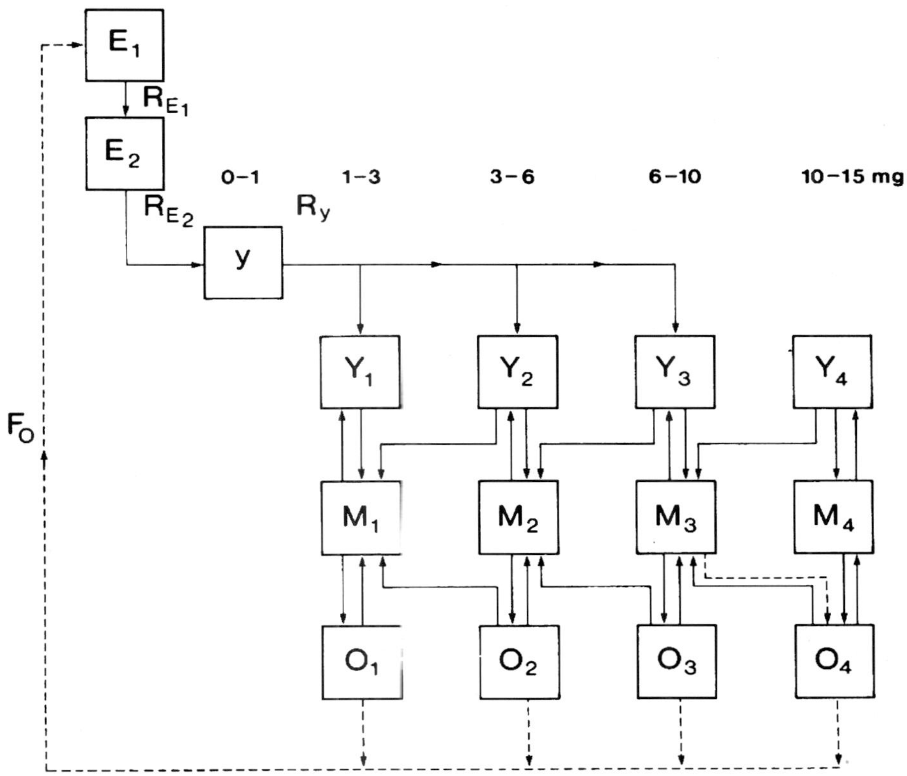


*Figure 2.* Bi-dimensional model describing egg development, growth, maturation, egg deposition and regression in *Tubifex tubifex* (from Bonomi & Di Cola, 1980). $E_1$: eggs and early embryos; $E_2$: advanced embryos; y: juveniles, individuals under 1 mg; Y: individuals without a distinct clitellum, over 1 mg (subscripts indicate the weight class); M: individuals with a distinct clitellum; O: individuals with clitellum and visible oocytes.

## *Matrix models*

Matrix population models were proposed by Lewis (1942) and Leslie (1945, 1948) and are a well-established and widely used tool in the general field of population dynamics. They are extensively discussed in Caswell (2001), where a complete presentation of the theory and several examples of applications to different kinds of organisms can be found. On the other hand, we are aware of only three papers which describe the application of this approach to aquatic oligochaetes.

In comparison with other approaches to mathematical modelling of population dynamics, these models are conceptually simple, and even

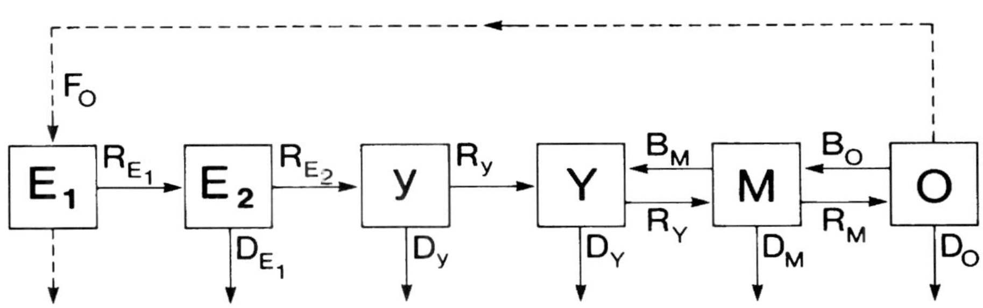


*Figure 3.* Simplified model describing egg development, growth, maturation, egg deposition and regression in *Tubifex tubifex* (from Bonomi & Di Cola, 1980). $E_1$: eggs and early embryos; $E_2$: advanced embryos; y: juveniles, individuals under 1 mg; Y: individuals without a distinct clitellum, over 1 mg; M: individuals with a distinct clitellum; O: individuals with clitellum and visible oocytes; R: transfer rates; D: death rates, B: regression rates; F: fertility.

ecologists with a limited mathematical background can understand at least their basic elements. Nevertheless, they provide a realistic and flexible quantitative description of complex life cycles. While they were first formulated as a tool for dealing with population age structure, they were further developed to deal with population structure defined by any other suitable descriptor, such as size or stage (Lefkovitch, 1965). Moreover, an estimate of the parameters is relatively easy to obtain. These features make these models especially suited for dealing with actual populations, in conjunction with laboratory and field data, though they also have their own role in theoretical population biology.

Matrix models make two basic assumptions: (i) The individuals in the population can be arranged in a number of discrete classes defined by age, instar, stage, size or other criteria or combination of criteria; (ii) time is a discrete variable, i.e., it assumes only values that are exact multiples of a basic unit, the projection interval (this is equivalent to saying that the population is observed and enumerated at regular time intervals and that what happens in between is not important). Because of their assumptions, matrix models are an obvious choice for populations with clearly distinct stages and seasonal breeding; however, the advantages of using them mean that they have also been widely applied to populations without these characteristics.

A simple hypothetical example for a population of a tubificid will be used to illustrate the basic principles of these models.

The first decision to be made when establishing a matrix model is the definition of the classes. When, as in oligochaetes, development is a continuous process and there are no obvious morphological differences between phases of the life cycle, this choice is somewhat arbitrary. However, using the development of the reproductive organs we can identify four stages (this is, of course, one of several alternatives):

- stage 1: eggs and embryos (the individuals inside the cocoons);
- stage 2: immature (individuals without a distinct clitellum);
- stage 3: mature (individuals with a distinct clitellum);
- stage 4: ovigerous (individuals with clitellum and visible oocytes).

The second decision involves the projection interval. When matrix models are applied to long-lived organisms with clearly defined seasonal reproduction, the projection interval is usually one year. For oligochaetes with reproduction that is usually continuous, albeit variable during the year according to environmental conditions, shorter intervals are more appropriate. The actual choice of interval is usually dictated by logistic considerations (frequency of sampling or observation). For the present example an interval of one week will be adopted.

The matrix model can be written as a system of finite difference equations:

$$\begin{aligned} n_1(t+1) &= P_1 \cdot n_1(t) + F_4 \cdot n_4(t) \\ n_2(t+1) &= G_1 \cdot n_1(t) + P_2 \cdot n_2(t) \\ n_3(t+1) &= G_2 \cdot n_2(t) + P_3 \cdot n_3(t) \\ n_4(t+1) &= G_3 \cdot n_3(t) + P_4 \cdot n_4(t) \end{aligned} \tag{1}$$

where: $n_1(t)$ is the number of individuals in stage 1 at any time $t$ and $n_1(t+1)$ the number of individuals in stage 1 at time $t+1$, i.e., 1 week later; the meanings of $n_2(t)$, $n_2(t+1)$, $n_3(t)$, $n_3(t+1)$, $n_4(t)$ and $n_4(t+1)$ are similar for stages 2, 3 and 4 respectively;

$P_1$ is the probability that an individual in stage 1 will survive for a projection interval and stay in the same stage, i.e., the probability of an egg or embryo inside the cocoon surviving for one more week without leaving the cocoon;

$F_4$ refers to the fertility of stage 4, i.e., the average number of eggs laid in one week by one ovigerous individual;

$G_1$ is the probability of an individual in stage 1 surviving for a projection interval and growing to stage 2, i.e., the probability of an egg or embryo inside the cocoon surviving for one more week and leaving the cocoon;

$P_2$ is the probability of an individual in stage 2 surviving for a projection interval and staying in the same stage, i.e., the probability of an immature surviving for one more week and not becoming mature;

$G_2$ is the probability of an individual in stage 2 growing to stage 3, i.e., the probability of an immature becoming mature within one week;

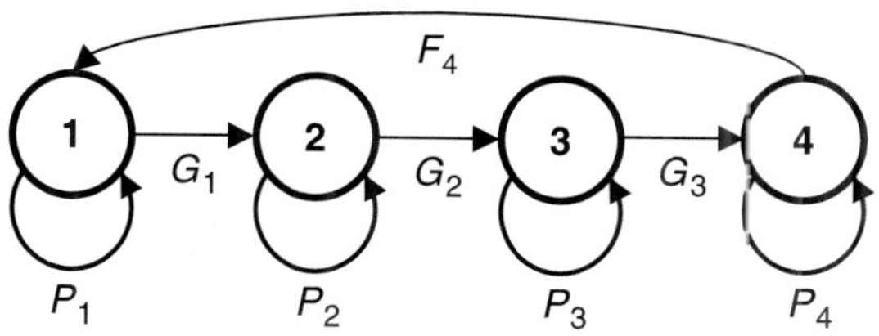


*Figure 4.* Life cycle graph for a tubificid. Numbers refer to life cycle stages; 1: eggs and embryos; 2: immature individuals; 3 mature individuals; 4: ovigerous individuals. Capital letters refer to rates; $P$ is the probability of an individual surviving and staying in the same stage; $G$ is the probability of an individual surviving and growing to the next stage; $F$ is fertility.

The meanings of $P_3$, $G_3$, $P_4$ are similar for stage 3 and 4.

Figure 4 shows the life cycle graph corresponding to the model. The model can also be written as an equation among matrices and vectors, which is completely equivalent to equation 1:

$$\begin{pmatrix} n_1(t+1) \\ n_2(t+1) \\ n_3(t+1) \\ n_4 t+1 \end{pmatrix} = \begin{pmatrix} P_1 & 0 & 0 & F_4 \\ G_1 & P_2 & 0 & 0 \\ 0 & G_2 & P_3 & 0 \\ 0 & 0 & G_3 & P_4 \end{pmatrix} \cdot \begin{pmatrix} n_1(t) \\ n_2(t) \\ n_3(t) \\ n_4(t) \end{pmatrix} \quad (2)$$

or:

$$\mathbf{n}(t+1) = \mathrm{A} \cdot \mathbf{n}(t) \quad (3)$$

The difference equation form (equation 1) may be clearer for biologists with no background in matrix algebra; however, one advantage of the matrix notation is that **A**, the projection matrix, displays all the parameters of the model in a concise and ordered form. In general, each coefficient on the $i$-th row and on the $j$-th column of the projection matrix represents the contribution of stage $j$ to stage $i$. This generic element may be referred to as $a_{ij}$ and may represent reproductive output (like $F_j$ in equation 2) or transition probabilities (like $P_j$ and $G_j$ in equation 2).

Matrix models were first used to deal with populations of aquatic oligochaetes by Pasteris et al. (1996). The model proposed in this paper was developed from the bi-dimensional model of Bonomi & Di Cola (1980), with the aim of estimating the transfer rates between the different compartments of the population and evaluating the effectiveness of weight and maturation stage to define population structure in *Tubifex tubifex* (matrix models are in fact very similar to compartmental models, the main difference being the discrete-time nature of the former).

In the laboratory experiment used to parameterise the model, some of the classes identified by Bonomi & Di Cola (1980) were not observed (for instance, immature individuals over 3 mg were never observed); as a result nine classes were included in the model.

The projection interval was set to 4 weeks, the same time lapse as between two successive weighings in the experiment.

The resulting projection matrix was rather more complicated than that of the previous example, because of the bi-dimensional classification of the individuals and the rather long projection unit.

The potential long-term population growth rate can be calculated from the projection matrix as $\lambda$, the finite growth rate. Given the highly unrealistic assumptions required to define $\lambda$, it is generally of little value in predicting or describing actual population growth. However, it is a very effective measure of the potential of a population, since it takes into account individual growth survival and fecundity over the whole life cycle, and integrates in a single number the complex information contained in the projection matrix.

An important part of the analysis of a matrix model is investigating how $\lambda$ would change in response to changes in the element of the projection matrix. In Pasteris et al. (1996) a very simple sensitivity analysis was performed to assess the effectiveness of the classification of the individuals adopted for the model. Estimates of the elements in the matrix were affected by two sources of error: (i) random variability; (ii) differences due to the effect of age that are not accounted for in the model (individuals with the same weight and maturation stage may have different ages but are treated as equal in the model). As a consequence the estimate of $\lambda$ is also affected by an error. The authors computed the upper and lower limits of the range of variation for their estimate of $\lambda$, and assumed this as a measure of the effectiveness of the adopted classification of the individuals and of the effects of not taking age into account in the model. The upper and lower limits of the range were 2.24 and 3.50. Even taking into account the fact that the range of variation is a crude, maximal measure of dispersion, this represents a substantial variability. The authors concluded that the adopted classification was not completely satisfactory, but that a demography based on individual weight and stage for aquatic oligochaetes was a reasonable working hypothesis worthy of further investigation.

A subsequent study (Pasteris et al., 1999) applied matrix models to laboratory cultures of *Limnodrillus hoffmeisteri*. The aim of this study was to compare the relative effectiveness of age, maturation stage and weight as criteria to define population structure. The data from the experimental cultures were used to parameterise four models, each based on a different classification of the individuals.

The experimental design adopted in this second study allowed the estimation of a confidence interval for the $\lambda$ value of each model. The width of confidence interval was very similar for the age- and the stage-based model, while the precision of the estimate was much lower for the weight-based model, suggesting that stage is a better substitute for age, which is not measurable in field populations.

Matrix models were also applied to marine naidid populations by Nilsson et al. (1997). The general aim of this study was to investigate genetic and demographic differences within a population of *Paranais littoralis* from an intertidal mudflat in Flax Pond, New York, USA.

In particular, the authors compared the finite growth rates of subpopulations of *P. littoralis* from different patches of the mudflat and assessed the influence of different kind of sediments and the seasonal variability on the growth rate. The adopted models were age-based and the duration of age classes and the projection interval was 4 days; 24 age classes were considered. The studied population reproduced exclusively by asymmetric agamic scission; the anterior worm was regarded as the parent and the posterior worm as an offspring with age 0. A distinct model was established for each of six subpopulations in June and October, using data from laboratory cultures, and $\lambda$ was computed for each model.

Significant differences were found both among subpopulations and among seasons; however, the subpopulations with the highest growth rates were not the same in June and October. In addition, there were no significant differences in growth rate between worms cultured in sediment collected from the patches where they originated and those cultured in sediment collected from other patches.

Sensitivity analysis was performed by calculating the elasticity of $\lambda$ with respect to each $a_{ij}$, i.e., the sensitivity of the growth rate to a proportional change in the elements of the matrix:

$$e_{ij} = \frac{a_{ij}}{\lambda} \cdot \frac{\partial \lambda}{\partial a_{ij}} \tag{4}$$

Comparison of elasticity values indicated that $\lambda$ was very little influenced by changes in survival and fecundity beyond 30 days of age: the difference in growth rate between June and October was due mainly to differences in early reproduction.

*Dynamic energy budget models*

A completely different approach to population dynamics modelling is represented by dynamic energy-budget (DEB) models (Nisbet et al., 2000; Kooijman, 2001). Matrix models may be regarded as merely a form of accounting: they supply a method for computing population growth, taking into account inputs to and outputs from all of the classes. However, matrix models do not attempt to explain what mechanisms control the values of the fertility and transfer rates.

DEB models, on the other hand, are more ambitious in so far as they are designed to assess the ecological consequences of physiological effects: population parameters are modelled as explicit functions of metabolic processes.

Biologically speaking, the idea behind DEB models is so simple as to sound almost trite: part of the energy ingested with the food is assimilated. A certain amount of this energy has to be spent on maintenance. The energy excess can be stored, spent on individual growth or spent on reproduction. However, growth controls body size and this in turn controls ingestion, assimilation and how assimilated energy is partitioned into storage, growth and reproduction. In addition, body size affect mortality and the input of energy through ingestion. Population growth is the result of reproduction and mortality. While this may sound obvious, dealing with these concepts in quantitative terms and translating them into mathematical equations is not, and DEB models are typically very complex from a mathematical standpoint.

As far as we know, only Ratsak et al. (1993) applied DEB models to aquatic oligochaetes. Their study was designed to analyse in quantitative terms the asexual propagation of the naidid *Nais elinguis* in a wastewater treatment plant, and was part of a program to model energy transduction in sewage treatment plants in relation to the optimisation of operating conditions. Like other metazoans dwelling in sewage treatment plants, naidids feed on bacteria, and it is hypothesised that they can reduce surplus sludge production and thus the high costs involved in sludge disposal.

The experimental data used were body length and time between divisions. Individuals from a single clone were reared in the laboratory, fed on activated sludge particles. After division, growth of the anterior and posterior daughters was followed individually and their length repeatedly measured until the next division.

The proposed DEB model described an individual with two state variables: volume and storage density (stored energy per unit volume). Food ingestion, assimilation, energy usage, individual growth and population growth were modelled, producing a set of fairly complex equations. The values for some of the parameters of the model (energy investment ratio, maintenance rate coefficient) were deduced from the average values in other organisms. Other parameters (position of division, length at division, energy conductance) were estimated by fitting the equation for the individual growth to the experimental data. This second group of parameters could be estimated once values for the first group had been fixed. In reality, estimating parameters of PED models can be a very difficult task; we think this is a significant disadvantage of this kind of model compared with matrix models.

The parameterised model fitted well with the experimental data for individual growth, even though differences among parameters estimated on different individuals were not trivial. This study represents a remarkable attempt to apply a sophisticated modelling method to the population dynamics of an aquatic oligochaete in the framework of an applied study, but the approach has not been developed further, at least as far as experimental work is concerned. However, El Houssif (2001) has explored some mathematical aspects of the model in greater depth.

## Concluding remarks

Most of the papers published on population dynamics of aquatic oligochaetes are accounts of how population number and population structure has varied over time. Relatively few studies have made an effort to apply to aquatic oligochaetes tools that are regarded as standard in the field of population dynamics at large. These tools include cohort and life table analysis and mathematical models, in particular matrix models. We think that the study of population dynamics of aquatic oligochaetes would benefit from an increased use of these tools.

Interactions with applied disciplines would be mutually beneficial. As illustrated above, possibly the most sophisticated mathematical model ever applied to population dynamics of oligochaetes was formulated as a tool to deal with a population of *Nais elinguis*, in an effort to reduce surplus sludge production in wastewater treatment plants. One of the basically unresolved core problems in ecotoxicology and ecological risk assessment is how to translate the results of laboratory toxicity tests into effects on field populations, which is actually an applied population dynamics problem.

In addition to providing funding opportunities, more connection with applied research would help to maintain the focus of population dynamics studies on finding answers to specific questions of major importance.

## References

Adreani, L., C. Bonacina, G. Bonomi & C. Monti, 1984. Cohort cultures of *Psammoryctides barbatus* (Grube) and *Spirosperma ferox* Eisen: a tool for a better understanding of demographic strategies in tubificids. Hydrobiologia 115: 113–119.

Archipova, N. R., 1976. Peculiarities of the biology and production in *Limnodrilus hoffmeisteri* Clap. (Oligochaeta, Tubificidae) on grey muds in the Rybinskoe reservoir. Proceedings of the Reservoirs Biology Institute 34: 5–15 [in Russian].

Argentesi, F., R. De Bernardi & G. Di Cola, 1974. Mathematical models for the analysis of population dynamics in species with continuous recruitment. Memorie dell' Istituto Italiano di Idrobiologia 31: 245–275.

Argentesi, F., R. De Bernardi & G. Di Cola, 1981. Analysis of population dynamics in species with continuous recruitment and unobservable age structure. Memorie dell' Istituto Italiano di Idrobiologia 39: 161–176.

Armendariz, L. C., 2000. Population dynamics of *Stylaria lacustris* (Linnaeus, 1767) (Oligochaeta, Naididae) in Los Talas, Argentina. Hydrobiologia 438: 217–226.

Begon, M. & M. Mortimer, 1986. Population Ecology, A Unified Study of Animals and Plants (2nd edn.). Blackwell Scientific Publication, Oxford, 220 pp.

Bonacina, C., G. Bonomi & C. Monti, 1986. Oligochaete cocoon remains as evidence of past lake pollution. Hydrobiologia 143: 395–400.

Bonacina, C., G. Bonomi & C. Monti, 1987. Progress in cohort cultures of aquatic Oligochaeta. Hydrobiologia 155: 163–169.

Bonacina, C., A. Pasteris, G. Di Cola & G. Bonomi, 1996. Production and population dynamics of *Tubifex tubifex* in the profundal zone of a freshwater reservoir in N. Italy. Hydrobiologia 334: 141–146.

Bonomi, G., 1979. Ponderal production of *Tubifex tubifex* Müller and *Limnodrilus hoffmeisteri* Claparède (Oligochaeta, Tubificidae) benthic cohabitants of an artificial lake. Bollettino di Zoologia 46: 153–161.

Bonomi, G. & G. Di Cola, 1980. Population dynamics of *Tubifex tubifex* studied by means of a new model. In Brinkurst, R. O. & D. G. Cook (eds), Aquatic Oligochaete Biology. Plenum Publishing Corporation, New York: 185–203.

Caswell, H., 2001. Matrix Population Models: Construction, Analysis and Interpretation (2nd edn.). Sinauer Associates, Sunderland, 722 pp.

Diaconu, I., A. Vadineanu & G. Risnoveanu, 1993. Changes of the structure and functioning of the benthic oligochaete communities from the Danube delta aquatic ecosystems (1976–1982) 1. Dynamics of the structure of the communities and their dominant populations. Revue Roumaine de Biologie, Série de Biologie Animale 38: 171–179.

Dumnicka, E., 1984. Laboratory studies on biology of cave-dwelling Enchytraeidae: *Enchytraeus dominicae* Dumnica and *Fridericia bulbosa* (Rosa). Mémoires de Biospéologie 11: 199–205.

Edmondson, W. T., 1960. Reproductive rates of rotifers in natural populations. Memorie dell' Istituto Italiano di Idrobiologia 12: 21–77.

El Houssif, N. E., 2001. A deterministic size-structured population model for the worm *Nais elinguis*. Journal of Mathematical Biology 42: 424–438.

Erséus, C. & R. J. Diaz, 1989. Population dynamics of *Tubificoides amplivasatus* (Oligochaeta, Tubificidae) in the Oresund, Denmark. Hydrobiologia 180: 167–176.

Finogenova, N. P., 1984. Growth of *Stylaria lacustris*. Hydrobiologia 115: 105–107.

Hentschel, B. T. & P. A. Jumars, 1994. In situ chemical inhibition of benthic diatom growth affects recruitment of competing, permanent and temporary meiofauna. Limnology and Oceanography 39: 816–838.

Jenderedjian, K., 1994. Population dynamics of *Potamothrix alatus paravanicus* Poddubnaya & Pataridze (Tubificidae) in different areas of Lake Sevan. Hydrobiologia 278: 281–286.

Jonasson, P. M. & F. Thorhauge, 1976. Growth and life cycle of *Potamotrix hammoniensis* (Tubificidae, Oligochaeta) in the profundal of Lake Esrom. A field and laboratory study. Archiv für Hydrobiologie 78: 71–85.

Juget, J., V. Goubier & D. Berthelemy, 1989. Intrinsic and extrinsic variables controlling the productivity of asexual population of *Nais* spp. (Naididae, Oligochaeta). Hydrobiologia 180: 177–184.

Kooijman, S. A. L. M., 2001. Quantitative aspects of metabolic organization; a discussion of concepts. Philosophical Transactions of the Royal Society of London Series B - Biological Sciences 356: 331–349.

Kosiorek, D., 1974. Development cycle of *Tubifex tubifex* Müller in experimental culture. Polskie Archiwum Hydrobiologii 21: 411–422.

Lefkovitch, L. P., 1965. The study of population growth in organisms grouped by stages. Biometrics 21: 1–18.

Leslie, P. H., 1945. On the use of matrices in certain population mathematics. Biometrika 33: 183–212.

Leslie, P. H., 1948. Some further notes on the use of matrices in population mathematics. Biometrika 35: 213–245.

Lewis, E. G., 1942. On the generation and growth of a population. Sankhya: The Indian Journal of Statistics 6: 93–96.

Lindegaard, C., 1992. Zoobenthos ecology of Thingvallavatn: vertical distribution, abundance, population dynamics and production. Oikos 64: 257–304.

Lindegaard, C., K. Hamburger & P. C. Dall, 1994. Population dynamics and energy budget of *Marionina southerni* (Cernosvitov) (Enchytraeidae, Oligochaeta) in the shallow littoral of Lake Esrom, Denmark. Hydrobiologia 278: 291–301.

Löhlein, B., 1996. Seasonal dynamic of aufwuchs Naididae (Oligochaeta) on *Phragmites australis* in a eutropic lake. Hydrobiologia 334: 115–123.

Marchese, M., 1994. Population dynamics of *Narapa bonettoi* Righi and Varela, 1983 (Oligochaeta: Narapidae) from the main channel of the middle Parana River, Argentina. Hydrobiologia 278: 303–308.

Monti, C., 1986. Population regulation in *Psammoryctides barbatus* Grübe (Oligochaeta: Tubificidae). Memorie dell' Istituto Italiano di Idrobiologia 44: 223–241.

Nilsson, P., J. P. Kurdziel & J. S. Levinton, 1997. Heterogeneous population growth, parental effects and genotype-environment interactions of a marine oligochaete. Marine Biology 130: 181–191.

Nisbet, R. M., E. B. Muller, K. Lika & S. A. L. M. Kooijman, 2000. From molecules to ecosystems through dynamic energy budget models. Journal of Animal Ecology 69: 913–926.

Pasteris, A., G. Bonomi & C. Bonacina, 1996. Age, stage and size structure as population state variables for *Tubifex tubifex* (Oligochaeta, Tubificidae). Hydrobiologia 334: 125–132.

Pasteris, A., M. Vecchi & G. Bonomi, 1999. A comparison among different population models for *Limnodrilus hoffmeisteri* Claparède (Oligochaeta, Tubificidae). Hydrobiologia 406: 183–189.

Pasteris, A., M. Vecchi, T. B. Reynoldson & G. Bonomi, 2003. Toxicity of copper-spiked sediments to *Tubifex tubifex*

(Oligochaeta, Tubificidae): a comparison of the 28-day reproductive bioassay with a 6-month cohort experiment. Aquatic Toxicology 65: 253–265.

Poddubnaya, T. L., 1980. Life cycle of mass species of Tubificidae (Oligochaeta). In Brinkurst, R. O. & D. G. Cook (eds), Aquatic Oligochaete Biology. Plenum Publishing Corporation, New York, 175–184.

Ratsak, C. H., S. A. L. M. Kooijman & B. W. Kooi, 1993. Modelling the growth of an oligochaete on activated sludge. Water Research 27: 739–747.

Risnoveanu, G. & A. Vadineanu, 2002. Observations on the population dynamics of *Potamothrix hammoniensis* (Michaelsen, 1901) (Tubificidae, Oligochaeta) in Lake Isacova in the Danube Delta. Hydrobiologia 479: 23–30.

Stites, D. L., 1987. Population and production dynamics of an enchytraeid worm in a subtropical blackwater river. Canadian Journal of Fisheries and Aquatic Sciences 44: 1469–1474.

Thorhauge, F., 1975. A field and laboratory study on the reproduction of *Potamotrix hanmmoniensis* (Tubificidae, Oligochaeta). Archiv für Hydrobiologie 74: 449–474.

Timm, T., 1974. On the life cycles of aquatic Oligochaeta in aquaria. Academy of Sciences of the Estonian Sovietic Socialist Republic, Institute of Zoology and Botany. Biology of the Estonian Fresh-water Organisms. Hydrobiological Investigations 6: 97–118 [in Russian, with English summary].

Timm, T., 1980. Methods of culturing aquatic Oligochaeta. In Belyaev G. M., G. G. Vinberg, N. S. Gaevskaya, V. I. Zhadin, L. A. Zenkevich, N. I. Kashkin, O. G. Reznichenko & A. P. Shcherbakov (eds), Aquatic Oligochaeta Worms. Taxonomy, Ecology and Faunistic Studies in the USSR. Proceedings of the Symposium on Aquatic Oligochaeta Held in Tartu, 1967. Translated from Russian, Amerind Publishing Co., New Delhi, 119–131.

Timm, T., 1984. Potential age of aquatic Oligochaeta. Hydrobiologia 115: 101–104.

Wilson, E. O. & W. H. Bossert, 1971. A Primer of Population Biology. Sinauer Associates, Sunderland, 192 pp.

Hydrobiologia (2006) 564:73–86
P.F.M. Verdonschot, H. Wang, A. Pinder & R. Nijboer (eds), Aquatic Oligochaete Biology IX
DOI 10.1007/s10750-005-1709-9


# Asexual reproduction and segmental regeneration, but not morphallaxis, are inhibited by boric acid in *Lumbriculus variegatus* (Annelida: Clitellata: Lumbriculidae)

Veronica G. Martinez, Prashant K. Reddy & Mark J. Zoran*
*Department of Biology, Texas A&M University, College Station, TX, 77843, USA*
*(*Author for correspondence: Tel.: +1-979-845-8099; Fax: +1-979-845-2891; E-mail: zoran@mail.bio.tamu.edu)*

*Key words:* morphallaxis, asexual reproduction, fission, regeneration, boric acid, escape reflex

## Abstract

Body fragmentation, in some animal groups, is a mechanism for survival and asexual reproduction. *Lumbriculus variegatus* (Müller, 1774), an aquatic oligochaete worm, is capable of regenerating into complete individuals from small body fragments following injury and reproduces primarily by asexual reproduction. Few studies have determined the cellular mechanisms that underlie fragmentation, either regenerative or asexual. We utilized boric acid treatment, which blocks regeneration of segments in amputated fragments and blocks architomic fission during asexual reproduction, to investigate mechanistic relationships and differences between these two modes of development. Neural morphallaxis, involving changes in sensory fields and giant fiber conduction, was detected in amputated fragments in the absence of segmental regeneration. Furthermore, neural morphallactic changes occurred as a result of developmental mechanisms of asexual reproduction, even when architomy was prevented. These results show that fragmentation in *L. variegatus*, during injury or asexual reproduction, employs developmental and morphallactic processes that can be mechanistically dissociated by boric acid exposure. In regeneration following injury, compensatory morphallaxis occurred in response to fragmentation. In contrast, anticipatory morphallaxis was induced in preparation for fragmentation during asexual reproduction. Thus, various forms of regeneration in this lumbriculid worm can be activated independently and in different developmental contexts.

## Introduction

The freshwater oligochaete worm, *Lumbriculus variegatus* (Müller, 1774), possesses a remarkable capacity for the regeneration of lost body parts and reproduces asexually by architomy. These developmental processes involve two general patterns of regeneration: epimorphosis and morphallaxis. Epimorphic regeneration is characterized by the differentiation of a blastema and the *de novo* generation of body parts or tissues; in this case, the regeneration of head or tail body segments. Morphallaxis, on the other hand, is a pattern of regeneration that involves the transformation of existing body parts or tissues into newly organized structures. This reorganization of extant tissues does not require the proliferation of new cells. During oligochaete morphallaxis, original body segments undergo anatomical and physiological transformations to match their new positional identity along the animal's body axis (Drewes & Fourtner, 1990; Myohara et al., 1999). Since a limited number of head segments are regenerated in the lumbriculid, *L. variegatus*, and the enchytraeid, *Enchytraeus japonensis*, posterior segments often become relocated anteriorly, requiring dramatic changes in the anatomy and physiology of these segments as they change positional identity.

Drewes & Fourtner (1990) demonstrated the functional reorganization of escape reflexes and their neural correlates in *L. variegatus* during morphallactic regeneration. Rapid escape behaviors in this worm are mediated by specific giant fiber pathways. A conserved feature of virtually all oligochaetes is the presence of three giant fibers, located in the dorsal region of the ventral nerve cord, that mediate rapid escape reflexes (Zoran & Drewes, 1987). These through-conducting pathways have been described as a chain of segmentally-arranged giant axons that run the length of the nerve cord (Günther & Walther, 1971). The medial giant fiber (MGF) mediates head specific escape behaviors such as rapid anterior shortening and body reversal, while the paired lateral giant fibers (LGF) govern tail withdrawal and activation of swimming bouts (Drewes & Fourtner, 1989; Drewes, 1999). Morphallaxis of these neural pathways during changes in segmental position, as in hydra regeneration (Wolpert et al., 1971), may result from the counterbalance of morphogenic influences localized within anterior and posterior activation centers at the ends of regenerating body fragments.

Recent studies have characterized several paradigms for the investigation of cellular and molecular events associated with morphallaxis (Martinez et al., 2005; Lesiuk & Drewes, 2001a, b). Specifically, body segments isolated from posterior regions, which acquire a new, more anterior segmental position, exhibit transformations in touch sensory fields, giant fiber conduction velocity, axonal diameter, and other physiological properties appropriate for the fragment's new positional identity. Lesiuk & Drewes (2001b) demonstrated the formation of ectopic heads in *L. variegatus* as a result of ventral nerve cord ablation. Following ectopic head formation, morphallactic changes occur within the giant fiber system, but only in 10–15 segments posterior to the newly formed ectopic head. These studies suggest that formation of a new head, *via* epimorphosis, may be required for activation of developmental mechanisms that regulate (neural) morphallaxis.

Boric acid, the most prevalent form of environmental boron, has been shown to be toxic in many species through its impact on reproductive and developmental events (Shomron & Ast, 2003), though its actions are not thought to be carcinogenic or mutagenic (Moore, 1997). It has been suggested that boric acid might disrupt epimorphosis (segmental regeneration) in lumbriculid worms (Drewes, 2003). Here we utilized boric acid treatment to test whether disruption of segmental regeneration would abrogate mechanisms of neural morphallaxis in *L. variegatus* (Martinez et al., 2005). In addition, we have set out to test whether morphallaxis is dependent on concomitant epimorphosis, employing boric acid to disrupt epimorphic production of new head and tail buds. We demonstrate that fragments do not undergo head or tail formation in the presence of active concentrations of boric acid. Nonetheless, morphallaxis was still detected in these fragments, suggesting that morphallactic mechanisms are not dependent on ongoing epimorphosis. Furthermore, boric acid suppressed architomic fission in worms, but did not disrupt morphallaxis. Therefore, developmental mechanisms requiring morphallaxis are independent of the processes governing segment formation (epimorphosis) and architomy.

## Materials and methods

### *Animals and maintenance*

Worms of *L. variegatus* were purchased from Flinn Scientific, Inc. (Batavia, IL). They were housed in moderately aerated Ozarka spring water, at a constant temperature of about 16 °C in the dark. Worms were provided brown paper towel clippings for substrate and were fed *Spirulina* powder and Tetramin staple flakes twice weekly.

### *Generation of regenerating body fragments*

To obtain experimental body fragments, worms were briefly anesthetized in 0.25 mM Nicotine in spring water. Segmental amputations were made at intersegmental boundaries with microdissecting scissors. Body fragments consisted of approximately 30 segments from either the anterior third or the posterior third of the worm. Regenerating body fragments were maintained individually in containers of spring water at 16 °C. Although animals regularly fragment by asexual fission in laboratory cultures, the rate of fission is depressed in animals maintained with sufficient aeration and substrate at cool temperatures. To induce production of body fragments by asexual fission, worms were exposed to an environmental shift. Animals were moved to room temperature (22 °C) for 3–4 days and then returned to 16 °C culture conditions that lacked paper substrate and aeration. Following this shift in environmental quality, ninety percent of animals typically fragment in 3–4 weeks.

*Boric acid treatment*

Whole animals, or body fragments, were immersed in spring water containing boric acid (99.9% purity; 100 pM–50 mM) for 2 weeks. Animal survival and segmental regeneration was monitored daily. Newly regenerated head and tail buds were visually inspected for the presence of defined segmental characteristics such as segmental boundaries and organization of the vasculature. Segments were counted as true segments based on the presence of defined intersegmental boundaries. Measurements of fragment weight, sensory fields, and giant fiber conduction velocities were calculated weekly. At the conclusion of exposure and testing periods, animals were dissected and fixed for histological analysis.

*Electrophysiological recording*

Impulse conduction along giant nerve fibers was studied using non-invasive electrophysiological recordings (O'Gara et al., 1982; Drewes & Fourtner, 1990). Touch stimuli were delivered by a hand-held plastic probe. Medial giant fiber (MGF) and lateral giant fiber (LGF) action potential waveforms were distinguished based on previously reported spike characteristics (Drewes & Fourtner, 1989, 1990; Rogge & Drewes, 1993). Extracellular voltage recordings were obtained using a printed-circuit-board grid of electrode pairs (1 mm space between positive and negative electrodes) and electrical signals were preamplified using a pair of differential recording amplifiers (100 × gain, AC-coupled differential inputs). These spike recordings were digitized with a Powerlab A–D conversion system (ADInstruments, Inc.) and were analyzed on a G4 Macintosh computer (Apple, Inc.) using the Powerlab Chart software.

Spike conduction time between two pairs of recording electrodes (5 mm distance between the electrode pairs) was measured from peak-to-peak of giant fiber spike waveforms recorded on the corresponding input channels. Conduction velocity (m $s^{-1}$) was obtained by dividing the conduction distance by the spike conduction time. Individual means, based on 3–5 measurements per animal, were used in calculating group means. Non-invasive recording grids were also used to map giant fiber sensory fields. Segments of specific identity (e.g. segment number 50) were marked with a spot of water-insoluble ink (Sharpie). Individual segments were then touched with a probe and giant fiber responses were monitored electrophysiologically.

*Histology*

Whole mount preparations were first pinned out on sylgard blocks while immersed in 0.25 mM nicotine in spring water, which blocks reflexive muscle movements and prevents autotomy (Lesiuk & Drewes, 1999). The VNC was exposed by removing the digestive tract, nephridia, and ventral blood vessel. Fragments were fixed in 4% paraformaldehyde, washed in PBS, stained with a dilute solution of toludine blue, rinsed with water and mounted on slides for light microscopy. All fixative and buffer solutions were maintained at ±4 °C and at pH 7.2.

Preparations were imaged using an Olympus inverted microscope, DIC optics, and a Hamamatsu CCD camera. Images were captured with Simple PCI software (C-Imaging, Inc.) and imported into Adobe Photoshop 6.0. Image analysis was conducted to select equivalent regions of interest from MGF and LGF dorso-longitudinal profiles along the length of the ventral nerve cord. GF diameters were measured for both MGF and LGF from whole mount preparations. Multiple measurements were made from each preparation along the length of a fragment and group means were then calculated.

*Statistics*

Two-tailed student's *t*-tests (Microsoft Excel) or ANOVA (Statistica, Inc.) were used for statistical analysis. Data are presented as mean plus or minus standard deviation (SD) or standard error of the mean (SEM) as indicated. Statistical significance was $p < 0.05$.

## Results

### *Effects of boric acid on segmental regeneration*

Following amputation, body fragments of 30 segments always regenerate a head of 8 segments and a tail of variable length, depending upon the fragment's original mass and positional identity (Fig. 1a). Exposure of fragments to 10 mM boric acid (BA) for 2 weeks disrupted segment formation without resulting in animal lethality (Fig. 1b). Lethality dose–response curves show that higher concentrations of BA (25–50 mM) produced significant reductions in fragment survival, while concentrations of 1 mM, 1 nM, 100 pM, and lower caused no mortality in worm fragments. However, concentrations of BA less than 10 mM also failed to disrupt segmental regeneration completely. Fragments exposed to 10 mM BA exhibited normal wound healing, but abnormal blastema formation. Consequently, initial bud formation was delayed until 4–5 days following amputation; whereas, in lower concentrations of BA, head and tail buds were visible 1–2 days post-amputation. In 10 mM BA, both anterior-third and posterior-third fragments had severe developmental defects in the number of head and tail segments produced as compared to untreated control fragments from similar axial regions of origin (Fig. 1c–e). Fragments treated with boric acid produced head buds that were reduced in size, from 8 well-defined segments to 3–5 segments, and lacked structural organization and a distinct prostomium (Fig. 1f, g). The impact of boric acid was most pronounced in tail regions of fragments. Tail buds produced by fragments of both anterior and posterior axial origin failed to develop any well-defined segments (Fig. 1g).

### *Analysis of neural morphallaxis in the absence of segmental regeneration*

Giant fiber mediated pathways are activated by sensory field inputs that vary along the anterior to posterior length of the animal. In our control animals, activation of touch receptors in the anterior 1/3 region of the animal's body wall (44 ± 2.1 segments, ~29% of total segments) triggered only the medial giant fiber (MGF), resulting in a head withdrawal (Fig. 2a). In the posterior 2/3 region of the worm (93.2 ± 2.5 segments, ~62% of the total segments), tactile stimulation elicited only lateral giant fiber (LGF) spiking, which resulted in a tail withdrawal. MGF and LGF sensory fields coexist in only 12.8 ± 2.3 segments of sensory field overlap, the equivalent of 9% of body segments (Fig. 2a). Therefore, fragments of 30 segments removed from anterior and posterior regions possessed exclusively MGF and LGF sensory fields, respectively, at the time of amputation. During neural morphallaxis, fragments removed from posterior regions transformed their escape reflex circuitry (i.e. gained MGF sensory activation), thus responding with behaviors appropriate for their new (anterior) positional identity.

Fragments treated in boric acid exhibited changes in giant fiber sensory fields similar to those of normal regenerating fragments (Fig. 2b). Specifically, 88% of posterior fragments treated in boric acid produced head withdrawal responses following 2 weeks of regeneration. Therefore, transformation of giant fiber sensory fields was not affected by BA-induced inhibition of segmental regeneration. In addition, BA-treated fragments exhibited changes in giant fiber conduction velocity that were appropriate for the fragment's new position along the length of the animal. Prior to amputation and BA exposure, MGF and LGF spikes were propagated through the posterior-third of the worm's body with similar conduction velocities (Fig. 2c, e). Following amputation and acute exposure to BA, tactile stimulation within the posterior-third of the worm's body did not result in activation of the medial giant fiber but LGF spikes were conducted at approximately 6 m $s^{-1}$ (Fig. 2e). However, a gradual increase in MGF conduction velocity was detected in these fragments over a subsequent 2 weeks in BA, such that conduction velocities eventually attained values similar to those of anterior control segments. These data indicate that neural morphallaxis occurred in posterior fragments, even in the absence of segmental regeneration.

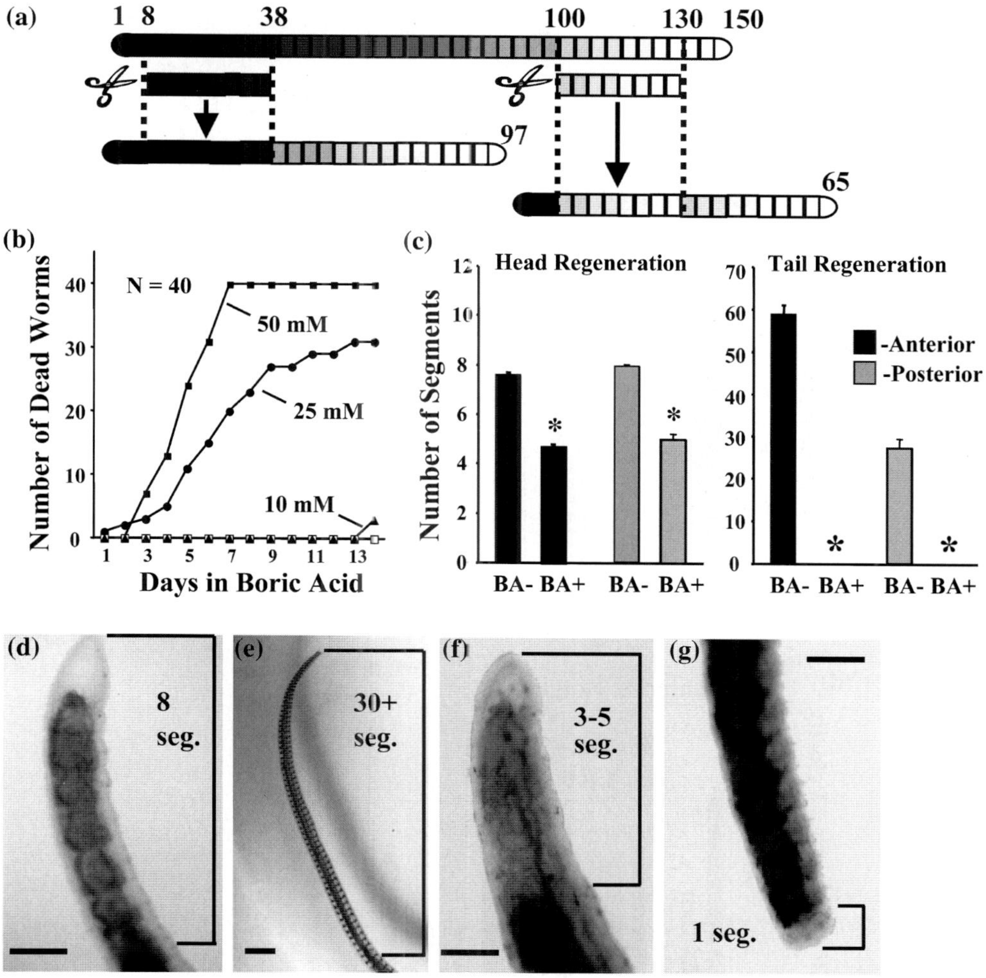


*Figure 1.* Effects of boric acid on segmental regeneration. (a) Segmental regeneration in *L. variegatus*. Whole animals of approximately 130 segments in length were cut into pieces, there by generating two populations of fragments: one from anterior (segment ~8–38) and one from posterior (segment ~100–130) regions of the worm. Amputated fragments always regenerated 7–8 segments of head and variable lengths of tail. Posterior fragments, which originally comprised segments 100–130, became more anteriorly positioned following segmental regeneration of the short head and longer tail. Numbers represent segmental numbers and shading implies anterior–posterior diffusion in original segment positional identity. (b) Lethality dose–response of boric acid. Graph showing lethality of concentrations of boric acid on *L. variegatus*. Concentrations ranged from 0 (open squares) to 50 mM (as indicated). Experimental number equals 40 animals per group. (c) Boric acid treatment of segmentally regenerating fragments. Head and tail regeneration in both anterior ($n = 39$) and posterior ($n = 36$) fragments was significantly reduced by 10 mM boric acid (*, $p < 0.001$). Bars = group mean. Error bars give SD (d–g). Regenerated head and tail segments on control fragments cultured in spring water (d, e) and experimental fragments cultured in 10 mM boric acid (f, g). Although some segmental regeneration of head segments occurred during one week in BA, disorganized structure of these segments (note the abnormal vasculature) was evident (f). Tail buds, even one week after BA treatment, were virtually nonexistent. All images were captured at 3 weeks after amputation. Scale bars = 2 μm (d, f, g) and 0.6 μm (e).

The diameters of the three giant fibers (Fig. 3a) vary along the length of the ventral nerve cord. MGFs have diameters that are much larger in the anterior region of the worm and then taper in size posteriorly. MGF and LGF diameter in posterior fragments was significantly changed following 9 weeks of morphallaxis (Fig. 3b). GF diameters in anterior fragments remained relatively

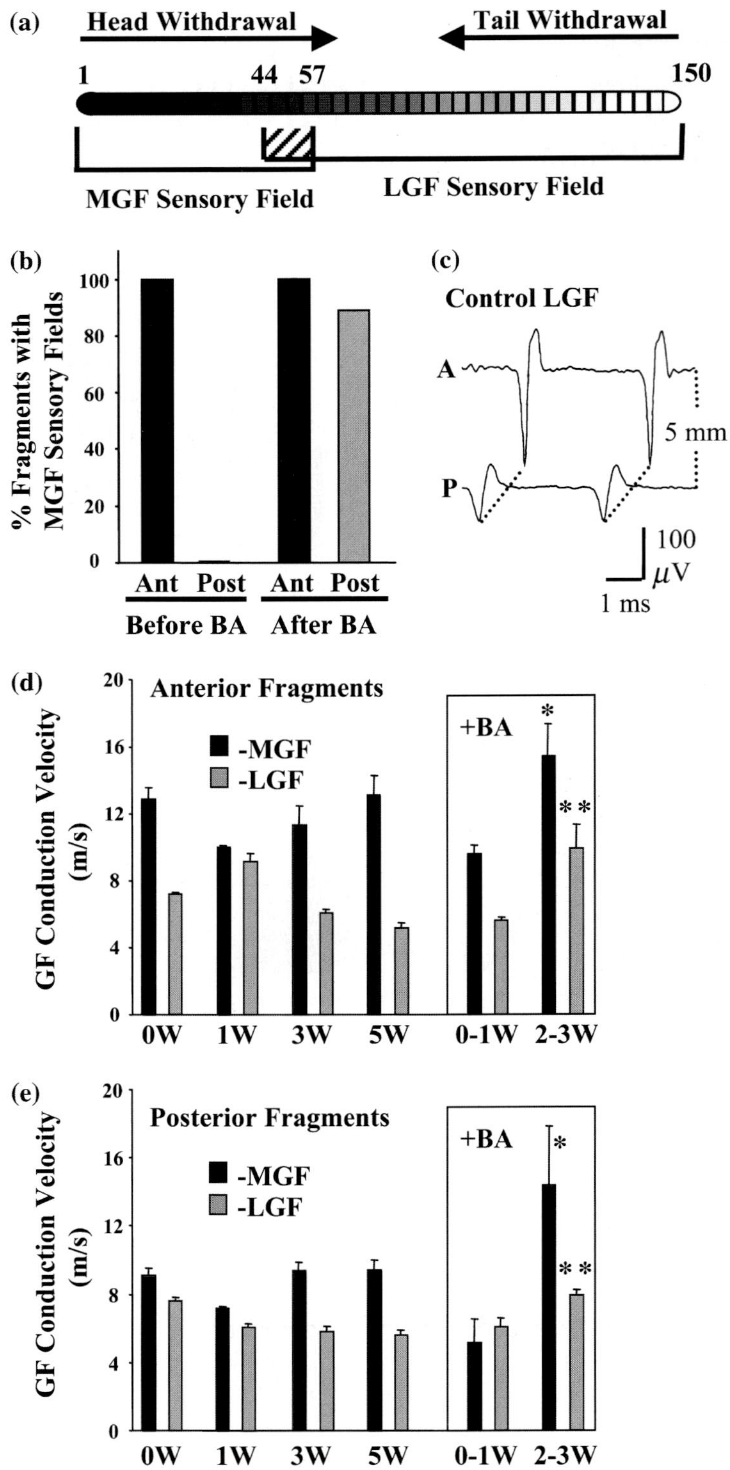
(a)
Head Withdrawal
Tail Withdrawal
1
44 57
150
MGF Sensory Field
LGF Sensory Field
(b)
% Fragments with MGF Sensory Fields
100
80
60
40
20
0
Ant Post
Ant Post
Before BA
After BA
(c)
Control LGF
A
P
5 mm
100 μV
1 ms
(d)
Anterior Fragments
GF Conduction Velocity (m/s)
20
16
12
8
4
0
-MGF
-LGF
+BA
0W 1W 3W 5W 0-1W 2-3W
(e)
Posterior Fragments
GF Conduction Velocity (m/s)
20
16
12
8
4
0
-MGF
-LGF
+BA
0W 1W 3W 5W 0-1W 2-3W

*Figure 2.* Boric acid does not effect changes in sensory field or conduction velocity during neural morphallaxis. (a) Giant fiber sensory field map for control whole animals. Head and tail rapid escape withdrawal was recorded following tactile stimulation to body segments. Specifically, stimulation of the anterior 1/3 region (segment 1–43) activated only a medial giant fiber (MGF) spike resulting in a head withdrawl. Stimulation of the posterior 2/3 region (segment 58–150) activated only lateral giant fiber (LGF) spiking causing a tail withdrawal. An area where both MGF and LGF spikes were activated (sensory field overlap) was detected, on average, between segments 44–57 ($n = 6$). (b) MGF sensory fields in fragments before and 3 weeks after boric acid treatment. Prior to boric acid (10 mM) exposure, newly amputated anterior fragments possessed medial giant fiber sensory fields in 100% of the population ($n = 34$). Medial giant fiber sensory fields were never detected in posterior fragments prior to boric acid treatment ($n = 39$). Following BA treatment, greater than 85% of posterior fragments possessed MGF sensory activation in a portion of the original 30 segments. (c) Representative giant fiber electrophysiological recording from a control fragment. To measure conduction velocities, MGF spikes or LGF spikes (as in this example recording) were recorded from two different electrode pairs set 5 mm apart. Velocities were measured by dividing the conduction distance by the peak-to-peak conduction time (as indicated by dotted lines). A and P equal anterior and posterior recording. Thus, this LGF spike is conducting in a posterior-to-anterior direction. (d) Giant fiber conduction velocity in anterior fragments during segmental regeneration. Prior to amputation (0 W), MGF conduction velocities measured $12.9 \pm 0.7$ m $s^{-1}$ and LGF conduction velocities measured $7.2 \pm 0.1$ m $s^{-1}$ in the anterior region of the worm ($n = 8$). Following amputation, MGF and LGF conduction velocities had equalized to approximately 10 m $s^{-1}$ at 1 week (1 W) and then returned to normal rates after 5 weeks of segmental regeneration. Fragments treated in boric acid (boxed region of histograms) demonstrated similar, relative changes in conduction velocity between MGFs and LGFs during segmental regeneration ($n = 9$), although both MGF and LGF velocities were significantly greater than controls at 3 weeks (*, $p < 0.05$; **, $p < 0.0001$). (e) Giant fiber conduction velocity in posterior fragments during segmental regeneration. Prior to amputation (0 W), MGF conduction velocities measured $9.1 \pm 0.1$ m $s^{-1}$ and LGF conduction velocities measured $7.6 \pm 0.4$ m $s^{-1}$ in the posterior region of *L. variegatus* ($n = 8$). Following amputation, MGF conduction velocities increased while LGF velocities decreased, indicative of the fragment's new anterior position. Even when treated in boric acid, relative changes in giant fiber conduction velocity emerged in posterior fragments by 2–3 weeks post amputation ($n = 8$). MGF and LGF conduction velocity in BA treated fragments were significantly greater than those of control fragments at 3 weeks (*, $p < 0.05$; **, $p < 0.0001$). Bars = group mean. Error bars = SD.

unchanged during regeneration. Fragments treated with 10 mM BA, both anterior and posterior, exhibited no significant differences in GF diameter, as compared to untreated control fragments. Thus, morphallactic changes in giant fiber size were not impacted by BA exposure.

### *Asexual reproduction in* Lumbriculus variegatus

Many oligochaetes, including *L. variegatus*, have marked regenerative capabilities and reproduce by asexual fission. Since boric acid blocked segmental regeneration without affecting neural morphallaxis, we tested the hypothesis that asexual fission is mediated by a developmental mechanism that can be dissociated from processes underlying morphallactic regeneration. To test this hypothesis, we developed an environmental shift paradigm (see Methods) to induce asexual reproduction by fragmentation.

Following environmental shift, controls animals in three separate experiments (involving a total of 76 worms) fragmented at a rate of 55% of the population over a 3-week period. In contrast, no worms fragmented in parallel control cultures treated with 10 mM BA, even though they had experienced the environmental shift protocol (Fig. 4a). Thus, BA inhibits asexual reproduction in this *L. variegatus* paradigm. Asexually reproducing lumbriculid worms formed an initial zone of architomy (Fig. 4b) that was condensed over several days into an architomy site (Fig. 4c).

Fragmentation by asexual reproduction resulted in the generation of two, zooid fragments. Interestingly, we never observed more than two fragments produced by any one animal using this protocol. Furthermore, the architomic fission site was predictable. Fragmentation resulted in the production of an anterior fragment of $48 \pm 10$ segments (mean ± SD) and a posterior fragment of $92 \pm 14$ segments (Fig. 4d). The two zooids produced by asexual fission in these experiments each formed a blastema at the site of fragmentation, which then differentiated appropriately into head or tail buds. Additionally, wet-weight measurements were collected for each fragment produced via asexual fragmentation. Anterior clones, on the day of fragmentation, measured $5.7 \pm 1.6$ mg (mean ± SD) and posterior clones measured $5.0 \pm 1.0$ mg ($n = 23$; Fig. 4e). Because there is no significant difference between the masses of anterior and posterior clones, we conclude that asexual fragmentation, also generating two fragments of significantly different segment

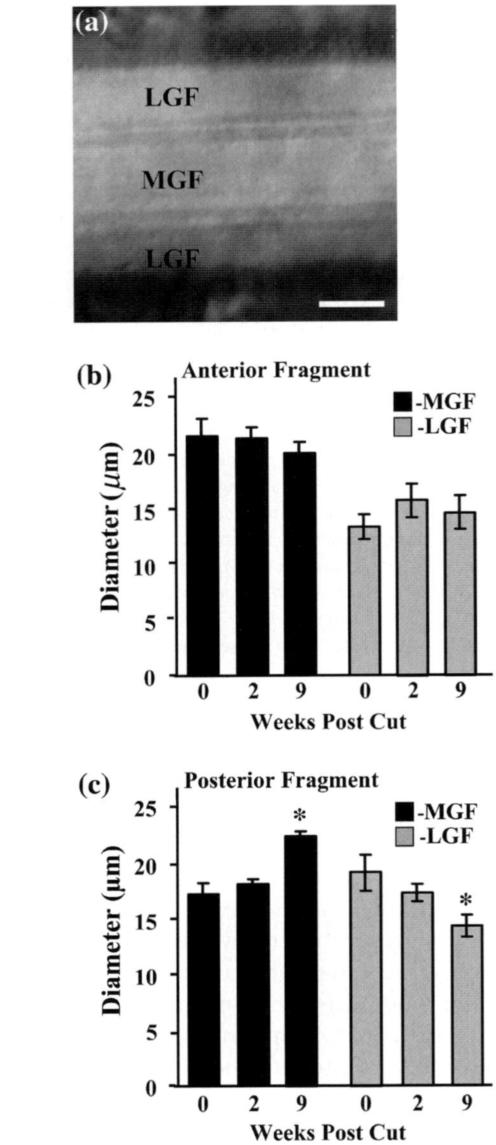


*Figure 3.* Boric acid does not effect changes in giant fiber size during neural morphallaxis. (a) Photomicrograph of an anterior fragment using Nomarski optics. Using similar images collected for all experimental groups, giant fiber diameters were measured by selection of equivalent regions of interest (using SimplePCI software) from giant fiber dorsal-longitudinal profiles along the length of the nerve cord. Axonal diameters were measured for both LGF and MGF in regions of interest. Scale bar = 15 $\mu$m. (b) Changes in giant fiber diameter during neural morphallaxis. No significant change in giant fiber diameter was detected in anterior fragments following 0, 2, and 9 weeks of regeneration. However, posterior fragments showed a marked decrease in LGF diameter after 9 weeks of regeneration ($p < 0.02$, $n = 9$). Significant changes in MGF size were also detected in posterior fragments after 9 weeks of regeneration ($p < 0.002$, $n = 9$).

numbers ($p < 0.0001$), resulted in the production of clones of equitable mass ($p = 0.1004$).

While neural morphallaxis has been described in fragments that were produced *via* injury, its involvement during asexual fragmentation has not been investigated. Our combined studies of giant fiber sensory fields and emerging architomy sites during asexual reproduction demonstrated that, on average, the predictable fission (architomy) site was located virtually at the center of the zone of MGF/LGF sensory field overlap. In some cases where animals had regenerated more recently, as suggested by differences in body wall pigmentation, new fission sites were located within 2–3 segments of the previous site. As the fission site developed, sensory field mapping detected a significant expansion of this area of sensory field overlap. Specifically, an increase in the MGF sensory field was observed, largely in segments posterior to the fission site. The zone of sensory field overlap expanded from 12 ± 2 segments (mean ± SD) to 29 ± 8 segments following fission site formation (Fig. 5a). Furthermore, an area of exclusively MGF sensory field of 10–15 segments just adjacent and posterior to the architomy site was formed.

*Asexual fragmentation is blocked by boric acid with no effect on neural morphallaxis*

To test the effects of boric acid on neural morphallaxis during asexual reproduction, we recorded morphallactic changes in giant fiber conduction velocity and axonal diameter following BA treatment. We measured conduction velocities specifically within segments 40 and 70 in asexually fragmenting (BA−) and non-fragmenting (BA+) animals. This region represented the segments with expanding MGF sensory fields, that is, segments posterior to the nascent fission site. Medial giant fiber conduction velocity recordings from BA-treated animals measured $9.5 \pm 0.8$ m $s^{-1}$(mean ± SD), as compared to MGF velocities of $9.3 \pm 0.6$ m $s^{-1}$ in non-fragmenting (BA−) animals (Fig. 5b). Thus, boric acid inhibition of fragmentation during asexual reproduction did not significantly effect alterations in GF conduction velocity associated with neural morphallaxis. Giant fiber size (axonal diameter) in asexually regenerating animal was not significantly

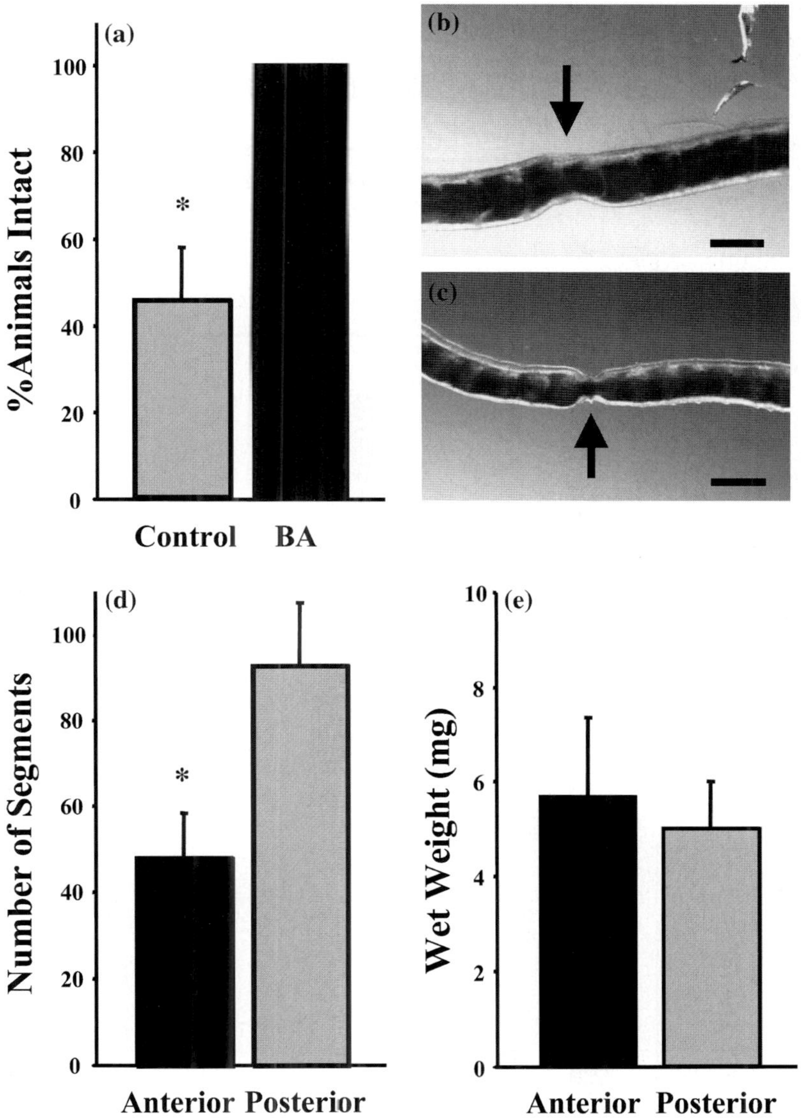


*Figure 4.* Asexual reproduction in *Lumbriculus variegatus.* (a) Asexual reproduction was induced in whole animals by an environmental shift in culture conditions (see Methods). Fragmentation, and consequently asexual reproduction, was blocked in animals treated with boric acid (10 mM). Following three weeks (two of these weeks being in BA-treated spring water), 100% of the animals had not fragmented or produced autotomy sites ($n = 76$). In control populations, 50% of animals fragmented by week 3 ($n = 76$). (b, c) *L. variegatus* reproduces asexually through architomic fission. During asexual reproduction, fission zones are formed (arrow in b, early fission). Obvious autotomy sites (arrow in c, late fission) are present prior to fission. No new segmental differentiation was seen in segments adjacent to the fission site, a feature characteristic of architomy. Scale bars = 2 $\mu$m. (d) Architomy sites are predictable in *L. variegatus.* In each animal fragmented during asexual reproduction, the numbers of head segments produced following fission were significantly less than tail segments. Architomy sites were consistently produced at segment 48 ± 10 in animals of approximately 140 segments ($p < 0.0001$; $n = 37$). (e) Cloned fragments produced following fission were of equal mass. Anterior fragments possessed wet weights that were not significantly different ($p = 0.11$, $n = 22$) from posterior fragments, even though they possessed approximately 1/3 of the total segments of the whole animal prior to fission.

different than that of animals treated with boric acid (Fig. 5c), again suggesting that the cellular correlates of neural morphallaxis are resistant to disruption by boric acid. Additionally, these GF size relationships did not differ from untreated control animals (not reproducing asexually).

Taken together, these results suggest that neural changes associated with morphallactic regen-

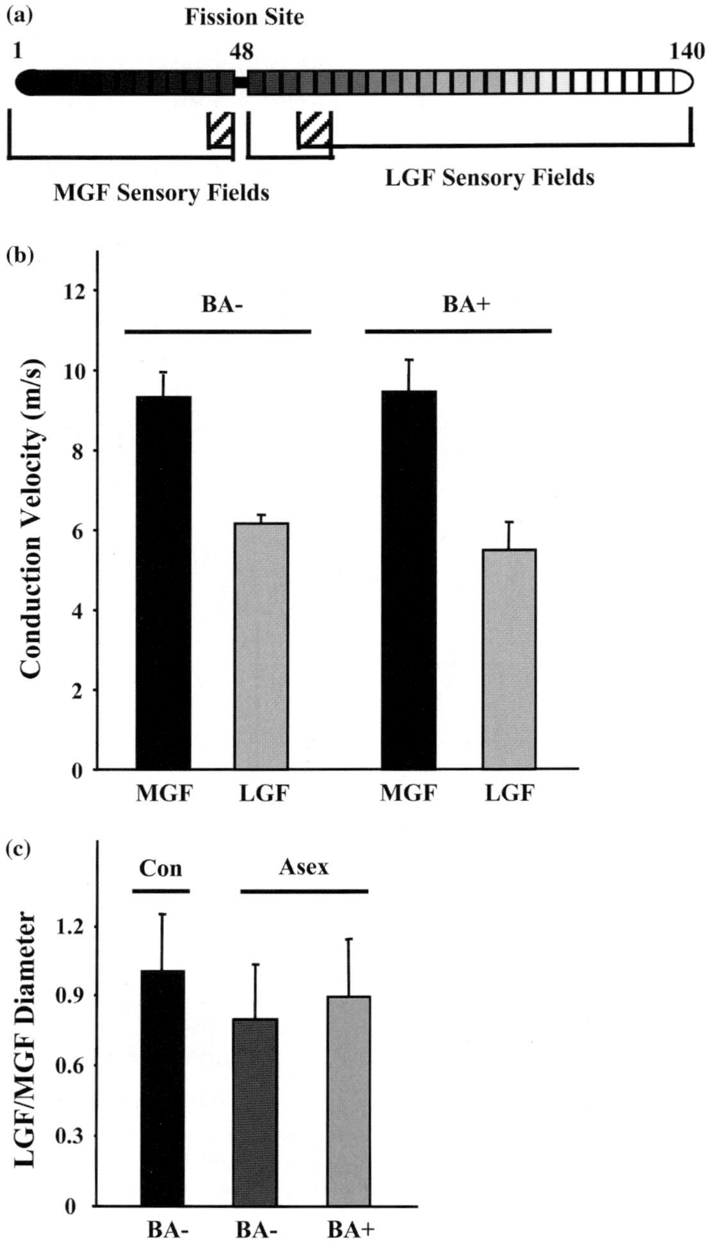


*Figure 5.* Boric acid does not affect neural morphallaxis during asexual reproduction. (a) Giant fiber sensory fields expanded prior to fragmentation during asexual reproduction. Fission sites developed at approximately segment 48, which was within the zone of MGF/LGF sensory field overlap in these animals. Sensory field mapping of asexually reproducing animals (w/fission sites) demonstrated an increase in the MGF sensory field in segments posterior to the fission site. Sensory field overlap expanded from 12 segments in controls to 29 segments in animals exposed to the environmental shift protocol for induction of asexual reproduction. In 10–15 segments posterior to the autotomy site, only MGF spikes could be evoked by tactile stimulation ($n = 6$). (b) Boric acid did not affect changes in giant fiber conduction velocity during asexual reproduction. MGF and LGF conduction velocities, measured within nascent or extant fission zones in both boric acid treated and control animals respectively ($n = 5$ per group) where not significantly different. (c) Changes in giant fiber diameter ratio were not effected by boric acid during asexual reproduction. There was no significant difference between giant fiber diameter ratios of animals treated with boric acid or spring water ($n = 5$ per group), following induction of asexual reproduction (Asex). Giant fiber diameter changes where comparable to those measured in control animals, which were not asexually reproducing (Con).

eration occurred in an anticipatory fashion (prior to architomy) during asexual reproduction.

## Discussion

Asexual reproduction and regeneration of lost segments are found in several groups of annelid worms, including several families of oligochaetes (Brinkhurst & Jamieson, 1971). Those oligochaetes that reproduce asexually typically possess the ability to regenerate head and tail segments following body transection (Morgulis, 1907; Berrill, 1952; Christensen, 1959; Myohara et al., 1999; Bely & Wray, 2001), with some exceptions (Bely, 1999). The ability of some oligochaetes, but not others, to regenerate body fragments, may be a consequence of the environmental niches in which they have evolved (Drewes & Zoran, 1989; Bely, 1999). *Lumbriculus variegatus* lives in shallow margins of ponds, where rapid environmental change and frequent predatory attack may be common. Asexual reproduction by architomic fission and regeneration following body injury are both common aspects of the life history of *L. variegatus* (Fig. 6a). These annelids are also hermaphrodites capable of reproducing sexually, however, they reproduced only by asexual reproduction in normal laboratory conditions.

Asexual reproduction in *L. variegatus*, cultured under controlled environmental conditions, involves a predictable fission site and reproductive outcome. Because fragmentation occurs prior to significant differentiation of head or tail buds, we conclude that fission in *L. variegatus* was by architomy. This form of asexual fission is rare among oligochaetes, where paratomic fission is the more common mode of reproductive fragmentation in those groups that have been extensively studied (Drewes & Fourtner, 1991; Bely, 1999; Bely & Wray, 2001). In paratomy, substantial head and tail segmental regeneration occurs prior to the actual separation of fragments (Giese & Pearse, 1975). The process of bud formation following architomic fission was much slower than bud formation following injury and wound healing (segmental regeneration). Buds with readily visible segmental boundaries exist at 3–4 days following body transection, while such buds are not evident on fragments produced by asexual fission until 8–11 days post-fragmentation. The delay in transition from blastema to bud in asexual reproduction may be slowed by the need for dedifferentiation of body wall tissue covering the initial blastema of asexual buds. Architomy sites formed at the anterior-to-posterior, one-third/two-third segmental boundary. This highly predictable location suggests that developmental mechanisms work to predetermine the site of asexual fission in this species. In contrast, in other annelids, for example, some enchytraeids, each asexual reproduction event leads to the generation of 5–10 fragments from a single animal (Yoshida-Noro et al., 2000).

Both asexual reproduction and segmental regeneration in *L. variegatus* utilize blastema formation and epimorphosis to add additional body segments to the fragment. A similar process exists in *Enchytraeus japonensis* (Myohara et al., 1999). This developmental process of epimorphosis is a form of compensatory regeneration found in oligochaetes. Another example of compensatory regeneration occurs in the tubificid worm, *Branchiura sowerbyi*, where gill filaments are produced on adult mid-body segments following loss of the gill filaments specific to posterior segments in adult worms (Drewes & Zoran, 1989). Both asexual fission and regeneration in *L. variegatus* involve morphallaxis, where adult segments transform the fragment to match their new positional identity. Previous studies characterized physiological, anatomical, and molecular changes that occur in the lumbriculid central nervous system during neural morphallaxis (Drewes & Fourtner, 1990; Martinez et al., 2005). Since fragments regenerate only a small number of new head segments, the anterior most segment of a posterior fragment becomes more anteriorly positioned following segmental regeneration. Since sensory systems and escape behaviors are mediated by mechanisms that vary along the anterior-posterior axis (Drewes & Fourtner, 1989; Drewes, 1999), neural morphallactic changes are especially dramatic in *L. variegatus*.

Many oligochaetes that reproduce by asexual fission also have significant regenerative capabilities. For this reason, it has been suggested that regenerative mechanisms might play an essential role during asexual fragmentation and may have contributed to the evolution of reproductive mechanisms such as paratomic fission (Bely, 1999;

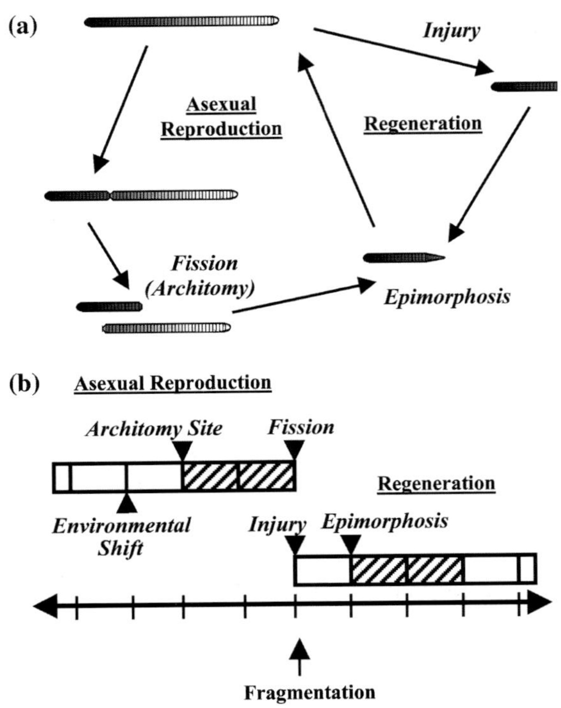


*Figure 6*. Asexual reproduction and segmental regeneration involve different sequences of developmental events. (a) *Lumbriculus variegatus* generated new individuals from fragments of body segments by asexual reproduction and regeneration. During asexual reproduction, whole animals produce architomic fission sites and following fragmentation the original animal separates into two body fragments. The formation of new segments by epimorphosis (bud formation) leads to the development of a complete adult worm. During regeneration following injury (or experimental transection), fragments form new buds by epimorphosis at the transection site. (b) Neural morphallaxis (changes in giant fiber sensory fields, area, and conduction velocity) occurs during both asexual reproduction and regeneration, however, the order of regenerative events is arranged differently over time during these two processes. Changes associated with neural morphallactic regeneration (filled squares) were detected prior to fragmentation events during asexual reproduction, appearing 2 weeks before fission. During regeneration, in contrast, morphallactic changes were not detectable until 1–2 weeks after injury. Thus, morphallaxis of the neural anatomy and physiology of original fragment segments were produced by anticipatory regenerative mechanisms in asexual reproduction; whereas, morphallaxis was retroactive during regeneration induced by injury.

Bely & Wray, 2001). We have demonstrated that the developmental processes of segmental regeneration and architomic fission both involve neural morphallactic regeneration. Additionally, the induction of epimorphosis and fission were compromised by boric acid treatment, but the developmental mechanisms that initiated and maintained morphallaxis were spared. Although we cannot determine whether the regenerative process of morphallaxis has been co-opted for asexual reproduction from segmental regeneration, it is shared, yet separable from each.

In the absence of regenerative epimorphosis or reproductive fragmentation, significant neural morphallaxis persisted. These results indicate that at least one regenerative process, neural morphallaxis, has been incorporated in asexual reproduction and although the mechanism of action of boric acid in these studies is unknown, it has been implicated in developmental defects in other systems including vertebrate axial development (Price et al., 1996; Fort et al., 1998). Boric acid is not metabolized in humans or animals (Murray, 1995) and at low consumption is reported to aid in wound healing possibly by increasing RNA synthesis of tumor necrosis factors (Benderdour et al., 1998), vascular endothelial growth factors and transforming growth factors (Dzondo-Gadet et al., 2002). Whether or not environmental levels of boric acid ever attain concentrations capable of impacting annelid asexual reproduction remains to be studied.

Anatomical and physiological correlates of neural morphallaxis in fragments produced by transection and asexual fission demonstrate that this form of regeneration is induced at different times relative to segmental regeneration by epimorphosis (Fig. 6b). In asexual reproduction, morphallactic regeneration is recruited in anticipation of architomy 1 week before the initiation of epimorphosis. In contrast, following injury-induced transection, morphallaxis occurred after the onset of epimorphosis. Thus, worms reproducing by asexual fission initiate developmental mechanisms in appropriate body regions (anticipatory morphallactic regeneration) in preparation for the future fragmentation event. Following fragmentation by injury, worms compensate for ongoing changes in axial position (compensatory morphallactic regeneration) coincident with the addition of new segments by epimorphosis. This process of anticipatory morphallaxis may not be peculiar to architomic fission, since changes is neural function likely occur prior to paratomic fission in the naidid worm, *Dero digitata*, even though the expression of these changes is inhibited prior to the fragmentation event (Drewes & Fourtner, 1991).

Not all annelids capable of segmental regeneration undergo asexual fission (Berrill, 1952), even though many of the developmental events that govern them are similar, if not co-opted. The processes of anticipatory and compensatory morphallaxis are critical elements of the developmental programs that regulate asexual reproduction and regeneration in *L. variegatus*. Still, the cellular and molecular mechanisms that govern morphallaxis are unclear. Based on similarities to developmental mechanisms in other invertebrates (Wolpert, 1971; Struhl, 1989), it has been hypothesized that formation of head and tail activation centers during these processes may regulate the establishment of axial gradients in GF sensory fields and escape behaviors (Drewes & Fourtner, 1990). If this is the case in *L. variegatus*, these organizing centers must be competent to form in the absence of ongoing segmental regeneration, as they occur in fragments where epimorphosis has been inhibited and during asexual reproduction prior to architomy. Nonetheless, we have demonstrated that the segmental zone of MGF/LGF sensory field overlap is virtually identical to that of the zone of fragmentation during asexual reproduction. It is likely that the developmental programs that establish sensory field axial gradients are the same events that direct the locus for architomic fission.

Clearly fragmentation in oligochaetes need not be located at zones of MGF/LGF sensory field overlap. Examples of fragmentation outside this region are occasionally observed in our *L. variegatus* cultures. In *Dero*, paratomic fission sites are typically located in segments at mid-LGF sensory field levels (Drewes & Fourtner, 1991). Whether or not these differences represent systematic disparity between architomy and paratomy remains to be determined. One common aspect of the regeneration of naidid (Drewes & Fourtner, 1991), lumbriculid (Lesiuk & Drewes, 2001b), and enchytraeid worms (Yoshida-Noro et al., 2000) is that the central nervous system likely plays a mechanistic role in this developmental process. The formation of a regeneration blastema often requires ongoing nerve fiber outgrowth in annelids (Dinsmore & Mescher, 1998). Our studies of morphallaxis during asexual fission and regeneration in *L. variegatus* provide new tools for addressing aspects of the relationship between these two developmental events and the potential role of neural influences. However, insight into the co-option of developmental mechanisms, the role of the nervous system in these processes, and the evolution of these events will not be clear until the genes that regulate them and their molecular evolutionary relationship are determined.

**Acknowledgements**

We thank Mr. Andres Mendoza for technical assistance in some of our studies and the late Dr Charles Drewes for many helpful suggestions and for kindly providing grids for non-invasive electrophysiological recording. This work was supported, in part, by an American Psychological Association, Minority Fellowship Program Grant T32 MH-18882 (V.G.M). Image analysis was conducted in the Cellular Physiology and Molecular Imaging Laboratory at Texas A&M University, supported by NINDS Grant PO1 NS-39546 (M.J.Z.).

**References**

Bely, A. E., 1999. Decoupling of fission and regenerative capabilities in an asexual oligochaete. Hydrobiologia 406: 243–251.

Bely, A. E. & G. A. Wray, 2001. Evolution of regeneration and fission in annelids: insights from engrailed- and orthodenticle-class gene expression. Development 128: 2781–2791.

Benderdour, M., K. Hess, M. Dzondo-Gadet, P. Nabet, F. Belleville & B. Dousset, 1998. Boron modulates extracellular matrix and TNFa synthesis in human fibroblasts. Biochemical and Biophysical Research Communications 246: 746–751.

Berrill, N. J., 1952. Regeneration and budding in worms. Biological Reviews 27: 401–438.

Brinkhurst, R. O. & B. G. Jamieson, 1971. Aquatic Oligochaeta of the World . University of Toronto Press, Toronto 860 pp.

Christensen, B., 1959. Asexual reproduction in the Enchytraeidae (Olig.). Nature 184: 1159–1160.

Dinsmore, C. E. & A. L. Mescher, 1998. The role of the nervous system in regeneration. In Ferretti, P. & J. Geraudie (eds), Cellular and Molecular Basis of Regeneration. Wiley, New York, NY, 79–108.

Drewes, C. D. & C. R. Fourtner, 1989. Hindsight and rapid escape in a freshwater oligochaete. Biological Bulletin 177: 363–371.

Drewes, C. D. & M. J. Zoran, 1989. Neurobehavioral specializations for respiratory movements and rapid escape from predators in posterior segments of the tubificid *Branchiura sowerbyi*. Hydrobiologia 180: 65–71.

Drewes, C. D. & C. R. Fourtner, 1990. Morphallaxis in an aquatic oligochaete, *Lumbriculus variegatus*: reorganization of escape reflexes in regenerating body fragments. Developmental Biology 138: 94–103.

Drewes, C. D. & C. R. Fourtner, 1991. Reorganization of escape reflexes during asexual fission in an aquatic oligochaete, *Dero digitata*. Journal of Experimental Zoology 260: 170–180.

Drewes, C. D., 1999. Helical swimming and body reversal behaviors in *Lumbriculus variegatus* (Annelida: Clitellata: Lumbriculidae). Hydrobiologia 406: 263–269.

Drewes, C. D., 2003. A toxicology primer for student inquiry: biological smoke detectors. Kansas School Naturalist, Emporia State University 50: 1–15.

Dzondo-Gadet, M., R. Mayap-Nzietchueng, K. Hess, P. Nabet, F. Belleville & B. Dousset, 2002. Action of boron at the molecular level: effects on transcription and translation in an acellular system. Biological Trace Element Research 85: 23–33.

Fort, D., T. Propst, E. Stover, P. Strong & F. Murray, 1998. Adverse reproductive and developmental effects in *Xenopus* from insufficient boron. Biological Trace Element Research 66: 237–259.

Giese, A. C. & J. S. Pearse, 1975. Reproduction of Marine Invertebrates, Annelids, and Echiurans, Vol. 3. Academic Press, New York 343 pp.

Günther, J. & J. B. Walther, 1971. Funktionelle Anatomie der dorsalen Riesenfaser-Systeme von *Lumbricus terrestris*. Zeitschrift fur Morphologie der Tiere 70: 253–280.

Lesiuk, N. M. & C. D. Drewes, 1999. Autotomy reflex in a freshwater oligochaete, *Lumbriculus variegatus* (Clitellata: Lumbriculidae). Hydrobiologia 406: 253–261.

Lesiuk, N. M. & C. D. Drewes, 2001a. Behavioral plasticity and central regeneration of locomotor reflexes in freshwater oligochaete, *Lumbriculus variegatus*. I: Transection studies. Invertebrate Biology 120: 248–258.

Lesiuk, N. M. & C. D. Drewes, 2001b. Behavioral plasticity and central regeneration of locomotor reflexes in the freshwater oligochaete, *Lumbriculus variegatus*. II. Ablation studies. Invertebrate Biology 120: 259–268.

Martinez, V. G., G. J. Menger III & M. J. Zoran, 2005. Regeneration and asexual reproduction share common molecular changes: upregulation of a neural glycoepitope during morphallaxis in *Lumbriculus*. Mechanisms of Development 122: 721–732.

Moore, J., 1997. An assessment of boric acid and borax using the IEHR evaluative process for assessing human developmental and reproductive toxicity of agents. Reproductive Toxicology 11: 123–160.

Morgulis, S., 1907. Observations and experiments on regeneration in *Lumbriculus*. Journal of Experimental Zoology 4: 549–574.

Murray, F., 1995. A human health risk assessment of boron (boric acid and borax) in drinking water. Regulatory Toxicology and Pharmacology 22: 221–230.

Myohara, M., C. Yoshida-Noro, F. Kobari & S. Tochinai, 1999. Fragmenting oligochaete *Enchytraeus japonensis*: a new material for regeneration study. Development, Growth, and Differentiation 41: 549–555.

O'Gara, B., E. P. Vining & C. D. Drewes, 1982. Electrophysiological correlates of rapid escape reflexes in intact earthworms, *Eisenia foetida*. I. Functional development of giant nerve fibers during embryonic and postembryonic periods. Journal of Neurobiology 13: 337–353.

Price, C., M. Marr, C. Myers, J. Seely, J. Heindel & B. Schwetz, 1996. The developmental toxicity of boric acid in rabbits. Fundamental and Applied Toxicology 34: 176–187.

Rogge, R. & C. D. Drewes, 1993. Assessing sublethal neurotoxicity effects in the freshwater oligochaete, *Lumbriculus variegatus*. Aquatic Toxicology 26: 73–90.

Shomron, N. & G. Ast, 2003. Boric acid reversibly inhibits the second step of pre-mRNA splicing. Federation of European Biochemical Societies Letters 552: 219–224.

Struhl, G., 1989. Differing strategies for organizing anterior and posterior body pattern in *Drosophila* embryos. Nature 338: 741–744.

Wolpert, L., 1971. Positional information and pattern formation. Current Topics in Developmental Biology 6: 183–224.

Yoshida-Noro, C., M. Myohara, F. Kobari & S. Tochinai, 2000. Nervous system dynamics during fragmentation and regeneration in *Enchytraeus japonensis* (Oligochaeta, Annelida). Development Genes and Evolution 210: 311–319.

Zoran, M. J. & C. D. Drewes, 1987. Rapid escape reflexes in aquatic oligochaetes: variations in design and function of evolutionarily conserved giant fiber systems. Journal of Comparative Physiology A 161: 729–738.

Hydrobiologia (2006) 564:87–93
P.F.M. Verdonschot, H. Wang, A. Pinder & R. Nijboer (eds), Aquatic Oligochaete Biology IX
DOI 10.1007/s10750-005-1710-3



# Seasonal vertical migration and aestivation of *Rhyacodrilus hiemalis* (Tubificidae, Clitellata) in the sediment of Lake Biwa, Japan

Tetsuya Narita
*8-25-9, Sakamoto, Otsu, 520-0113, Shiga, Japan*
*(Fax: +81-77-5790057; E-mail: narita8259@ybb.ne.jp)*

*Key words:* *Rhyacodrilus hiemalis*, vertical distribution, seasonal vertical migration, aestivation, Tubificidae, Lake Biwa

**Abstract**

The vertical distribution of the tubificid worm *Rhyacodrilus hiemalis* Ohtaka, the numerically dominant species of oligochaete in the littoral of Lake Biwa, was studied with special reference to seasonal vertical migration in the lake sediment. Monthly collections of lake sediment cores were made using PVC tubes. Core sections of sample sediments ranged from 76 to 117 cm. The vertical distribution of the worms showed no diurnal variation; therefore diel vertical migration was not evident. Seasonal downward migration started in April, and upward migration started in October. From December to March, almost all worms remained in the near-surface sediment layer (surface to 30 cm deep), while from July to September almost all worms remained deeper than 30 cm. However, few individuals migrated deeper than 90 cm. No discontinuous layers were found in grain size composition, water content, loss on ignition, particulate carbon, nitrogen or phosphorus. In deep sediment there was no free oxygen, as evidenced by negative ORP values. For 4 months in summer, *R. hiemalis* aestivated, probably utilizing anaerobic respiration. It appeared that *R. hiemalis* moved deeper in the sediment in response to sediment temperature, because sediment temperatures in the deep layers seemed to converge at around 20–21 °C in the summer months. The life history traits of seasonal vertical migration and summer aestivation perhaps arose as an adaptation to the climatic conditions accompanying the geographical origin of *R. hiemalis*, and they also serve to minimize predation risk during summer when most invertebrate predatory fishes are active.

## Introduction

For benthic animals, vertical movement into lake sediment is advantageous to escape and hide from predators, and to avoid unfavourable environmental conditions. Disadvantages of this movement are that lake sediments are sometimes subject to oxygen deficit and a lower food availability, though tolerance to such features may differ between species (Nelson, 1991; Hamburger et al., 1998).

The vertical distribution of aquatic oligochaetes has been studied from the viewpoints of predation avoidance (Kajak & Dusoge, 1971; Milbrink, 1973; Newrkla & Mutayoba, 1987), food availability (Nalepa & Robertson, 1981), environmental condition (Newrkla & Wijegoonawardana, 1987) and life history (Lazim et al., 1989). Nalepa & Robertson (1981) discussed the seasonal change of vertical distribution in relation to the organic detritus thickness. However, few studies have considered the temporal changes in vertical distribution.

*Rhyacodrilus hiemalis* Ohtaka is a freshwater tubificid species endemic to Japan (Timm, 1997) and is commonly found in many Japanese lakes (Ohtaka, 1995). In Lake Biwa *R. hiemalis* is a dominant tubificid and distributed mainly in the littoral zone. Its reproductive cycle of breeding in mid-winter, when it deposits cocoons on hard substrata (Narita, 2001), is unique among the tubificids in Lake Biwa.

In this paper, as a part of the life history study of dominant oligochaetes in Lake Biwa, the seasonal vertical migration of *Rhyacodrilus hiemalis* in the sediment of Lake Biwa is examined from the viewpoint of life history traits.

## Study area, materials and methods

The study was conducted at a littoral site (400 m offshore, about 4 m deep) of the south basin of Lake Biwa from September 1983 to December 1984. The south basin is ca. 27 $km^2$ in area, and owing to its shallowness (max depth, ca. 6 m), does not show stable thermal stratification even in mid-summer. The sediment of the study site was predominantly silt-clay, and the sediment surface was partially covered with detritus; aquatic macrophytes were absent.

Core collections were made with PVC tubes (diameter, 12.3 cm; length, 100 or 125 cm) in the mornings of 5 September, 5 October, 1 November, 7 December in 1983, and 18 January, 29 February, 30 March, 16 April, 9 May, 7 June, 4 July, 7 August, 5 September, 9 October, 7 November and 7 December in 1984. Night collections were made at around midnight of 23 April and 23 July, 1984. Three core samples were collected with 100-cm tubes on each sampling date, except for 5 September 1983, when four cores were collected. The sediment depth that was actually penetrated was 76–80 cm in 14 cores, 80–85 cm in 22 cores, 86–90 cm in 15 cores, and 92 cm in 1 core, except in September 1984 when sediment cores of 111, 113 and 117 cm were collected with 125-cm tubes. There was no seasonal difference in the length of sediments collected. The core tube was pushed vertically into the sediment by a SCUBA-assisted diver. Before extracting the core tube from the sediment, a rubber stopper closed the upper opening. The core tube was then carefully extracted from the sediment by hand and, with the aid of a rope, pulled up from the boat. Before the core reached the water surface, another rubber stopper was pushed into the bottom opening of the core tube.

In September 1983, and April, May, July, August, October and December 1984, the sediment samples were sectioned on the boat, but in other months the samples were taken back to shore and then sectioned. The time delay from the sampling to the end of sectioning was about 40 min for the first core and about 90 min for the last core. Night samples of sediment were sectioned immediately on the boat.

The sample sediments were sectioned at 5 cm intervals from the sediment surface to 30 cm deep, and at 10 cm intervals below 30 cm, by gently extruding the mud upward from the bottom of the core tube to the open upper end using a rubber piston. Sectioned sediments were washed through a 0.5 mm screen, and sorted live by eye with a hand-held magnifying glass. Animals were preserved in 10% formalin solution for later identification and counting. Dubious oligochaete specimens were slide-mounted for identification by chaetal morphology.

To examine whether any large-scale vertical displacement of *R. hiemalis* took place in the cores after sampling, a short sediment sample (about 18 cm) taken by Lenz-type grab (14.2 cm square, 30 cm high) was partitioned immediately after sampling, since it was assumed that a large-scale displacement of *R. hiemalis* would not occur within a few minutes. From February to December 1984 one sample with a Lenz-type grab was collected on the same date as the long cores. Immediately after the grab was withdrawn onto the boat, the sediment in the grab was partitioned into six 3 cm thick layers by inserting metal plates. The time interval from the sampling to the insertion of the plates was usually only several minutes.

Oxidation–reduction potential (ORP) was examined by ORP meter (Horiba, 7-SD model) for the core sediments collected in different dates from benthos samples on 7 July and 18 July, 1984.

Grain size composition, water content, loss on ignition, contents of particulate carbon, nitrogen and phosphorus were examined in every 5 cm layer of the sediment collected on 25 June, 1984.

Water content was measured by drying sediments at 100 °C for several days until a stable weight was attained. Loss on ignition was examined for a portion of the dried and ground sediment (1–2 g) by igniting in an electric furnace at 550 °C for 3 h. Particulate carbon and nitrogen were measured by a CNH analyzer (Yanako model 2).

To examine sediment temperatures, an additional sediment core of about 80–90 cm long was collected with a transparent acryl tube (10 cm in diameter and 1 m long) in each month. The tube had small holes in the wall at 5 cm vertical intervals, through which thermometers were inserted into the contained sediment. To prevent the sediment extruding through the holes, a strip of adhesive vinyl tape was placed over the holes. After retrieving the core onto the boat, a small thermometers was immediately pushed through each hole by peeling off the tape, beginning from the lowermost hole.

## Results

*Species composition of oligochaetes*

Of 2246 individual oligochaetes collected, 84% belonged to the Family Tubificidae and 7.9% to the Family Naididae. The remainder comprised unidentifiable individuals of very small size. Among the total oligochaete numbers, 31% were *Rhyacodrilus hiemalis*, 26% were *Aulodrilus* spp. (*A. japonica*, *A. limnobios*, *A. pigueti*), 15% were *Teneridrilus mastix* and 8.6% were *Limnodrilus grandisetosus*. *Branchiura sowerbyi*, *Limnodrilus hoffmeisteri*, *Bothrioneurum vejdovskyanum* were also collected. For the Naididae, *Specaria josinae* was relatively common (5.2%), and *Stephensoniana trivandrana*, *Nais* sp. *Vejdovskyella simplex* occurred rarely. Among all oligochaetes, *R. hiemalis* was numerically dominant at the study site.

*Seasonal changes of vertical distribution*

*Rhyacodrilus hiemalis* did not show any detectable diel vertical migration, since there were no differences in the day and night pattern of vertical distribution, and no significant difference between day average density and night one in each sediment layer in April and July ($p > 0.1$, two sample *t*-test) (Fig. 1).

In September and October 1983, more worms were found deep in the sediment (>30 cm), but in November 1983 some worms were present in near surface layers (< 15 cm) (Fig. 2). In December 1983 to April 1984, almost all worms were found

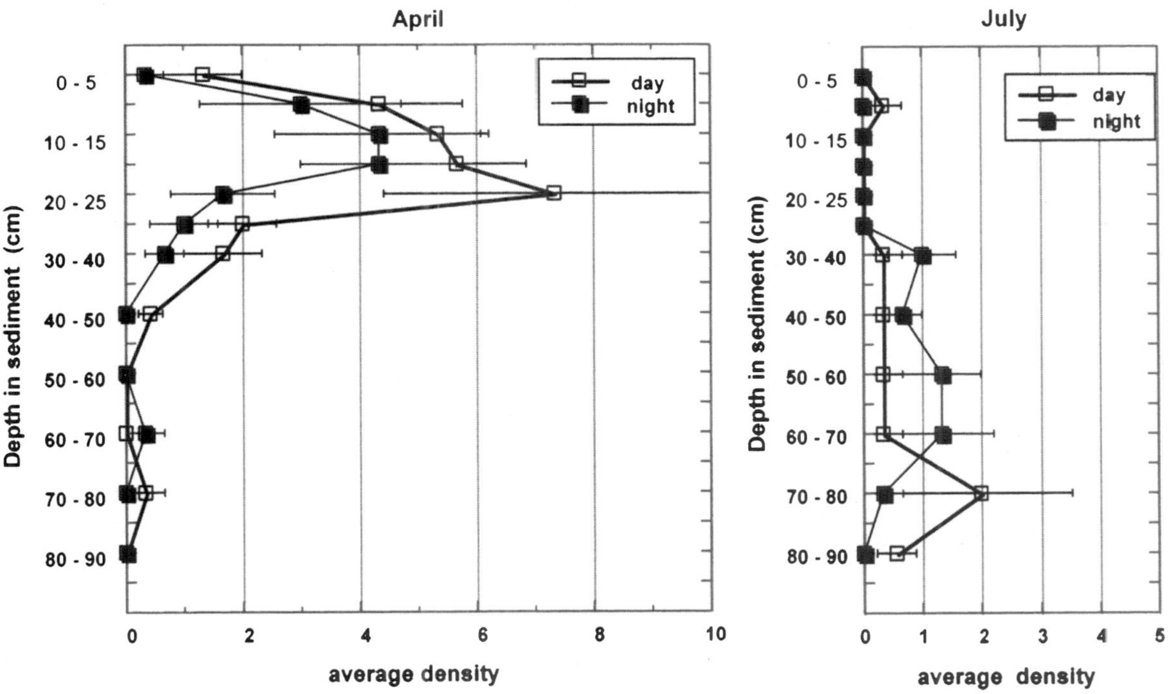


*Figure 1.* Vertical distributions of the average *R. hiemalis* density and SE in the day (open square) and night (filled square) cores in April (16 and 23) and July (4 and 23), 1984. Average density indicates the number in sediment of 119 cm$^2$ area of each section. Sediment samples were sectioned at 5 cm intervals to 30 cm, and at 10 cm intervals below 30 cm.

in near-surface and sub-surface layers, shallower than 30 cm, In April, a few worms were present in the middle (30–60 cm) and deep layers (>60 cm), and by May and June most worms were found in the middle and deep layers. In July to September worms were rarely found in near and sub-surface layers. In October, the worms were observed in sub-surface, middle and deep layers, indicating that they had begun seasonal upward migration.

In September 1984, long sediment cores were collected (111–117 cm deep), but no *R. hiemalis* were collected in the layers below 90 cm.

Mean vertical depth of *R. hiemalis* was large (42–67 cm) in summer (July to October), but small (11–20 cm) from November to April, except November 1983 (54 cm) (Table 1).

Calculated population densities of *R. hiemalis* were low during the summer months. Densities increased in November and December, and remained at a high level from December to April (Fig. 3).

*Table 1.* Mean vertical depth of distribution in *R. hiemalis*

| Date | Mean depth (cm) |
|---|---|
| 1983 Sep. 5 | 63.9 |
| Oct. 5 | 67.2 |
| Nov. 1 | 53.9 |
| Dec. 7 | 13.1 |
| 1984 Jan. 18 | 10.7 |
| Feb. 29 | 14.9 |
| Mar. 30 | 17.5 |
| Apr. 16 | 18.4 |
| Apr. 23 | 16.8 |
| May 9 | 42.7 |
| June 7 | 29.4 |
| July 4 | 61.9 |
| July 23 | 53.6 |
| Aug. 7 | 44.7 |
| Sep. 5 | 57.7 |
| Oct. 9 | 41.9 |
| Nov. 7 | 19.7 |
| Dec. 5 | 12.6 |

*Distribution near sediment surface*

Figure 4 shows the density of *R. hiemalis* near sediment surface in each month. From July to October no *R. hiemalis* were collected in the 18 cm sediment section collected by Lenz-type grab. In other months, vertical distribution pattern varied, but the distribution pattern obtained from Lenz-type grab collections differed only slightly from that seen in the near-surface layers of the cores. Statistical comparison to core sample could not be

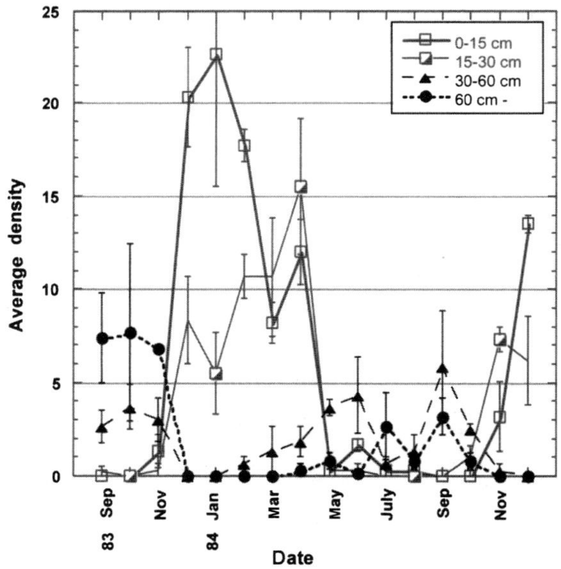


*Figure 2.* Seasonal change of the average *R. hiemalis* density and SE in near surface (0–15 cm), sub-surface (15–30 cm), middle (30–60 cm) and deep (> 60 cm) layers from September 1983 to December 1984.

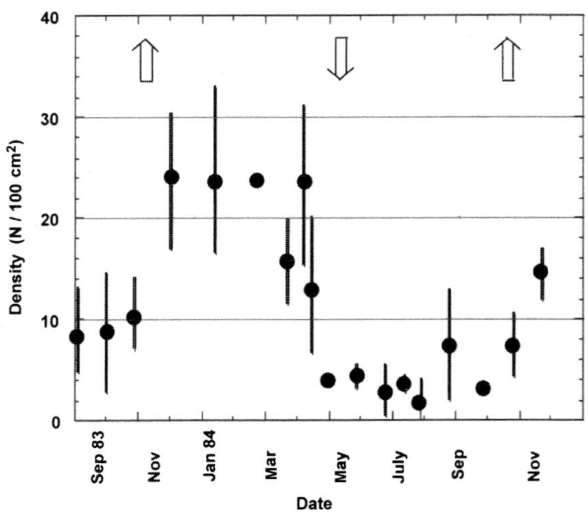


*Figure 3.* Calculated population density and SD of *R. hiemalis* in sediment column of 100 $cm^2$. Arrows indicate the period of upward and downward migration.

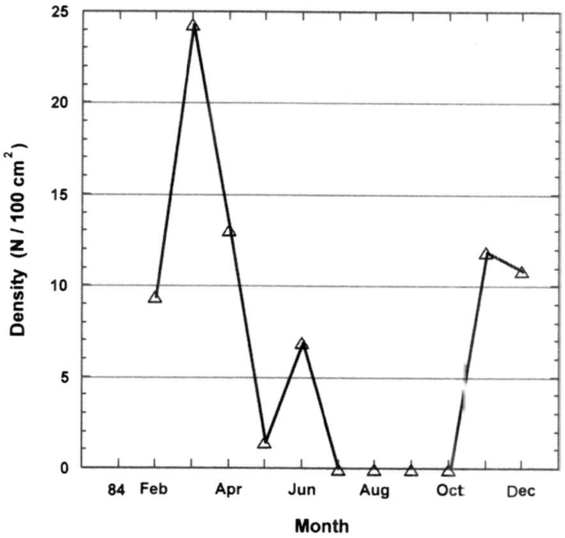


*Figure 4*. Seasonal change of the *R. hiemalis* density in sediment of 100 cm² collected by a Lenz-type grab. No worms were collected from July to October in the grab samples

adapted for one sample of Lenz grab, the pattern of seasonal change of the density by Lenz-type grab was similar to that of near surface density in core sample.

*Relation of worm density to environmental conditions of sediments*

The ORP values were negative at all sediment depths, with the minimum at around 10 cm deep, and were rather constant below 30 cm. This indicated that there was no free oxygen for aerobic respiration of animals. The colour of sediments was light grey in the deeper layers, indicating reductive conditions. There was no significant correlation between the worm density on 4 July or 23 July (night sample) and ORP value at each depth on 7 July or 18 July 1984 ($r = 0.134$ or 0.198, $p = 0.677$ or 0.147).

Regarding grain size distribution, ratios of larger size particles, being mostly organic debris especially >500 $\mu$m, were higher in the upper layers than in the deeper ones, except the uppermost layer.

The parameters ORP, grain size composition, water content, loss on ignition, and particulate carbon, nitrogen and phosphorus decreased gradually in the upper layers (to 50 cm depth), and then were relatively constant below 50 cm, and no zone of discontinuity was evident in any environmental parameter related to worm density (Fig. 5).

*Sediment temperatures*

The sediment temperature of the uppermost layer varied between 3.8 °C in January and 28.4 °C in September, and that of the deepest layer ranged from 8.0 °C in February to 21.9 °C in September (Fig. 6). The sediment temperatures changed gradually and continuously, decreasing with depth from May to October, increasing with depth from November to February, and being almost same at all layers in March and in October. There was no discontinuous layer in sediment temperature in any month. The sediment temperatures of the deepest layer converged to 20–21 °C in summer (July–October).

## Discussion

The aim of this study was to describe the life history of one of the dominant oligochaetes in Lake Biwa. Although *R. hiemalis* was a dominant littoral component during the cool season, this spe-

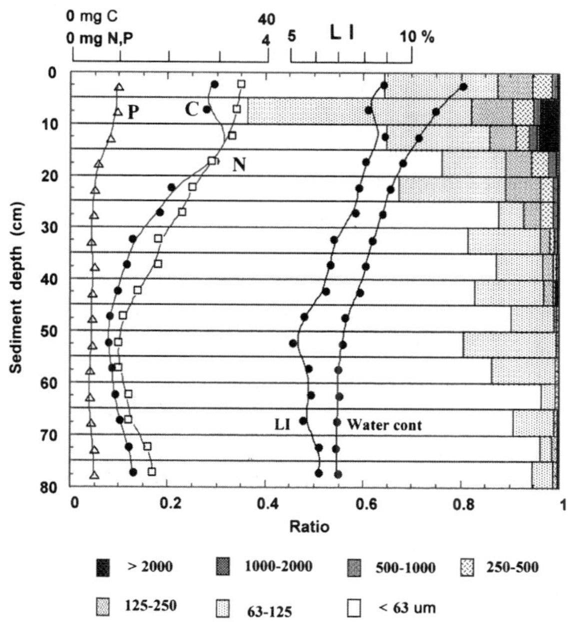


*Figure 5*. Vertical profile of grain size composition (histogram), water content, loss on ignition (LI), contents of particulate carbon (C), nitrogen (N) and phosphorus (P).

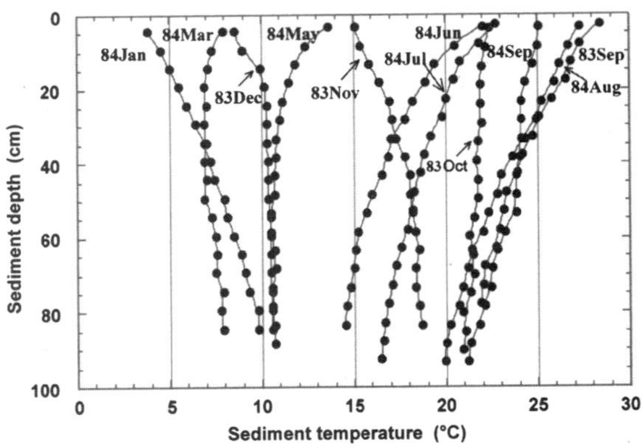


*Figure 6.* Vertical profiles of sediment temperatures examined at 5 cm intervals in each month.

cies disappeared from the upper sediment layers in summer, displaying a remarkable seasonal vertical migration in the sediment.

It is difficult to describe rigorously the vertical microdistribution of infaunal animals in sediment, since some animals may be displaced within cores following sampling. Several devices have been designed to study the vertical microdistribution of infauna (Milbrink, 1973; Fisher & Beeton, 1975; Nalepa & Robertson, 1981). However, these could penetrate not deeper than 20 cm, and since *R. hielmalis* often inhabited sediments deeper than 30 cm, other sampling devices were necessary to effectively study the microdistribution of *R. hiemalis*. The fact that the seasonal pattern in the rapidly sectioned Lenz grabs were similar to that in the near-surface layer of the cores suggested that large-scale vertical displacement did not occur in the cores after sampling. Though the microdistribution of *R. hiemalis* was not known in detail owing to the limitation of collecting methods, it was clearly evident that *R. hiemalis* migrated seasonally, and remained deep in the sediment with rather cool and constant temperature for 4 months during the summer, probably to avoid the high temperatures in the upper layer.

*Rhyacodrilus hiemalis* did not seem to penetrate deeper than 90 cm. The reason is not known. The sediment temperature at the deepest layer converged to around 20 °C maximum, and the sediment temperature below 90 cm may not change greatly even if the worms were to go deeper. The worms may orient towards the sediment depth with a cool and constant temperature, and upon reaching it, they may remain there and not proceed deeper.

Since the worms did not show diel vertical migration, and free oxygen which the animals could use for aerobic respiration was absent in the deep sediment layer, *R. hiemalis* must use anaerobic respiration when aestivating. The energy for anaerobic metabolism might be obtained through the glycolysis of glycogen (Hamburger et al., 1998), because *R. hiemalis* stored a whitish substance in the body in April and May (pers. obs.), which might be glycogen.

Regarding to other species collected in the study, small sized species, such as *Aulodrilus* spp., *Teneridrilus mastix, Limnodrilus hoffmeisteri* and naidid species, and small individuals of larger size species, such as *Branchiura sowerbyi* and *Limnodrilus grandisetosus*, were distributed mostly or exclusively in the upper part ($< 15$ cm) of near sediment surface layer. However, *B. sowerbyi* (8 out of 16 months) and *L. grandisetosus* (11 out of 18 months) were often caught in the sediment deeper than 30 cm. These species did not show any seasonal change in vertical distribution.

Population density of *R. hiemalis* was low in the summer months, increasing in November and December, when the worms migrated to the near-surface layers, and remained at a high level from December to April. The increase in the population density in November and December was not due to recruitment of a new generation, because cocoon deposition in *R. hiemalis* starts in January (Narita, 2001). The reason for the increment in population densities in December is not known, but it might be due to horizontal migration. Some *R. hiemalis* might have penetrated deeper than 90 cm, and it is possible, but unlikely, that there was also some recruitment from deeper layers of sediment, which the cores did not reach.

In terms of bio-geographical distribution, the genus *Rhyacodrilus* is an Amur-Indian component within Eurasia (Timm, 1997). The observed preference for cooler temperatures derived from its phylogenetic origin may have given rise to an adaptive life history trait of aestivation during the warm summer months. Similar behaviour of seasonal vertical migration and aestivation in summer has been observed also in the larvae of *Propsilocerus akamusi* (Tokunaga) (Orthocladiinae, Chironomidae) in the littoral zone of Japanese lakes (Yamagishi & Fukuhara, 1972). This chironomid species is endemic to Japan, and bio-geographically the genus is believed also to be originated from the Amur region (Sasa, 1978). In the process of extending their distribution southwest to the Japanese islands, several benthic animals originating from the cool temparate or sub-frigid zone, such as Amur region, seem to have evolved a life history strategy of summer aestivation. The long absence from the near surface sediment during the summer months may also be advantageous for *R. hiemalis* by avoiding from the predation by warm-water cyprinid fishes, which are abundant and active in the littoral zone of L. Biwa at this time.


## Acknowledgements

The author thanks T. Ueda, A. Kawabata and O. Kawai, Otsu Hydrobiological Station of Kyoto University, for their help in collecting samples, and K. Kawabata, Kanazawa University, and A. Rossiter, Lake Biwa Museum, for their critical reading of the manuscript.


## References


Fisher, J. A. & A. M. Beeton, 1975. The effect of dissolved oxygen on the burrowing behaviour of *Limnodrilus hoffmeeisteri* (Oligochaeta). Hydrobiologia 47: 273–290.

Hamburger, K., C. Lindegaard, P. C. Dall & I. Nilson, 1998. Strategies of respiration and glycogen metabolism in oligochaetes and chironomids from habitats exposed to different oxygen deficits. Verhandlungen Internationale Vereinigung fur theoretische und angewandete Limnologie 26: 2070–2075.

Kajak, Z. & K. Dusoge, 1971. The regularities of vertical distribution of benthos in bottom sediments of three Masurian lakes. Ekologia polska 19: 485–499.

Lazim, M. N., M. A. Learner & S. Cooper, 1989. The importance of worm identity and life history in determining the vertical distribution of tubificids (Oligochaeta ) in a riverine mud. Hydrobiologia 178: 81–92.

Milbrink, G., 1973. On the vertical distribution of oligochaetes in lake sediment. Report Institute Freshwater Research Drottningholm 54: 34–50.

Narita, T., 2001. Cocoon deposition of *Rhyacodrilus hiemalis* Ohtaka (Tubificidae) in Lake Biwa, Japan. Hydrobiologia 463: 141–148.

Nalepa, T. & A. Robertson, 1981. Vertical distribution of the zoobenthos in southeastern Lake Michigan with evidence of seasonal variation. Freshwater Biology 11: 87–96.

Nelson, D. R., 1991. Tardigrada. In Thorp, J. H. & A. P. Covich (eds), Ecology and Classification of North American Freshwater Invertebrates. Academic Press, London: 501–521.

Newrkla, P. & S. Mutayoba, 1987. Why and where do oligochaetes hide their cocoons? Hydrobiologia 155: 171–178.

Newrkla, P. & N. Wijegoonawardana, 1987. Vertical distribution and abundance of benthic invertebrates in profundal sediments of Mondsee, with special reference to olighochaetes. Hydrobiologia 155: 227–234.

Ohtaka, A., 1995. A new species of the genus *Rhyacodrilus* Bretscher (Oligochaeta, Tubificidae) from Japanese lakes. Zoological Science 12: 491–498.

Sasa, M., 1978. Taxonomical and biological notes on *Tokunagayusurika akamusi* (Tokunaga), with description of immature stages (Diptera, Chironomidae). Japanese Journal of Sanitary Zoology 29: 93–101.

Timm, T., 1997. Freshwater oligochaeta of some urban watercourses in the Russian Far East. Internationale Revue der gesamten Hydrobiologie 82: 407–467.

Yamagishi, H. & H. Fukuhara, 1972. Vertical migration of *Spaniotoma akamusi* larvae (Diptera: Chironomidae) through the bottom deposits of Lake Suwa. Japanese Journal of Ecology 22: 226–227.

Hydrobiologia (2006) 564:95–99
P.F.M. Verdonschot, H. Wang, A. Pinder & R. Nijboer (eds), Aquatic Oligochaete Biology IX
DOI 10.1007/s10750-005-1711-2



# Survival and reproduction in *Aeolosoma viride* (Annelida, Aphanoneura)

Rosanna Falconi, Tommaso Renzulli & Francesco Zaccanti*
*Department of Biology, University of Bologna, Via Selmi 3, I 40126 Bologna, Italy*
*(*Author for correspondence: Tel.: +39-051-2094177; Fax: +39-051-2094286; E-mail: francesco.zaccanti@unibo.it)*

*Key words:* paratomic fission, survival, intrinsic growth rate of population

## Abstract

We studied variability in the reproduction and survival of the zooids of *Aeolosoma viride* under controlled environmental conditions. Reproduction is by paratomic fission with pygidial budding. We collected data regarding the survival and reproduction of two generations of zooids (G0 and G1) raised in single cultures. Average survival was 69 days while reproductive activity varied according to the age of the specimen as well as to that of its parent at the time of fission.

## Introduction

Aeolosomatidae are small limnetic annelids that, for the most part, reproduce agametically. Sexual reproduction has, however, been reported in a few species. Only one species, *Aeolosoma singulare*, reproduces exclusively by sexual reproduction (Marotta et al., 2003). Agametic reproduction occurs by paratomic fission with pygidial budding that begins at different stages of growth in the different species (Herlant-Meewis, 1950, 1951).

In *Aeolosoma viride* Stephenson, 1913 which is known to reproduce agametically only, paratomy begins when the worm reaches a length of 9 setigerous segments and a first fission zone forms to isolate the pygidial segment. The morphogenesis of a new zooid begins at this point with the formation of new segments and the head forming before the zooid separates from the parent. Also during the morphogenesis of the first filial zooid, new fission zones form in the parent anteriorly to the first offspring (Herlant-Meewis, 1951).

There have not been many population dynamics studies of annelids that reproduce by paratomic fission reported in the literature (Christensen, 1984; Bely, 1999; Salminen & Haimi, 2001). In the case of Aeolosomatidae, quantitative data on reproductive strategies have been collected on *Aeolosoma japonica* (Kuwahara & Yamamoto, 1981), on *Aeolosoma hemprichi* (Kamemoto & Goodnight, 1956; Parker & Kamemoto, 1959; Suzuki & Kurihara, 1981; Wei et al., 2003) and on *Aeolosoma viride* (Zaccanti & Falconi, 2002).

In this paper we will consider the biological cycle and reproductive activity of *Aeolosoma viride* in order to construct a model to be used in the study of the physiological factors influencing morphogenesis, reproduction and survival in these animals: effects of neuromediators and neurohormones on the fissiparity and on the regenerative morphogenesis (Stagni et al., 1997) and measurement of the oxidative stress status in biological systems (Valgimigli et al., 2001).

## Materials and methods

The specimens of *Aeolosoma viride* used for this study came from a strain that we began breeding in cultures in the Biology Department's laboratory of zoology at the University of Bologna over 30 years ago. The newly formed worms, derived from a single clone, were placed in single cultures in Petri dishes with 2.5 ml of dechlorinated water which was kept at a constant temperature of 23 ± 1 °C. Photoperiods were divided into 12 h of artificial light and 12 h of darkness. The animals were fed finely chopped spinach. We changed the cultural medium every day as well as observed and recorded reproductive activity, isolating all newly formed zooids.

As fission occured, we immediately classified the worms as parental zooids (GO), zooids of the first filial generation (G1), and zooids of the second filial generation (G2). We kept track of the age in days from the original fission for G0 and G1 zooids (aG0, aG1); the survival in days for G0 and G1 zooids (sG0, sG1); and the number of G1 and G2 zooids produced (nG1 and nG2).

From the data collected, we were able to calculate survival rates and reproductive activity in terms of the cumulative number of filial zooids (N), number of offspring "born" each day (dN/dt) and the intrinsic growth rate "r", expressed by the Euler–Lotka function where $(\Sigma(S.dN/dt.e^{-r.t}) = 1$, with S = survival and t = age in days of the zooids.

## Results

A series of observations included 61 G0 zooids. Survival in this first group was found to be at 68.6 ± 15.9 days from fission to natural death.

We observed 1559 fissions with formation of G1 zooids. Figure 1(a) illustrates the cumulative number of G1 zooids produced in relation to the age of the G0 zooids at the time of fission. The progression follows a sigmoidal curve with a lower asymptote of −13.22 ± 3.74 (standard error), an upper asymptote of 25.88 ± 0.26, a center of 8.50 ± 2.77, and a slope of 11.10 ± 0.86. The mean daily reproductive activity for G0 zooids, expressed in terms of the number of G1 zooids produced per day [dN/dt (G1)] is 0.37 ± 0.22.

The graph in Figure 1(b) illustrates the distribution of daily reproductive activity [dN/dt (G1)] in relation to the progressive ages of the G0 zooids. The graph shows a short incremental phase peaking when G0 zooids are about 10 days old, followed by a long, increasingly less active phase during which fission notably and steadily decreases as the G0 zooids age.

We summarized reproductive activity by calculating the intrinsic growth rate "r" based on the model for population growth in which population density remains constant: $N(t) = N(o).e^{rt}$. The mean intrinsic growth rate "r (G0)" calculated according to the equation of Euler–Lotka was 0.377 ± 0.146.

We conducted an other series of observations on 58 G1 worms, all of which had separated from 3 G0 zooids. We kept track of the age of the G0 zooids at the time of each fission; i.e., the age of the G0 parents at the time of the fissions with formation of each new G1 zooid. We also observed 1452 fissions with formation of G2 zooids from their G1 parents.

The average survival rate of the second group of G1 zooids was 68.8 ± 16.2 days. The difference in the survival rate of this group compared to the first group observed is not statistically significant.

Average daily reproductive activity for G1 zooids (dN/dt (G2)) is 0.35 + 0.21 while the intrinsic growth rate "r (G1)" is 0.374 + 0.150.

We also noted that the reproductive capacity of the G1 zooids was affected by how old the G0 parent zooids were at the time of fission; i.e. how old they were when they "gave birth" to each G1 zooid.

In fact, the cumulative number of G2 zooids produced from G1 zooids (Fig. 1c), the daily reproductive activity (dN/dt (G2)) (Fig. 1d), and the intrinsic growth rate (r(G1)) (Fig. 1e) are significantly correlated ($p < 0.01$) to the age of the G0 zooids at the time each G1 zooid separated from its parent.

## Discussion

Results of our observations have allowed us to define the life cycle of *Aeolosoma viride* under

*Figure 1.* (a) The cumulative number of filial zooids (G1) in relation to the age of the G0 parental zooids. The average cumulative number is in bold face type. (b) Distribution of the daily reproductive activity (dN/dt) of filial zooids (G1) in relation to the progressive ages of the G0 parental zooids. (c) Distribution of the cumulative number of G2 zooids in relation to the age of G0 zooids at the time each G1 zooid separated from its parent. (d) Distribution of the daily reproductive activity (dN/dt (G2)) of G1 zooids in relation to the age of the G0 zooids at the time each G1 zooid separated from its parent. (e) Distribution of the intrinsic growth rates for G1 zooids in relation to the age of G0 zooids at the time each G1 zooid separated from its parent.

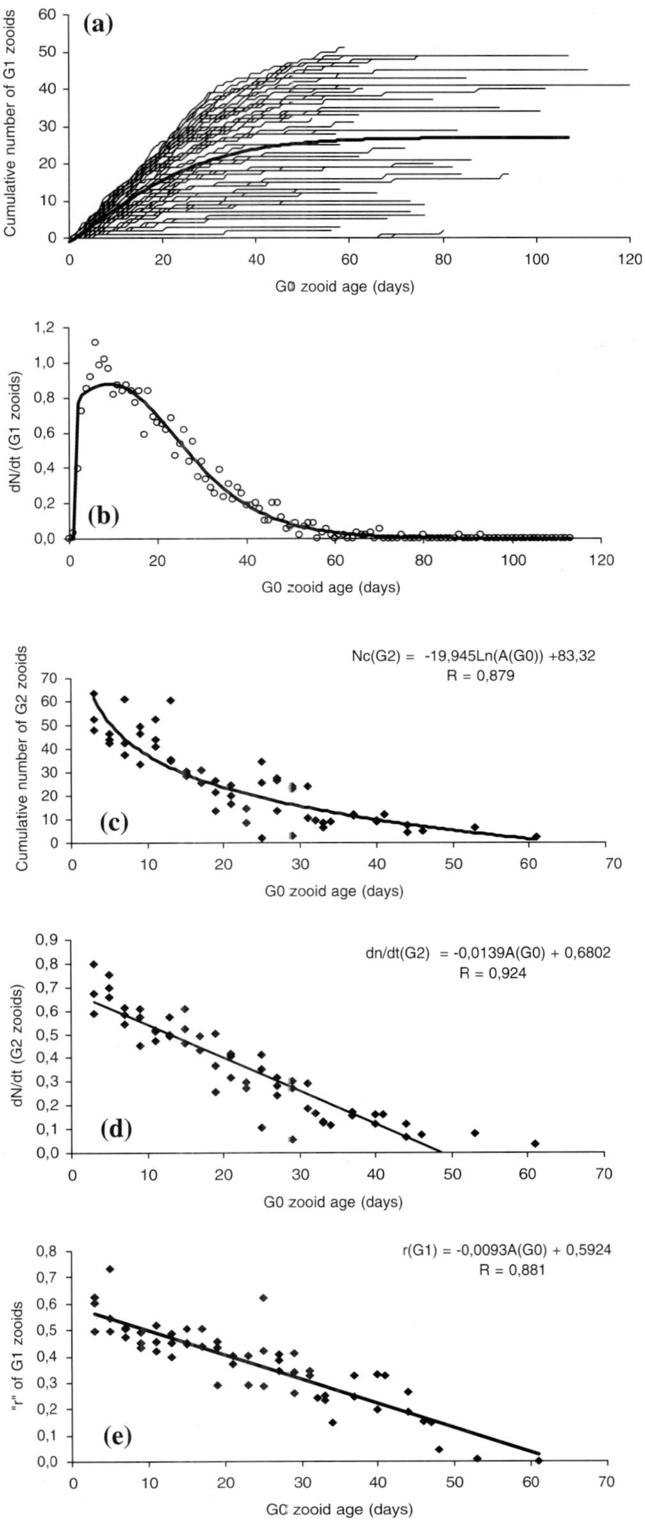
(a)
Cumulative number of G1 zooids
G0 zooid age (days)
(b)
dN/dt (G1 zooids)
G0 zooid age (days)
(c)
Cumulative number of G2 zooids
Nc(G2) = -19,945Ln(A(G0)) +83,32
R = 0,879
G0 zooid age (days)
(d)
dN/dt (G2 zooids)
dn/dt(G2) = -0,0139A(G0) + 0,6802
R = 0,924
G0 zooid age (days)
(e)
"r" of G1 zooids
r(G1) = -0,0093A(G0) + 0,5924
R = 0,881
G0 zooid age (days)

controlled conditions in terms of survival, cumulative number of zooids produced, daily reproductive activity, and intrinsic growth rate of the population. Our findings are not easily comparable with those of other authors regarding the population dynamics of other species of Aeolosomatidae, because of the different experimental conditions. The mean generation time is in the same order of magnitude, but longer, than that one of 1,2 days, as suggested for *Aeolosoma japonica* from Kuwahara & Yamamoto (1981).

Our results show that, while these animals reproduce solely by paratomic fission and were all kept under the same conditions in terms of temperature, light, space and food, survival rates and reproductive behavior were highly variable.

The high variability may have to do in part with the aging cycle of the individuals which is characterized by a short juvenile phase and longer mature and senescent phases.

We noted, furthermore, that the age of the parental zooids at the time of fission of each filial zooid also had a bearing on variability. In fact, the reproductive activity of zooids of young parents was much higher than in zooids of older parents.

If we take the estimated intrinsic growth rates for individuals, a considerable part of observed variability may be linked to the age of the parent animals. If we statistically factor out parental age, the variability among individuals decreases to 47.3% of total variability.

We hypothesize that residual variability may be due to uncontrolled environmental conditions and/or to the presence of genetic variation. Even in the case of agamic reproductive processes in there is a high rate of mitotic activity typical of schizogenesis and paratomic regeneration.

In conclusion, the reproductive behavior of *Aeolosoma viride* changed as individual aged. Furthermore, reproductive capacity of filial zooids was heavily affected by the age of the parental zooids when fission of new zooids occured. This kind of transmission of physiological states from the parental generation to the filial generation could be interpretatated in term of hypotheses the regarding overall aging and the neuroendocrine regulation of paratomic fission (Wolpert, 1969, 1971; Stagni & Vannini, 1977; Vannini, 1979; Zaccanti &Tognato, 1983).

We could also hypothesize the "possible" transmission of residual variability due to a certain degree of genetic mosaicism which could have been created by the accumulation of mutations linked to the high rate of mitotic activity which characterizes schizogenesis and paratomic regeneration (Lynch et al., 1993; Colato & Fontanari, 2001; Fontanari et al., 2003).

## References


Bely, A. E., 1999. Decoupling of fission and regenerative capabilities in an asexual oligochaete. Hydrobiologia 406: 243–251.

Christensen, B., 1984. Asexual propagation and reproductive strategies in aquatic Oligochaeta. In Bonomi, G. & C. Erséus (eds), Aquatic Oligochaeta: Proceedings of the 2nd International Symposium on Aquatic Oligochaeta Biology, held in Pallanza, Italy, September 21–24, 1982. Developments in Hydrobiology, 24. W. Junk Publishers, Dordrecht, The Netherlands, 91–95.

Colato, A. & J. F. Fontanari, 2001. Soluble model for the accumulation of mutations in asexual populations. Physical Review Letters 87: 238102.

Fontanari, J. F., A. Colato & R. S. Howard, 2003. Mutation accumulation in growing asexual lineages. Physical Review Letters 91: 218101.

Herlant-Meewis, H., 1950. Le leggi della scissiparità in *Aeolosoma hemprichi* (Ehrenberg). Revue Canadienne de Biologie Montréal 9: 123–198.

Herlant-Meewis, H., 1951. Les lois de la scissiparité chez les Aeolosomatidae: *Aeolosoma viride*. Annales de la Societé Royale Zoologique de Belgique 82: 231–284.

Hessling, R. & G. Purschke, 2000. Immunohistochemical (cLSM) and ultrastructural analysis of the central nervous system and sense organs in *Aeolosoma hemprichi* (Annelida, Aeolosomatidae). Zoomorphology 120: 65–78.

Kamemoto, F. I. & C. J. Goodnight, 1956. The effects of various concentrations of ions on the asexual reproduction on the oligochaete *Aeolosoma hemprichi*. Transactions of the American Microscopical Society 75: 219–228.

Kuwahara, M. & T. Yamamoto, 1981. Culture of Archioligochaeta, *Aeolosoma* species, found in activated sludge. Kagawa Daigaku Nogakubu Gakujutsu Hokoku 33: 75–80.

Lynch, M., R. Burger, D. Butcher & W. Gabriel, 1993. The mutational meltdown in asexual populations. Journal of Heredity 84: 339–344.

Marotta, R., M. Ferraguti & P. Martin, 2003. Spermiogenesis and seminal receptacles in *Aeolosoma singolare* (Annelida, Polychaeta, Aeolosomatidae). Italian Journal of Zoology 70: 123–132.

Parker, R. & F. J. Kamemoto, 1959. The effects of various concentrations of ions on the asexual reproduction of the oligochaete *Aeolosoma hemprichi.* Transactions of the American Microscopical Society 78: 207–211.

Salminen, J. & J. Haimi, 2001. The asexual enchytraeid worm *Cognettia sphagnetorum* (Oligochaeta) has increased Cu resistance in polluted soil. Environmental Pollution 113: 221–224.

Stagni, A. M. & E. Vannini, 1977. Neurosecrezione, "informazione di posizione" e gradiente morfogenetico in idre e planarie. Accademia Nazionale dei Lincei Classe Scienze Fisiche Matematiche Naturali 8(62): 839–847.

Stagni, A., M. L. Sini, R. Falconi, A. Osti, A. Quaglia & P. P. Giorgi, 1997. The effect of octopamine and mianserin on growth and schizogenesis in *Aeolosoma viride* (Annelida Aeolosomatidae). Animal Biology 6: 47–52.

Suzuki, T. & Y. Kurihara, 1981. Studies on the population dynamics of the aquatic oligochaete, *Aeolosoma hemprichi*, in continuous cultures. Japanese Journal of Ecology 31: 125–130.

Valgimigli, L., G. F. Pedulli & M. Paolini, 2001. Measurement of oxidative stress by EPR radical-probe technique. Free Radical Biology & Medicine 31: 708–716.

Vannini, E., 1979. Alcuni modelli di differenziamento sessuale in animali pluricellulari. Accademia Nazionale dei Lincei Classe Scienze Fisiche Matematiche Naturali 8: 465–468.

Wei, Y., R. T. van Houten, A. R. Borger, D. H. Eikelboom & Y. Fan, 2003. Comparison performances of membrane bioreactor and conventional activated sludge processes on sludge reduction induced by Oligochaete. Environmental Science and Technology 37: 3171–3180.

Wolpert, L., 1969. Positional information and spatial pattern of cellular differentiation. Journal of Theoretical Biology 25: 1–24.

Wolpert, L., 1971. Positional information and pattern formation. Current Topics in Developmental Biology 6: 183–224.

Zaccanti, F. & R. Falconi, 2002. Dinamica di popolazioni artificiali di *Aeolosoma viride* (Annelida Aphanoneura). Bollettino Accademia Gioenia Scienze Naturali 35: 549–558.

Zaccanti, F. & G. Tognato, 1983. Quantitative study on the reproduction of two strains of agamic planarians *Dugesia gonocephala* and *Dugesia tigrina*. Journal of Experimental Zoology 225: 149–154.

# Part Three
# Ecology and Toxicology

Hydrobiologia (2006) 564:103–116
P.F.M. Verdonschot, H. Wang, A. Pinder & R. Nijboer (eds), Aquatic Oligochaete Biology IX
DOI 10.1007/s10750-005-1712-1



# Ecology and biology of marine oligochaeta – an inventory rather than another review

Olav Giere
*Zoologisches Institut und Zoologisches Museum, Universität Hamburg, Martin-Luther-King-Platz 3, D-20146, Hamburg, Germany*
*(E-mail: olav.giere@zoologie.uni-hamburg.de)*

*Key words:* marine oligochaeta, life cycle, population dynamics

## Abstract

Instead of updating the1982-review by Giere & Pfannkuche, it was deemed more useful to highlight some recent and relevant studies given the limited number of new and pertinent publications. In addition, new data on population ecology, some recent results on symbiotic associations and descriptions of adaptations to oxygen deficiency and hydrogen sulfide will be commented on. The scarcity of new ecological and biological studies on marine Oligochaeta, an animal group that often attains key positions in important shore areas, is alarming. Future research efforts on this ecologically relevant annelid group are seriously needed.

## Introduction

In an earlier review (Giere & Pfannkuche, 1982), we described the pertinent literature up to 1980. The original aim of this contribution was to provide an update with more recent important data. However, a literature research showed only some sixty references directed specifically at oligochaetes over the past 20 years. Some of these publications were of more general interest and/or point to directions of future relevance. Those will be briefly mentioned. Therefore, this compilation represents more an inventory rather than another review. If relevant, selected studies of brackish water and freshwater oligochaetes have also been included. For comparison, the topics in this update are the same as those in our previous review.

## Topics considered

### *Methods*

In field, culture and experimental techniques, studies on marine oligochaetes do not need novel or specific approaches other than those established methods for the general benthos or previously described. The main changes have been in the development of mathematical and statistical evaluation methods. These have been summarized by Burd et al. (1990) and they describe examples explicitly referring to oligochaetes. Burd et al. (1990) first evaluate sampling patterns and efforts to analyze data matrices. Classical univariate descriptors such as various diversity indices and their application are covered. The use of biomass/size spectra is also discussed as well as the value of multivariate statistics such as Multidimensional Scaling and other ordination techniques.

### *Faunistics, distribution, zoogeography*

There are numerous papers in this field that are of local faunistic relevance only. Focus is given here only to those with wider than regional aspects or with generalizable conclusions. Erséus (2002) summarized the knowledge on oligochaetes from mangrove habitats. This ecotone harbors more than 50 species. The majority (38 spp), aquatic tubificids, occur in more marine habitats, while the more terrestrial enchytraeids (16 spp.) dominated in the upper reaches of the mangrove belt. This pattern conforms to the distribution ascribed to meiobenthic oligochaetes (Fig. 1) on marine shores in general (Giere, 1993).

Brackish-water tolerance appears to be a prerequisite for colonizing mangrove biotopes where typical taxa like *Paranais*, *Ainudrilus*, *Rhizodrilus*, *Monopylephorus* occur. The second dominant factor for mangrove benthos is the absence of an interstitial system. This is the reason for the absence of the mostly interstitial phallodriline Tubificidae, a group that is species-rich in sandy habitats in warm water regions (e.g., Erséus, 2003). The absence of oxygen in the sediment is not mentioned as an important factor in worm distribution by Erséus (2002).

Mangroves are ecologically important but rather exotic habitats. The same can be said for *Spartina* salt marshes along the North American Atlantic coast, a rapidly disappearing habitat. Here, biological diversity and productivity is of prime importance and seems well assessed. Examination of the air passages in *Spartina* stems has revealed this to be a preferred habitat for enchytraeid species (Healy & Walters, 1994), which has generally been disregarded. At the present time the trophic basis of these rich oligochaete populations (*Marionina* spp, *Monopylephorus parvus*) is unclear. However, this one habitat alone can increase estimates of oligochaete abundance by an order of magnitude from 1150 to 10,250 ind. $m^{-2}$.

High ecological relevance is also attributed to another unusual habitat, the drift zone of beaches, where debris from plants and animals tends to accumulate and decomposing organisms prevail. Here, Koop & Griffiths (1982) studying the meiofauna of South African beaches, found accumulations of meiobenthic oligochaetes (taxa unidentified). They suggest that this wrack zone is a center of oligochaete distribution. As assumed in earlier studies from boreal shores (Giere, 1970), these detritivorous oligochaetes play an important role in the breakdown of larger organic particles thus facilitating bacterial decomposition. The exceptional role of this zone is emphasized by the fact that other beach habitats also contain considerable amounts of organic matter, yet are less attractive to the meiofauna (Giere, 1993).

## *Life history*

This general heading includes studies on development, reproductive biology, life cycles with the

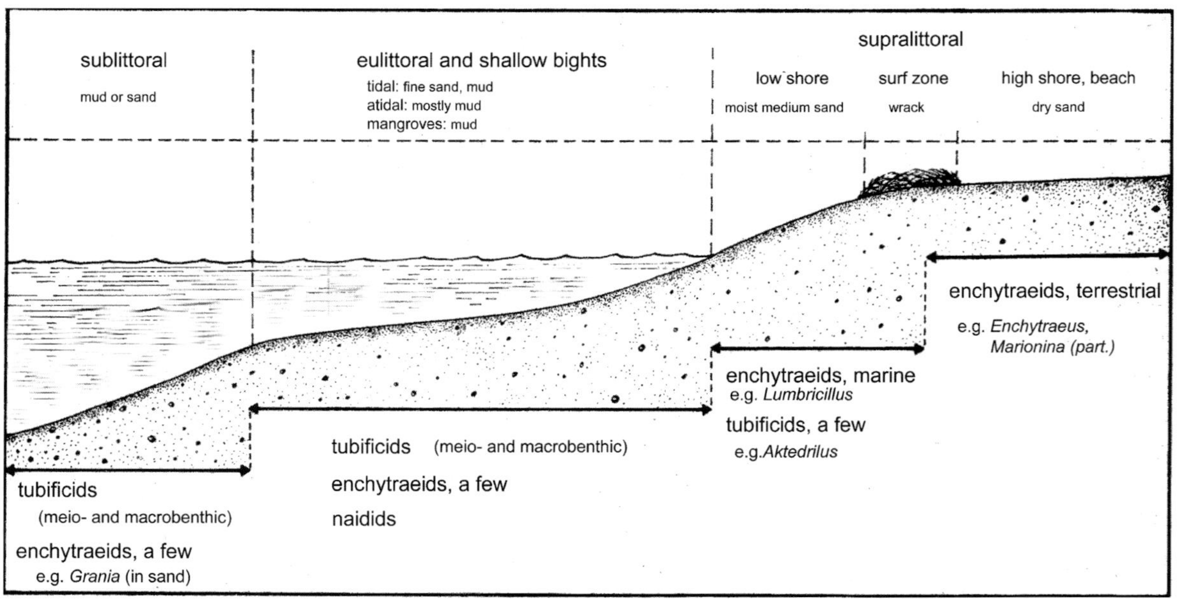


*Figure 1.* Generalized distribution pattern of marine oligochaetes (from Giere, 1993, modified).

annual fluctuations as well as patterns of symbiotic coexistence.

*General aspects*

Working on general developmental and genetic problems, Levinton's group chose the asexually reproducing *Paranais litoralis* (Naididae) as their animal model. Investigating the senescence of animals without a specific germ line, Martinez & Levinton (1992) found that asexually reproducing animals are also subject to aging and they reason that senescence is linked to somatic differentiation rather than germ line sequestration.

Nilsson et al. (1997) revealed that several clones belonging to subpopulations of *P. litoralis* from different environments showed marked differences in life span, production of offspring, growth etc. DNA-analysis revealed not only a genetic interclonal difference, but that the sequences were unstable and seemed to interact with the environment. The authors interpret this flexibility as adaptive, creating a mosaic of local reactions to a shifting environment. In a later study (Nilsson et al., 2000), observations from the field and from cultures showed that in times of resource depletion, the clones grew longer and thinner and become migratory, swimming up in the water column. Thus, areas of food scarcity may be avoided and migration helps the populations not only to persist but also to maintain a genetic flux and variability between metapopulations.

*Reproductive biology*

Oligochaeta, especially many Tubificidae, are characterized by their complex and elaborate reproductive organs which have become the basis for identification. Yet, we are far from understanding the reproductive biology of these animals and cannot even guess at the interrelations between the form and function of external genital structures in these hermaphroditic organisms. The paper by Cuadrado & Martinez-Ansemil (2001), although dealing with limnetic tubificid species, gives an example about differentiation of structures required for copulation and sperm exchange. Our lack of knowledge in this field is apparent in those species where number and position of genital organs and openings deviates from normal (e.g., mid-dorsal opening of spermathecae).

Takashima & Mawatari (1998) concluded that the function of the genital structures and, thus, the mode of sperm transfer remains mostly enigmatic. Accordingly, the autapomorphic character of many genital structures cannot be relied on from extrapolation of old observations on earthworms and we cannot consider the megadriles to be representative for all the oligochaeta. What is required in this somewhat mundane, but much needed, research field is patience, appropriate culture techniques and modern documentation methods, best characterized by the film of Westheide & Müller (1995) on the reproductive behavior of terrestrial enchytraeids.

*Symbioses*

A remarkable biological and ecological feature of some marine oligochaetes is the existence of complex and obligatory symbioses with bacteria. Occurring in two phallolodriline genera (Tubificidae). These associations parallel the symbiotic trends in deep-sea hydrothermal vents, but differ in the diversity of symbionts participating in the relationships (Giere & Langheld, 1987; Giere & Krieger, 2001). Since the discovery of two gutless species in coralline sands of Bermuda (Giere, 1979) some 100 species of the monophyletic genera *Olavius* und *Inanidrilus* completely lacking digestive and excretory organs have been described (Erséus, 2003) (Fig. 2).

The bacteria–symbiotic worms occur in shallow sulfidic sediments of warm water areas from all major seas. At present approximately 30 spp. have been studied in detail. They all have incorporated a fairly thick layer of extracellular bacteria beneath the cuticle while the wide body cavity is devoid of an intestine (Giere et al., 1995). Only in two species from the Mediterranean island of Elba, are the bacteria mostly found enclosed by the epidermal cells, thus often attaining an intracellular position (Giere & Erséus, 2002). The symbiotic bacteria in the gutless worms belong to different morphotypes. Large, oval ones are frequently combined in dual symbiosis with smaller, rod-shaped ones (Fig. 3). However, in some cases the picture becomes complicated by additional morphologically divergent types of different

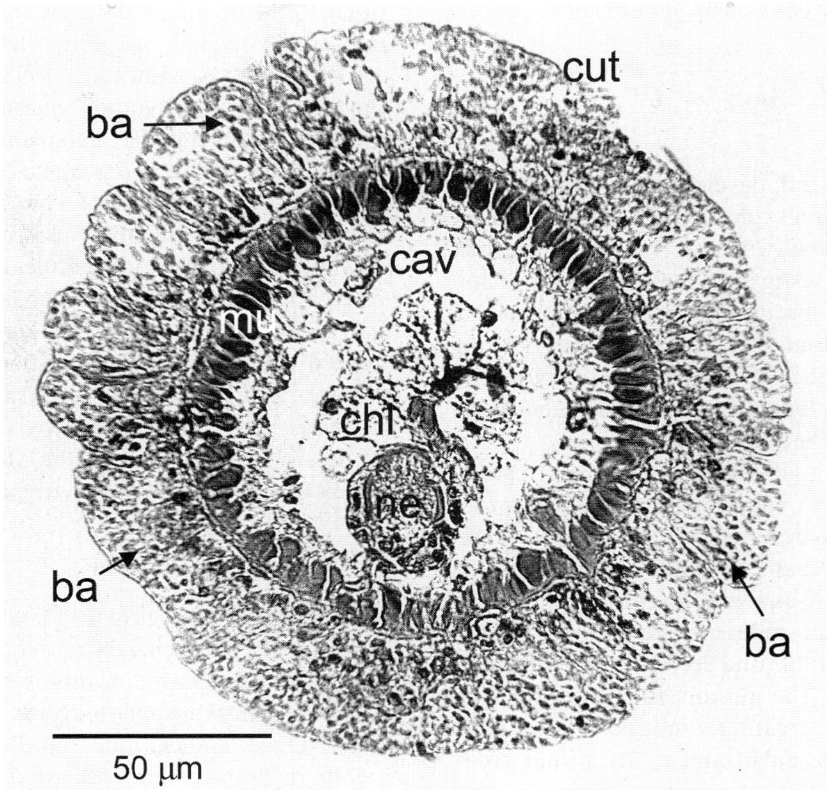


*Figure 2.* Cross-section through *Inanidrilus leukodermatus* (Tubificidae, Oligochaeta) from Bermuda. Light micrograph. Abbreviations: ba = symbiotic bacteria, cav = body cavity, chl = chloragogue cells, cut = cuticle, mu = dermal musculature, ne = ventral nerve cord.

shapes and sizes, including long, twisted filiforms (Giere & Krieger, 2001; Fig. 4).

It was not till molecular methods were developed that the function of some of the various symbionts could be revealed. The large oval symbionts belonging to the gamma-Proteobacteria have been shown to be sulfide oxidizers (Dubilier et al., 1995), while the function of the smaller ones, often alpha-Proteobacteria, still remains unclear. In *Olavius algarvensis*, from Elba, close functional cooperation between the two bacterial types has been documented (Dubilier et al., 2001). Here sulfide, oxidized by the larger gamma-Proteobacteria into sulfate, was then reduced again into sulfide by the small rod-shaped alpha-Proteobacteria (Fig. 5).

This unique bacterial 'cyclotrophism' of directly adjoining symbionts enables the gamma-Proteobacteria to obtain sulfide (as electron donor) even in sand poor in, or temporally devoid, of free sulfide. In this case, the gutless hosts, obligately dependent on their bacteria, are not any longer strictly bound to sulfidic/oxic interfaces (Giere et al., 1991) but can remain in the oxic layer. Further molecular studies have recently revealed that the two tubificid species found in Elba harbor even a set of up to five bacterial phylotypes (Dubilier et al., in press). Other bacterial phylotypes, including Spirochaeta, have been identified by molecular methods as symbionts in various gutless species from Australian, Caribbean and Peruvian regions.

Obligate reliance of the hosts on their bacterial partners, the internal position of the bacteria, and the complicated mode of reproduction in oligochaetes (see above) suggest vertical pathways of bacterial transmittance, i.e. from the parental generation directly to their offspring. On the other hand, the divergence of bacterial associations in closely related worm species rather speaks for a horizontal uptake, i.e. uptake from the environment. Sometimes (e.g. in *O. algarvensis*), the symbionts differ even in separate populations of

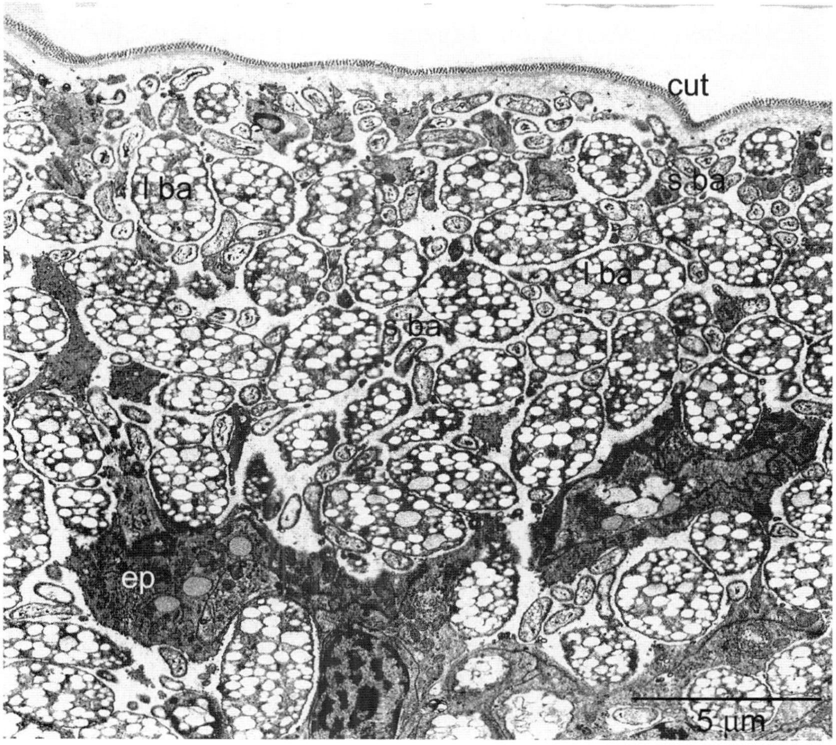


*Figure 3.* Cross-section through *Inanidrilus leukodermatus* (Tubificidae, Oligochaeta) from Bermuda showing subcuticular symbiotic bacteria of two different morphotypes. Transmission electron micrograph. Abbreviations: cut = cuticle, ep = epidermal cell, l ba = large γ-Proteobacteria, s ba = small α-Proteobacteria.

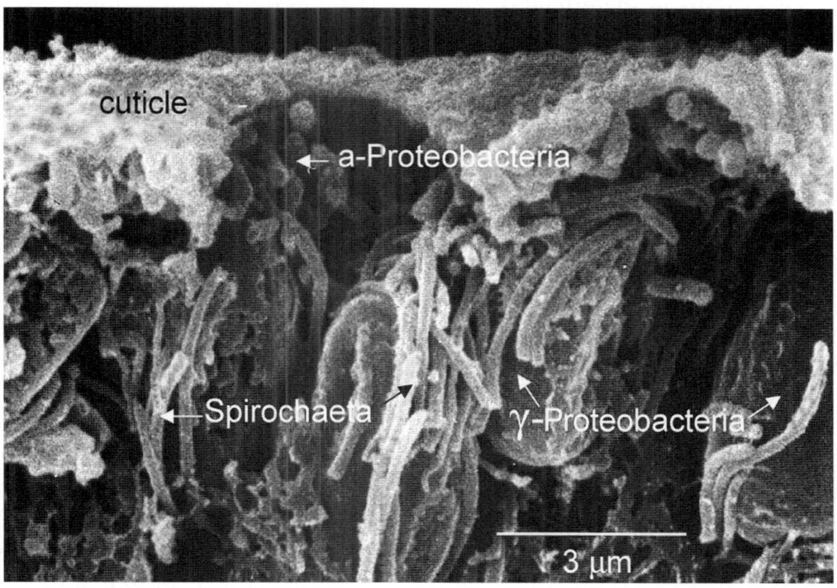


*Figure 4.* Longitudinal micropreparation of *Olavius crassitunicatus* (Tubificidae, Oligochaeta) from the shelf off Peru showing spirochete symbionts together with oval γ-Proteobacteria and small α-Proteobacteria. Scanning electron micrograph.

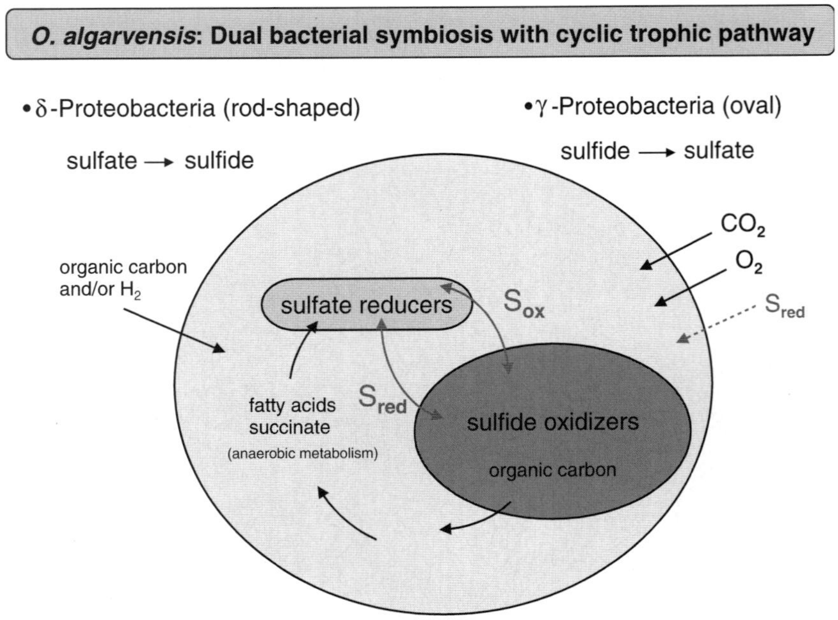


*Figure 5.* Functional diagram of symbiotic bacteria and their metabolic pathways in *Olavius algarvensis* from Elba (from Dubilier et al., 2001, modified).

the same host species. This also supports the view of environmental transmission of the bacteria. Our studies on *I. leukodermatus* from Bermuda have revealed that only the oval sulfide oxidizing gamma-Proteobacteria are vertically transmitted (Krieger, 2001), 'infecting' the egg at the moment of deposition (the sticky eggs in these gutless oligochaetes are layed singly onto the sediment grains and not deposited in cocoons as usual in oligochaetes). Hence, we have to assume that there is a combination of vertical and environmental transmission, at least in the species studied (Fig. 6).

Though there are many parallels that can be drawn between the bacteria–symbiotic associations in these gutless Oligochaeta and the deep-sea gutless tube worms and molluscs in sulfide-emanating hydrothermal vents, the complexity, variability, and diversity of the symbiotic pattern in the gutless tubificids seem unique. Their various bacterial associations, each perhaps with specific metabolic requirements, might be a clue to the syntopic occurrence of various gutless species (see below).

*Life cycle studies*

The assessment of life cycles has been a neglected, but is an ecologically important field of oligochaete research. This type of study requires patient long-term data collection with numerous sampling campaigns and/or tedious culturing to separate the 'noise' from the trend, so as to reveal the role of environmental factors (temperature, salinity) and demonstrate true life cycle attributes. This long-term approach was adopted in the comprehensive study by Erséus & Diaz (1989) on *Tubificoides amplivasatus* (Tubificidae) from the Kattegat. The subtidal species reproduces year round without clear cohort formation. It has a very patchy distribution with irregular peaks of abundance, which can be related to the warmer temperatures in late summer and autumn. The long time to maturity (200 days) infers a life span of several years. Recruitment in the warmer season and for prolonged periods is also reported for the congener *T.* cf. *brownae* in North American brackish water (Diaz, 1984).

A relationship with temperature was not found in *Tubificoides benedii* from the shallow Forth estuary (Bagheri & McLusky, 1982). Here, the erratic annual population curves with peaks also in the winter did not show any clear seasonality and no marked period of recruitment. This was in contrast to the other dominant oligochaetes in this brackish water habitat, the naidid species *Amphichaeta sannio* and *P. litoralis*. Their population

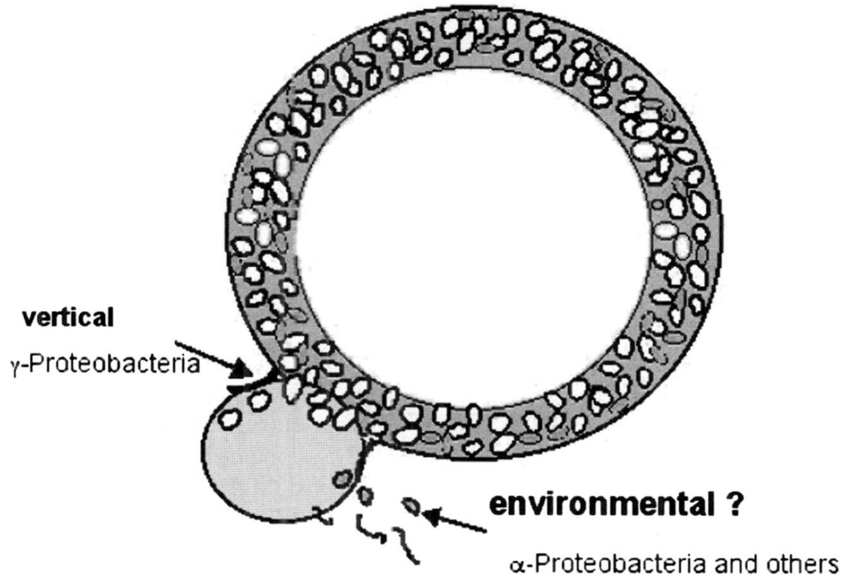


*Figure 6.* Schematic pathways of bacterial transmission in symbiotic oligochaetes.

peaks could be correlated to periods of algal blooms. The prevailing asexual reproduction by fission supports erratic population fluctuations as soon as algal blooms arise.

*Quantitative population data, production*

"Few estimates of production in aquatic oligochaetes have been made. Those, that have vary considerably in methodology, the units employed, the values obtained, and sometimes in the concept of production used." (R.O. Brinkhurst, 1st. Olig. Symposium, Sidney, Can., 1979)

By and large, this statement (Brinkhurst, 1980) is still valid, with the result that in the field of marine oligochaeta, we have to rely on a few isolated publications.

Using respiration values combined with biomass data, Bagheri & McLusky (1984) calculated production of *T. benedii* and the naidid species *A. sannio* and *P. litoralis* together with that of the small polychaete *Manayunkia aestuarina*. These species are dominant in the Forth estuary, Scotland, contributing about half and sometimes up to >90% of the total invertebrate production of this brackish water area.

*Tubificoides benedii* is a species which contradicts general ecological classification: This common oligochaete is dominant in coastal areas that are highly enriched in organic matter, and is often typified as 'opportunist' that is adapted to the rapid environmental fluctuations and harsh conditions in estuaries (see Bagheri & McLusky, 1982). The behavior of this oligochaete species does not conveniently fit into the concepts of, 'r' and 'K' strategies described in the ecological literature. Despite its huge population density ($1 \times 10^6$ m$^{-2}$) and production (14–27 g wet wt. m$^{-2}$ a$^{-1}$). *T. benedii* is not a typical r-strategist. Its long life span (a few years, see also *T. amplivasatus*, above), discontinuous breeding, internal fertilization with brooding and low P/B-ratio (0.85:1) contradict the classical 'r'-strategy. Among the aquatic groups, this contradiction refers mainly to the long-lived tubificids and to many larger terrestrial oligochaetes as well, so that oligochaetes do not fit the simplified pattern of r- and K-strategists.

The low P/B-ratio of *T. benedii* is in contrast to that of some enchytraeids, e.g., the small enchytraeid *M. southerni* (oligohaline to limnic species, P/B ratio of 2.5) which only has a 1-year life cycle (Lindegaard et al., 1994). Reliable calculations of annual P/B ratios in naidids are impaired by their excessive multiplication through asexual fission in periods of algal blooms.

Obtaining more reliable life history data for marine oligochaeta for calculating production will require both field sampling and laboratory cultures. The work of Bouguenec & Giani (1989) on *Enchytraeus variatus*, a terrestrial species with both sexual and asexual reproduction is an excellent example of the type of study required. Using cultures the relevant life history data for the complete calculation of growth and production were acquired including: embryonic development, post-embryonic, maturity period, laying period, individual life time, production, biomass, and growth curves of populations.

*Abiotic habitat factors*

The impact of habitat on distribution and life history is one of the basic ecological questions. Nevertheless, multifactorial interactions render its assessment difficult. For oligochaetes, this complexity is summarized by Diaz (1984) "... The complexity of species interactions... combined with climatic, or salinity, or substrate changes may overshadow any progressive tendencies in the benthos."

*Sediment*

One of the most puzzling features of marine oligochaete communities is the diversity of closely related species in seemingly uniform sediments and syntopic occurrence. An analysis of the speciose tubificid fauna in shallow water sediments of the Caribbean (Diaz & Erséus, 1994) explained the co-occurrence of species by strong niche partitioning and by differences in organic contents of the sediment. Using cluster analysis these authors classified species associations and related them to sediment granulometry and water depth.

*Oxygen/hydrogen sulfide*

Sediment structure and organic content are likely the main factors affecting community structure, especially in marine habitats: The supply of oxygen and exposure to hydrogen sulfide have often been found to be key factors in the distribution of marine oligochaetes (Giere & Pfannkuche, 1982). "To fully understand the distribution and ecology of aquatic plants and animals, sulfide must be taken into account along with temperature, salinity, oxygen, and other environmental factors" (Bagarinao, 1992). In the preferred coastal habitat of many marine oligochaetes, periods of oxygen deficiency and strong sulfide formation are frequently observed as a result of high enrichment with organic matter. The general importance of oxygen/free sulfide in affecting the distribution of marine oligochaetes is reflected in four pertinent reviews published in the 90s: although not restricted to oligochaetes, the conclusions of these reviews can be generalized and are of broad importance. Giere (1992) considered the tolerance of many specialized marine animals to hypoxia and sulfide from an ecological point of view, while Grieshaber et al. (1992), in a parallel paper, considered the physiological mechanisms involved in survival under sulfidic conditions. Grieshaber et al. (1994) supplemented this paper with a compilation of physiological and metabolic responses to hypoxia. A more ecophysiological view was taken in the paper by Bagarinao (1992) who also covered some of the symbioses developed in sulfide-rich environments. In the context of this paper, these review papers can be summarized:

1. Hypoxia and sulfide closely interact physiologically, the presence of hydrogen sulfide exacerbates the effects of hypoxia ("sulfide dependent anaerobiosis").
2. Different taxa use different mechanisms of detoxification and adaptation.
3. The main defense against hydrogen sulfide is the capacity for intermittent anaerobiosis and/or oxidation by the mitochondria. During anaerobiosis, succinate is accumulated, possibly in all aquatic oligochaetes, and can be used as an indicator of an anaerobic metabolism.
4. The frequently observed precipitation of metal-sulfides in various tissues of the body often interpreted as a mechanism for detoxification needs scrutiny, since it will not cope with the rapid sulfide influx and export pathways have never been reported (see also Dubilier et al., 1995).
5. Most enchytraeids and naidids are sensitive to hydrogen sulfide and hypoxia, while tubificids often are more resistant, some of them having special physiological adaptations (see below).
6. The ecological advantage of living in hypoxic, temporarily sulfidic sediments is the ample supply of organic matter and the reduced interspecific competition.

Lindegaard et al. (1994) in their study on *Marionina southerni* from Danish brackish waters, confirmed the limited capacity of many enchytraeids to regulate in low oxygen conditions. This species was a relatively good regulator but only at low temperatures. Between 10 and 20 °C, it reached its critical oxygen partial pressure ($P_{crit.}$) at 60% saturation (=12.6 kPa). This corresponds to the average field occurrence of marine enchytraeids, which are never encountered in severely hypoxic sediments (Giere, 1973).

In contrast, our own studies on *T. benedii* (Tubificidae), often a 'pioneer' in hypoxic and sulfidic environments (Giere et al., 1999), resulted in a $P_{crit.}$ of only 10% (=2.1 kPa) at 15 °C. This value is extremely low compared to other aquatic oligochaetes and polychaetes (Table 1) documenting that *T. benedii* is one of the best annelid oxyregulators, even under sulfidic conditions. Its extreme oxygen tolerance corresponds with its unusually low respiration rate, compared to limnetic oligochaetes and polychaetes of comparable size (Table 1).

*Salinity*

In coastal marine areas, especially in tidal flats and estuaries, salinity fluctuations also play an important role in the distribution of oligochaetes. This is especially evident from studies in long tidal estuaries such as the Rhine delta where Verdonschot (1981) and Verdonschot et al. (1982) could relate salinity to field distribution. It was shown that *Heterochaeta costata* (=*Tubifex costatus*) preferred shallow-water brackish stations avoiding areas of usually euhaline salinity, while *T. benedii* covered a range from meso- to euhaline stations.

To examine the salinity-dependent osmoregulatory potential of aquatic oligochaetes, comparative tolerance experiments have been conducted (Generlich & Giere, 1996). Using terrestrial and coastal marine populations of *Enchytraeus albidus* (Enchytraeidae) the osmolality of the coelomic fluid was measured after varying acclimation times in a range of salinities. The enormous regulatory potential of *E. albidus* was underlined not only by the survival of test animals in concentrations from freshwater to fully marine conditions (after long-term acclimation), but particularly by the continuous maintenance of a hyperosmotic coelomic fluid. The main difference in populations of *E. albidus* from the seashore and garden soil was the higher stability of the osmotic barrier inside/outside in the shore populations.

Rapid salinity change (hyperosmotic shock) can result in a similarly rapid adjustment of the internal fluid concentration to the changed external conditions. However, it took longer for the animals to regulate their body volume by adjustment of the water content. Molecular studies are required to show to what extent ecophysiological differences in terrestrial and marine populations of *E. albidus* relate to differences in their genetic background. Comparison with other annelids suggests that different regulatory pathways exist in the various test animals. In the tubificid

*Table 1.* Inter-annelid comparison of critical oxygen concentrations ($P_{crit.}$) and respiration at about 15 °C

| Species | Critical $PO_2$ (kPa) | Respiration rate, 15 °C (nmol $O_2$ $g^{-1}$ fwt $min^{-1}$) | Reference |
|---|---|---|---|
| *Tubificoides benedii* | – | 37 | Bagheri & McLusky (1984) |
| *T. benedii* | 1 | | Dubilier et al. (1994) |
| | 2 | 63 | Giere et al. (1999) |
| *Limnodrilus hoffmeisteri* | 4 | 357 | Schneider (1994) |
| *Tubifex tubifex* | 4 | 218 | Schneider (1994) |
| *Marionina achaeta* | 3.2 | 570 | Lasserre (1976) |
| *M. southerni* | 12.6 | 169 | Lindegaard et al. (1994) |
| *Scoloplos armiger* | 2.1 | – | Schöttler & Grieshaber (1988) |
| *Capitella capitata* (large form) | 17 | | Gamenick (1997) |
| | | 125 | Gamenick et al. (1998) |
| *Arenicola marina* | 16 | – | Hauschild & Grieshaber (1997) |

*Heterochaeta costata*, regulation of water loss in a hyperosmotic medium was more limited.

*Biotic habitat factors*

*Bioturbation*

For endobenthos bioturbative reworking of the sediment is one of the most effective environmental factors. In the marine realm most work in this field has been done with polychaetes, but the paper by Reible et al. (1996), although dealing with the limnetic tubificid *Limnodrilus hoffmeisteri*, may also parallel conditions in marine sediments. From experimental sediment arrays contaminated with pyrene it was concluded that the intense bioturbative activity of these tube-building worms would lead to an enhanced long-term diffusion of pore water into the overlying water column (about 370 times greater than in non-bioturbated sediment).

Given the frequent high abundance of common littoral oligochaetes (e.g., $10^5$–$10^6$ ind. $m^{-2}$, Giere & Pfannkuche, 1982), the burrows of these non-tube builders would likely cause a considerable 'secondary porosity' in muddy sediments which adds to that created by other marine burrowers (mainly polychaetes and bivalves). The exchange of solutes and pollutants at the water/sediment interface is further increased by the continuous reworking of sediment through the digging and feeding activity of most endobenthic fauna. Kaster et al. (1984) calculated for *Limnodrilus* that sediment up to several times its own body weight is brought to the surface every day *via* defecation. Grossly enhanced diffusion and reworking of sediment by the various littoral oligochaete species will substantially contribute to the overall exchange of solutes, an important factor in sediment biogeochemistry.

*Disturbance*

Another important effect of bioturbation is disturbance, particularly between similar functional groups, e.g., deposit feeders (Rhoads & Young, 1970) or between benthic adults and settling larvae (adult – larvae interaction, Woodin, 1976). McCann & Levin (1989) examined the disturbance associated with dense populations of a marine tubificid (*Monopylephorus evertus*). Surprisingly, the experiments showed that there was no measurable negative impact between the large populations of the oligochaete and the adults of another common surface deposit feeder, the polychaete *Streblospio benedicti*. Even negative 'adult-larval interaction' between the mobile burrower (*Monopylephorus*) and the settling planktonic larvae of the polychaete were not observed. But this absence of effect seemed stage-specific: *M. evertus* did inhibit the development and survival of newly settled juveniles of *S. benedicti*.

In another study, Commito (1987) used the marine tubificid *T. benedii* to test the general applicability of the 'trophic group amensalism' hypothesis (Woodin, 1976). Do dense populations of filter feeders (mussels) prevent colonization of *all* deposit feeding benthos, regardless of developing a planktonic or benthic larval/juvenile phase? He showed that population density of *T. benedii* was the same or even higher in mussel beds than in open mud areas. Absence of an inhibitory effect was also found by Bagheri & McLusky (1982) studying the interaction between the filter feeding polychaete *M. aestuarina* and the grazer/deposit feeder *P. litoralis*. Thus, amensalism between filter feeders and deposit feeders is not evident for holobenthic deposit feeding species, based on these data (and in fact amelioration may be true). The only demonstrated interactions are inhibitory effects by the filtering activity reducing the planktonic larvae of deposit feeders that settle onto the bottom.

The results of Bagheri & McLusky (1982) provide some explanation of disturbance effects. The naidid *P. litoralis* differs from *T. benedii* in its biology: It often occurs in the suprabenthic water layer actively swimming above the sediment/water interface and potentially interferes with the filtering structures of *Manayunkia*. In the absence of negative interaction by disturbance, explanations for the general ecological interactions are ascribed to predator/prey relationships. And these relationships are highly dependent on size classes and life history characteristics.

An example of the effect of size class on competition for food, between oligochaetes and polychaetes was given in experiments, which artificially inhibited the viability of diatoms (Hentschel & Jumars, 1994). The breakdown of diatom populations reduced the stocks of the brackish water naidid *Amphichaeta leydigi* and the polychaete *Hobsonia florida* only in the larger life stages. The smaller juveniles showed no reduction in abundance, indicating a different feeding strategy.

The lack of disturbance interactions, the extreme adaptation to hypoxic and sulfidic sediments, described above, and the predilection for centers of biodeposits would explain the particularly rich populations of *T. benedii* in the bottom sediment of mussel beds with their masses of feces and pseudofeces. Despite the intense predation by *Nereis diversicolor* on *T. benedii* in mussel beds, recruitment of this opportunistic oligochaete seems to compensate for its losses, while predatory pressure on another oligochaete of a similar ecological character, *Tubifex pseudogaster*, was much less (Dittmann, 1990).

*Oligochaetes as food*

The different fate of the two eulittoral tubificid species in mussel beds leads to consideration of the fate of the rich oligochaete production as a prey for higher trophic levels. Eulittoral tidal flats are known as nursery and over-wintering grounds for many marine fauna. Here, the rich biomass of marine oligochaeta results in them playing an important role in the coastal food web (Giere & Pfannkuche, 1982). This was confirmed in the experiments of Dittmann (1990) with the polychaete *N. diversicolor*. In our own studies (unpubl.) juvenile *Crangon crangon* and gobiid fish consume *T. benedii* in large amounts preferring it to polychaetes of similar size. For some migratory birds *T. benedii* is a 'significant part of the diet' (Bagheri & McLusky, 1984). Finally, the meiofauna also utilize marine oligochaetes. Watzin (1986) found severe depletion of oligochaetes (unidentified) in experimental boxes after the addition of two turbellarian species as predators. Virnstein (1977), however, showed that in the field and with natural reproduction, populations of *Tubificoides* spp. remained fairly stable and were not depleted by predation.

*Nutrition of oligochaetes*

Aquatic oligochaetes, similar to their terrestrial relatives, the 'earthworms,' are usually considered as non-selective detritivores or deposit feeders. Harper et al. (1981) found bacteria to be a major food item for *Nais variabilis* (Naididae) in freshwater sediments. Given the very large populations of naidids in sediments rich in organic matter the authors extrapolated that up to 48 times the bacterial stock in the sediment were continuously ingested by the worms. Thus, bacterivory seems to be the primary feeding strategy for these oligochaetes, which confirms earlier findings on tubificids (Wavre & Brinkhurst, 1971). More detailed studies, also on marine species, are required and feasible with evolving micro-ecological techniques and the application of molecular and isotopic methods. The indiscriminate terms 'detritus feeders' or 'deposit feeders' are probably not appropriate for oligochaetes (they are useful general terms cf. carnivore or herbivore).

Another food source that should be investigated is the uptake of dissolved organic substances, particularly sugars and amino acids. All degraded organic matter, each bacterial cell and biofilm, releases or excretes rich amounts of solutes of high nutritive value. The relatively large body surface area to volume ratio of oligochaetes with their thin cuticle renders them well adapted to uptake of nutrients across the body wall. Petersen et al. (1998) found in the brackish water species *Nais elinguis* (a salinity tolerant freshwater species) that amino acids alone could contribute up to 1/4 of its food requirement. Uptake of dissolved substances is obligatory in the gutless tubificids (*Inanidrilus leukodermatus*) (Liebezeit et al., 1983). Uptake rates of glucose were comparable to those found in gutless pogonophorans where this feeding mode was estimated (Southward & Southward, 1980) to cover 30% of the metabolic needs.

Despite the absence of other detailed studies on marine oligochaetes, there is a high probability that transepidermal and transintestinal uptake of dissolved substances are significant trophic pathways especially for the smaller, meiobenthic oligochaetes.

## Conclusions

The marine environment plays a dominant economic, recreational and protective role in our society. Bagheri & McLusky (1984) suggest that in estuarine waters "oligochaetes ... form an absolutely vital, sometimes dominant component of secondary production ...". This view can probably be generalized for all eutrophic littoral sediments. It has also been shown that 'oligochaetes are capable of major modifications of their physical environment' (McCann & Levin, 1989). Thus, their ecological role in eulittoral ecotones requires at least as much intense scientific attention as do polychaetes. In fact McCann and Levin (1989), who mostly work on polychaetes have suggested with reference to oligochaetes. 'Future research efforts on the ecology and biology of this ecologically relevant annelid group are seriously needed'.

The lack of new ecological and biological studies on marine oligochaeta, since the 1982 review, is disturbing. Perhaps our focus on marginal aspects, relevant to only a small group of specialists and their symposia, and published only in specialized journals of small influence, may be contributory to this lack of new research? I have tried to show in this review that it is not only specialists in marine oligochaetes who have brought this group of organisms into main stream marine research. These biologists likely see the values of these organisms in a larger context. Beyond just coastal ecology, marine oligochaetes can serve as valuable models for solving problems of general biological relevance. Topics such as tissue senescence, genetic strategies of populations, or patterns and evolution of symbioses can and should be tackled using marine Oligochaeta as convenient tools and examples.

## References


Bagarinao, T., 1992. Sulfide as an environmental factor and toxicant: tolerance and adaptations in aquatic organisms. Aquatic Toxicology 24: 21–62.

Bagheri, E. A. & D. S. McLusky, 1982. Population dynamics of oligochaetes and small polychaetes in the polluted forth estuary ecosystem. Netherlands Journal of Sea Research 16: 55–66.

Bagheri, E. A. & D. S. McLusky, 1984. The oxygen consumption of *Tubificoides benedeni* (Udekem) in relation to temperature and its application to production biology. Journal of Experimental Marine Biology and Ecology 78: 187–197.

Bouguenec, V. & N. Giani, 1989. Biological studies upon *Enchytraeus variatus* Bougenec & Giani 1987 in breeding cultures. Hydrobiologia 180: 151–165.

Brinkhurst, R. O., 1980. The production biology of the Tubificidae (Oligochaeta). In Brinkhurst, R. O. & D. G. Cook (eds), Aquatic Oligochaete Biology. Plenum Press, New York, London, 205–209.

Burd, B. J., A. Nemec & R. O. Brinkhurst, 1990. The development and application of analytical methods in benthic marine infaunal studies. Advances in Marine Biology 26: 169–247.

Commito, J. A., 1987. Adult – larval interactions: predictions, mussels and cocoons. Estuarine Coastal and Shelf Sciences 25: 599–606.

Cuadrado, S. & E. Martinez-Ansemil, 2001. External structures used during attachment and sperm transfer in tubificids (Annelida, Oligochaeta). Hydrobiologia 463: 107–113.

Diaz, R. J., 1984. Short term dynamics of the dominant annelids in a polyhaline temperate estuary. Hydrobiologia 115: 153–158.

Diaz, R. J. & C. Erséus, 1994. Habitat preferences and species associations of shallow-water marine Tubificidae (Oligochaeta) from barrier reef ecosystems off Belize, Central America. Hydrobiologia 278: 93–105.

Dittmann, S., 1990. Mussel beds – amensalism or amelioration for intertidal fauna?. Helgoländer Meeresuntersuchungen 44: 335–352.

Dubilier, N., A. Blazejak & C. Rühland, 2006. Symbioses between bacteria and gutless marine oligochaetes. In Overmann, J. (ed.), Molecular Basis of Symbiosis. Springer, Berlin, Heidelberg, New York (in press).

Dubilier, N., O. Giere, D. L. Distel & C. M. Cavanaugh, 1995. Characterization of chemoautotrophic bacterial symbionts in a gutless marine worm (Oligochaeta, Annelida) by phylogenetic 16S rRNA sequence analysis and *in situ* hybridization. Applied and Environmental Microbiology 61: 2346–2350.

Dubilier, N., O. Giere & M. K. Grieshaber, 1994. Concomitant effects of sulfide and hypoxia on the aerobic metabolism of the marine oligochaete *Tubificoides benedii*. Journal of Experimental Zoology 269: 287–297.

Dubilier, N., O. Giere & M. K. Grieshaber, 1995. Morphological and ecophysiological adaptations of the marine oligochaete *Tubificoides benedii* to sulfidic environments. American Zoologist 35: 163–173.

Dubilier, N., C. Mülders, T. Ferdelmann, D. de Beer, A. Pernthaler, M. Klein, M. Wagner, C. Erséus, F. Thiermann, J. Krieger, O. Giere & R. Amann, 2001. Endosymbiotic sulphate-reducing and sulphide-oxidizing bacteria in an oligochaete worm. Nature 411: 298–302.

Erséus, C., 2002. Mangroves and marine oligochaete diversity. Wetlands Ecological Management 10: 197–202.

Erséus, C., 2003. The gutless Tubificidae (Annelida: Oligochaeta) of the Bahamas. Meiofauna Marina 12: 59–84.

Erséus, C. & R. J. Diaz, 1989. Population dynamics of *Tubificoides amplivasatus* (Oligochaeta, Tubificidae) in the Öresund, Denmark. Hydrobiologia 180: 167–176.

Gamenick, I., 1997. Ökophysiologische und enzymatische Differenzierung verschiedener Geschwisterarten des *Capitella capitata* – Komplexes (Annelida, Polychaeta). Shaker, Aachen 112 pp.

Gamenick, I., M. Abbiati & O. Giere, 1998. Field distribution and sulphide tolerance of *Capitella capitata* (Annelida: Polychaeta) around shallow water hydrothermal vents off Milos (Aegean Sea). A new sibling species?. Marine Biology 103: 447–453.

Generlich, O. & O. Giere, 1996. Osmoregulation in two aquatic oligochaetes from habitats with different salinity and comparison to other annelids. Hydrobiologia 334: 251–261.

Giere, O., 1970. Untersuchungen zur Mikrozonierung und Ökologie mariner Oligochaeten im Sylter Watt. Veröffentlichungen des Instituts für Meeresforschung Bremerhaven 12: 491–529.

Giere, O., 1973. Oxygen in the marine hygropsammal and the vertical microdistribution of oligochaetes. Marine Biology 21: 180–189.

Giere, O., 1979. Studies on marine Oligochaeta from Bermuda, with emphasis on new *Phallodrilus*-species (Tubificidae). Cahiers de Biologie Marine 20: 301–314.

Giere, O., 1992. Benthic life in sulfidic zones of the sea – ecological and structural adaptations to a toxic environment. Verhandlungen der Deutschen Zoologischen Gesellschaft 85(2): 77–93.

Giere, O., 1993. Meiobenthology. The Microscopic Fauna in Aquatic Sediments. Springer-Verlag, Heidelberg, Berlin 328 pp.

Giere, O., N. M. Conway, G. Gastrock & C. Schmidt, 1991. 'Regulation' of gutless annelid ecology by endosymbiotic bacteria. Marine Ecology Progress Series 68: 287–299.

Giere, O. & C. Erséus, 2002. Taxonomy and new bacterial symbioses of gutless marine Tubificidae (Annelida, Oligochaeta) from the Island of Elba (Italy). Organism Diversity and Evolution 2: 289–297.

Giere, O. & J. Krieger, 2001. A triple bacterial endosymbiosis in a gutless oligochaete (Annelida): ultrastructural and immunocytochemical evidence. Invertebrate Biology 120: 41–49.

Giere, O. & C. Langheld, 1987. Structural organisation, transfer and biological fate of endosymbiotic bacteria in gutless oligochaetes. Marine Biology 93: 641–650.

Giere, O., C. Nieser, R. Windoffer & C. Erséus, 1995. A comparative structural study on bacterial symbioses of Caribbean gutless Tubificidae (Annelida, Oligochaeta). Acta Zoologica (Stockholm) 76: 281–290.

Giere, O. & O. Pfannkuche, 1982. Biology and ecology of marine Oligochaeta, a review. Oceanography and Marine Biology: An Annual Review 20: 173–308.

Giere, O., J.-H. Preusse & N. Dubilier, 1999. *Tubificoides benedii* (Tubificidae, Oligochaeta) – a pioneer in hypoxic and sulfidic environments. An overview of adaptive pathways. Hydrobiologia 406: 235–241.

Grieshaber, M., I. Hardewig, U. Kreutzer, A. Schneider & S. Völkel, 1992. Hypoxia and sulfide tolerance in some marine invertebrates. Hypoxie- und Sulfidtoleranz einiger mariner Invertebraten. Verhandlungen der Deutschen Zoologischen Gesellschaft 85(2): 55–76.

Grieshaber, M. K., I. Hardewig, U. Kreutzer & H.-O. Pörtner, 1994. Physiological and metabolic responses to hypoxia in invertebrates. Reviews on Physiology and Biochemical Pharmacology 125: 43–147.

Harper, R. M., J. C. Fry & M. A. Learner, 1981. A bacteriological investigation to eludicate the feeding biology of *Nais variabilis* (Oligochaeta: Naididae). Freshwater Biology 11: 227–236.

Hauschild, K. & M. K. Grieshaber, 1997. Oxygen consumption and sulfide detoxification in the lugworm *Arenicola marina* (L.) at different ambient oxygen partial pressures and sulfide concentrations. Journal of Comparative Physiology, B 167: 378–388.

Healy, B. & K. Walters, 1994. Oligochaeta in *Spartina* stems: the microdistribution of Enchytraeidae and Tubificidae in a salt marsh, Sapelo Island, USA. Hydrobiologia 278: 111–123.

Hentschel, B. T. & P. A. Jumars, 1994. *In situ* chemical inhibition of benthic diatom growth affects recruitment of competing, permanent and temporary meiofauna. Limnology and Oceanography 39: 816–838.

Kaster, J. L., J. V. Klump, M. Meyer, J. Krezoski & M. E. Smith, 1984. Comparison of defecation rates of *Limnodrilus hoffmeisteri* Claparède (Tubificidae) using two different methods. Hydrobiologia 111: 181–184.

Koop, K. & C. L. Griffiths, 1982. The relative significance of bacteria, meio- and macrofauna on an exposed sandy beach. Marine Biology 66: 295–300.

Krieger, J., 2001. Funktion und Übertragung endosymbiontischer Bakterien bei darmlosen marinen Oligochaeten. PhD-thesis, Univ-Hamburg, Shaker, Aachen.

Lassèrre, P., 1976. Metabolic activities of benthic microfauna and meiofauna. Recent advances and review of suitable methods of analysis. In McCave, I. N. (ed.), The Benthic Boundary Layer. Plenum Publ. Corp, New York, NY, 95–142.

Liebezeit, G., H. Felbeck, R. Dawson & O. Giere, 1983. Transepidermal uptake of dissolved carbohydrates by the gutless marine oligochaete *Phallodrilus leukodermatus* (Annelida). Oceanis 9: 205–211.

Lindegaard, C., K. Hamburger & P. C. Dall, 1994. Population dynamics and energy budget of *Marionina southerni* (Cernosvitov) (Enchytraeidae, Oligochaeta) in the shallow littoral of Lake Esrom, Denmark. Hydrobiologia 278: 291–301.

Martinez, D. E. & J. S. Levinton, 1992. Asexual metazoans undergo senescence. Proceedings of the National Academy of Science USA 89: 9920–9923.

McCann, L. D. & L. A. Levin, 1989. Oligochaete influence on settlement, growth and reproduction in a surface-deposit-feeding polychaete. Journal of Experimental Marine Biology and Ecology 131: 233–253.

Nilsson, P., J. P. Kurdziel & J. S. Levinton, 1997. Heterogeneous population growth, parental effects and genotype–environment interactions of a marine oligochaete. Marine Biology 130: 181–191.

Nilsson, P. G., J. S. Levinton & J. P. Kurdziel, 2000. Migration of a marine oligochaete: induction of dispersal and microhabitat choice. Marine Ecology Progress Series 207: 89–96.

Petersen, S., G. Arlt, A. Faubel & K. R. Carman, 1998. On the nutritive significance of dissolved free amino acids uptake for the cosmopolitan oligochaete *Nais elinguis* Müller (Naididae). Estuarine Coastal and Shelf Science 46: 85–91.

Reible, D. D., V. Popov, K. T. Valsaraj, L. J. Thibodeaux, F. Lin, M. Dikshit, M. A. Todaro & J. W. Fleeger, 1996.

Contaminant fluxes from sediment due to tubificid oligochaete bioturbation. Water Research 30: 704–714.

Rhoads, D. C. & D. K. Young, 1970. The influence of deposit-feeding organisms on sediment stability and community trophic structure. Journal of Marine Research 28(2): 150–178.

Schneider, A., 1994. Die Sulfidtoleranz der Schlammröhrenwürmer *Limnodrilus hoffmeisteri* (Cl.) und *Tubifex tubifex* (M.). Shaker, Aachen, 122 pp.

Schöttler, U. & M. Grieshaber, 1988. Adaptation of the polychaete worm *Scoloplos armiger* to hypoxic conditions. Marine Biology 99: 215–222.

Southward, A. J. & E. C. Southward, 1980. The significance of dissolved organic compounds in the nutrition of *Siboglinum ekmani* and other small Pogonophora. Journal of the Marine Biological Association of the United Kingdom 60: 1005–1034.

Takashima, Y. & S. F. Mawatari, 1998. *Mitinokuidrilus excavatus* n.g., n.sp., a marine tubificid (Oligochaeta) with a unique mode of reproduction. Zoological Sciences 15: 593–597.

Verdonschot, P. F. M., 1981. Some notes on the ecology of aquatic oligochaetes in the Delta region of the Netherlands. Archiv für Hydrobiologie 92: 53–70.

Verdonschot, P. F. M., M. Smies & A. B. J. Sepers, 1982. The distribution of aquatic oligochaetes in brackish inland waters in the SW Netherlands. Hydrobiologia 89: 29–38.

Virnstein, R. W., 1977. The importance of predation by crabs and fishes on benthic infauna in Chesapeake Bay. Ecology 58: 1199–1217.

Watzin, M. C., 1986. Larval settlement into marine soft-sediment systems: interactions with the meiofauna. Journal of Experimental Marine Biology and Ecology 98: 65–113.

Wavre, M. & R. O. Brinkhurst, 1971. Interactions between some tubificid oligochaetes and bacteria found in the sediments of Toronto Harbour, Ontario. Journal of the Fishery Research Board of Canada 28: 335–341.

Westheide, W. & M. C. Müller, 1995. Organisation und Fortpflanzung von Enchytraeiden (Oligochaeta). Film C 1821. Publikationen ZV Wissenschaftlichen Filmen, Biologie 22: 153–170.

Woodin, S. A., 1976. Adult-larval interactions in dense infaunal assemblages: patterns of abundance. Journal of Marine Research 34: 25–41.

Hydrobiologia (2006) 564:117–126
P.F.M. Verdonschot, H. Wang, A. Pinder & R. Nijboer (eds), Aquatic Oligochaete Biology IX
DOI 10.1007/s10750-005-1713-0



# Habitat preferences of aquatic Oligochaeta (Annelida) in the Rokytná River, Czech Republic – a small highland stream

Jana Schenková* & Jan Helešic
*Institute of Botany and Zoology, Masaryk University, Kotlářská 2, 611 37 Brno, Czech Republic*
*(*Author for correspondence: Fax: +420541211214; E-mail: schenk@sci.muni.cz)*

*Key words:* Oligochaeta, habitat preferences, distribution pattern, temporal variability, multivariate analysis

## Abstract

Research was carried out to determine aquatic oligochaete habitat preferences in the Rokytná River (Thaya River basin), a sixth order highland stream in the Czech Republic during the period of April 1999–April 2001. Quantitative samples were collected and current velocity and basic physico-chemical variables were measured monthly in four typical habitats in the Rokytná River. During this study, 28, 842 individuals representing 44 oligochaete species or higher taxa were collected. Temporal variability of proportional occurrence of trophic groups found on selected habitats (gathering collectors and grazers) was analysed. Habitat preferences of oligochaete species were evaluated by Canonical Correspondence Analysis (CANOCO). Biochemical oxygen demand (BOD) and nitrate ($NO_3^-$) ion concentration were the most important variables explaining the distribution of Oligochaeta along the first axis. Current velocity (at 40% of the depth) and presence of oligochaetes associated with the habitat where gravel bars never formed were the most important variables along the second axis. Both axes were correlated with the temperature, reflecting the sampling in summer along the first axis and sampling in late spring along the second axis. The amount of organic matter (BOD) and concentrations of $NO_3^-$ ions represented both oligochaete food source and decomposition products contributing to the growth of algae. Current velocity and preferred habitat explained the spatial pattern of oligochaete distributions.

## Introduction

A wide range of environmental factors permanently influences aquatic ecosystems. Major physico-chemical factors are mostly defined by geology, geomorphology, and climate, and aquatic ecosystems can also be affected significantly by human activity. The oligochaete community composition reflects the states and changes in these variables. Even under natural conditions, a dynamic process of disturbance and change in the community composition occurs continuously (Hynes, 1970; Williams, 1980). The influence of stream hydrology, and physical and chemical factors on aquatic Oligochaeta (Annelida) have been studied by many authors (e.g., Korn, 1963; Dumnicka & Pasternak, 1978; Prenda & Gallardo, 1992; Martínez-Ansemil & Collado, 1996; Verdonschot, 2001).

A moderately polluted highland stream was selected to study habitat preferences within the aquatic oligochaete community. The first objective was to follow the temporal variability of the community composition over a 2-year study. Oligochaete species were divided according to their feeding habits (Moog, 1995) into two groups – grazers and gathering collectors, and temporal variability in their proportional occurrence was evaluated. Such changes were followed separately on four dominant habitats differing from hydrological point of view.

The second objective was to evaluate, by means of multivariate analysis, environmental variables significant for the oligochaete distribution and thus to establish if selected habitats play an important role in oligochaete community composition.

## Study area

The oligochaete fauna was studied in the Rokytná River (Thaya River basin, Czech Republic), a sixth order stream (Strahler, 1957). The Rokytná River is 89.3 km long, with a catchment area of 585.4 $km^2$; the mean annual discharge at the mouth is 1.27 $m^3 s^{-1}$. The present study was conducted in a 60 m reach of the river, located at river km 9, at an altitude of 220 m and a slope of 3.2 m $km^{-1}$.

The hydrology at the study reach is characterised by two (summer and winter) annual maximum discharge levels: 1999 (14.6 and 6.0 $m^3 s^{-1}$ respectively), 2000 (3.8 and 24.3 $m^3 s^{-1}$ respectively), and 2001 (2.2 and 7.5 $m^3 s^{-1}$ respectively). Highly variable discharges are often associated with agricultural landscapes – the catchment area is not able to absorb either summer rainstorms or winter snow-melt. The substrate of the stream within this study area consists of pebbles, gravel, and small amount of sand. For the studied river stretch, an alternating of the straight and meander parts is typical.

On the gravel substratum, periphyton developed in spring, dominantly represented by Bacillariophyceae (Marvan, 1998), but the character of the river bottom did not enable a growth of macrophyta.

## Materials and methods

### Habitat types

Four major habitat types on two cross-sections were sampled monthly, from April 1999 to April 2001. One sample was taken at cross-section A, situated in the streamline. This part of the river was chosen, as a representative of the natural straight part of the river, where gravel bottom does not form either gravel bars or islands. Three samples were taken in a meander at cross-section B to include all different habitats of this river stretch. One in the littoral, drying up part (depositional zone indicated as B littoral), one on the gravel bar (indicated as B gravel bar) and one in streamline (erosional zone: B streamline), respectively (Fig. 1). Water depth, current velocity at the bottom and at 40% of the depth characterised selected habitats.

"Freeze core" method (Bretschko & Klements, 1986) down to the depth of 0.1 m on each of the selected habitats was used for sediment structure evaluation twice – in October 2000 and June 2001. Detailed results of the sediment analysis including organic matter content down to the 0.7 m depth were published by Helešic et al. (2005). The sediment particles were sorted and percent representation of individual particle sizes was calculated. The roughness of the substrate of each habitat was expressed by the phi value (Cummins, 1962) for all selected habitats as a weighted average of phi values for each particle size. The smallest value (−3.5) and hence the highest roughness was found on B gravel bar, similar to that found on habitat A streamline (−3.2), habitats B littoral and B streamline had higher values (−2.4 and −2.5), all of them in October 2000. The bottom roughness almost did not change on habitats A streamline (−3.1), B gravel bar (−3.4) and B streamline (−2.6) while in B littoral the bottom roughness increased (−3.6) at the end of investigation in June 2001. Due to possible habitat damage it was refrained from more frequent sampling of the substratum.

Although the Rokytná River is quite a dynamic stream, the substratum did not change within a year except in one of the selected habitats. The substrate data were not included in analyses below, but were used to describe the characteristics of studied habitats.

### Environmental variables

The following physico-chemical variables were recorded during each sampling period: water temperature, hydrogen ion concentration (as pH), dissolved oxygen, conductivity, current velocity at 40% of depth, current velocity at bottom, water depth, biochemical oxygen demand (BOD), chemical oxygen demand (COD), ammonia and nitrate nitrogen, and phosphate phosphorus ($PO_4^{3-}$). The values of physico-chemical variables measured were very similar between habitats, however, the current velocity differed considerably. Mean daily

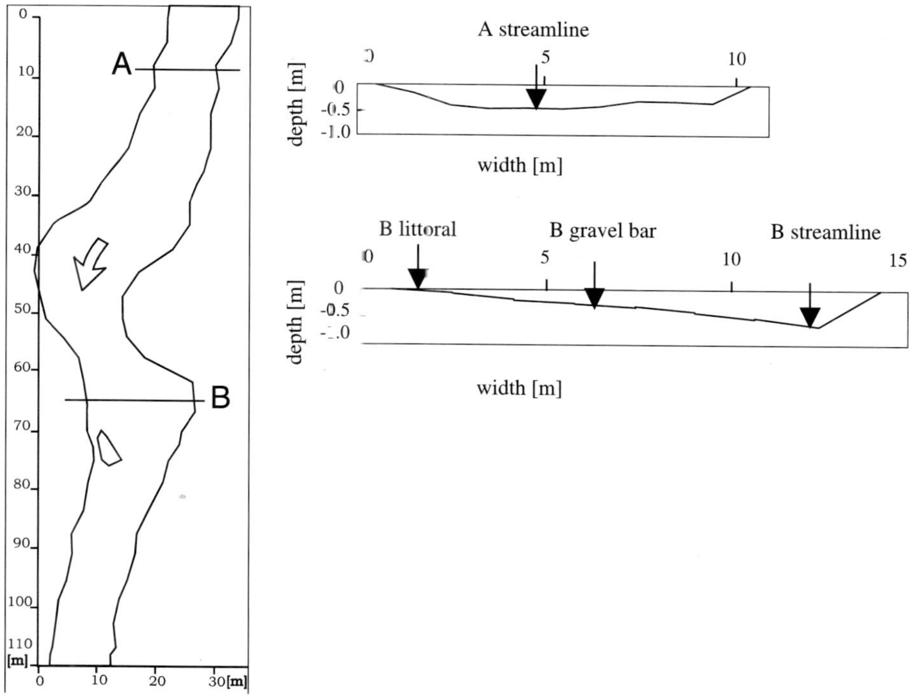


*Figure 1.* Cross-sections of habitats A and B on the Rokytná River, Czech Republic (5 April 2000), and the position of the sampling points: A streamline, B littoral, gravel bar and streamline (arrows).

discharge values of the sampling dates were obtained from the Czech Hydrometeorological Institute (Table 1). One value per sampling date was obtained, since the values did not differ among studied habitats.

*Oligochaeta*

The macrozoobenthic populations were sampled using a Kubíček sampler (0.1 m$^2$, mesh size 0.5 mm) (Helan et al., 1973). Oligochaete specimens were fixed in formalin (4%) in the field, and sorted under the stereomicroscope in the laboratory. All oligochaete species were mounted in Canada Balsam (Cranston, 1982) and determined using keys by Hrabě (1954, 1981), Brinkhurst & Jamieson (1971), Sperber (1950), Timm (1999) and Kathman & Brinkhurst (1999). Identification of *Bythonomus lemani* was confirmed by microscopical examination of mounted specimens that had been embedded in paraffin, sectioned longitudinally, then stained in haematoxylin (Jírovec, 1958).

After the Oligochaeta were divided into trophic groups (Moog, 1995), the temporal variability in proportional occurrence of trophic groups found (gathering collectors and grazers) and their occurrence on selected habitats was evaluated.

*Data processing*

Habitat preferences of oligochaete trophic groups on selected habitats were compared using the Wilcoxon Matched pair test, because the measured data did not have normal distributions (Kolmogorov–Smirnov test) and were dependent on sampling date. The correlation of two trophic groups and two environmental variables (temperature and daily discharge) were evaluated using the Spearman rank correlation coefficient (Zar, 1984).

Faunistic and environmental data were analysed by Canonical Correspondence Analysis (CCA), CANOCO (Ter Braak & Schmilauer, 2002). This method permits the construction of theoretical variables (ordination axes) that best fit the species data according to a unimodal method

*Table 1.* Minimum, maximum, mean and median values of environmental variables of the Rokytná River over the period April 1999–April 2001

| Variable | Min. | Max. | Mean | Median |
|---|---|---|---|---|
| Water temperature (°C) | −0.2 | 19.4 | 11.0 | 13.0 |
| pH | 4.4 | 9.0 | 7.6 | 7.7 |
| $O_2$ [mg $l^{-1}$] | 6.59 | 18.20 | 10.4 | 9.43 |
| Conductivity [$\mu$S $cm^{-1}$] | 453 | 725 | 615 | 615 |
| Current velocity at bottom [m $s^{-1}$] | 0 | 0.87 | 0.34 | 0.30 |
| Current velocity at 40% of depth [m $s^{-1}$] | 0 | 1.46 | 0.67 | 0.68 |
| Mean daily discharge [$m^3$ $s^{-1}$] | 0.13 | 3.95 | 1.36 | 1.08 |
| Depth [m] | 0.10 | 0.47 | 0.22 | 0.20 |
| BOD [mg $O_2$ $l^{-1}$] | 1.9 | 15.8 | 5.4 | 4.5 |
| COD [mg $O_2$ $l^{-1}$] | 26.1 | 51.1 | 27.8 | 32.2 |
| $NO_3^-$ [mg $l^{-1}$] | 12.0 | 57.5 | 27.8 | 20.0 |
| $NH_4^+$ [mg $l^{-1}$] | 0.25 | 1.10 | 0.60 | 0.60 |
| $PO_4^{3-}$ [mg $l^{-1}$] | 0.05 | 1.80 | 0.85 | 0.90 |

of ordination. In the canonical ordination these axes are a linear combination of environmental variables. The correlations between environmental variables and the ordination axes are derived and summarised in the eigenvalue of the ordination axis. The result of CCA is a set of scores for both sites and taxa, which can be presented as two-dimensional ordination diagram where points represent sites or taxa and lines outgoing from the origin expresses environmental variables (Ter Braak & Schmilauer, 2002). The length of the lines represents the importance of the respective variable.

The data were not transformed and species were not down-weighted. The statistical significance of the relationship between all species and all variables was tested by Monte Carlo permutation test using 9999 permutations (Ter Braak & Schmilauer, 2002).

## Results

### *Oligochaeta*

During this study, 28 842 individuals representing 36 species of the families Lumbriculidae, Naididae, Tubificidae, and Lumbricidae were collected and identified (Table 2). In addition, the family Enchytraeidae was represented by unidentified species in four genera. *Psammoryctides barbatus* and *Bothrioneurum vejdovskyanum* (Tubificidae) and *Nais elinguis* (Naididae) were the dominant species.

### *Trophic groups*

Most of the recorded species belong to two trophic groups, gathering collectors and grazers. Species with more trophic adaptations (Moog, 1995), the dominant feeding habit was used for evaluating. The third group, predators (only within the genus *Chaetogaster*) represented only about 0.1% of the total number of individuals and therefore was omitted in the analyses. Though the proportion of trophic groups (gathering collectors and grazers) varied among habitats, these differences were insignificant ($p > 0.05$) (Wilcoxon Matched pair test).

The occurrence of grazers was positively correlated (Spearman rank correlation coefficient, $p < 0.05$) with the discharge and negatively with temperature on habitat A streamline (Fig. 2a). The grazers were also negatively correlated with temperature habitat B streamline (Fig. 2b) and gathering collectors were positively correlated with temperature on habitat A streamline (Fig. 2c), (Spearman rank correlation coefficient, $p < 0.05$).

*Table 2.* List of oligochaete taxa recorded, taxon code, frequency of occurrence, percentage of occurrence and trophic group (gat – gathering collector, gra – grazer and pre – predator)

| Taxon | Abbreviation | No. of individuals | % | Trophic group |
|---|---|---|---|---|
| Lumbricidae | | | | |
| *Allolobophora chlorotica* Savigny, 1826) | Allo chl | 28 | 0.10 | gat |
| *Eiseniella tetraedra* (Savigny, 1826) | Eise tet | 26 | 0.09 | gat |
| Enchytraeidae | | | | |
| *Enchytraeus* g. sp. div. | Ench sp | 64 | 0.22 | gat |
| *Fridericia* g. sp. Div. | Frid sp | 54 | 0.19 | gat |
| *Cernosvitoviella* g. sp. div. | Cern sp | 3 | 0.01 | gat |
| *Cognettia* g. sp. div. | Cogn sp | 32 | 0.11 | gat |
| Tubificidae | | | | |
| *Aulodrilus pluriseta* (Piguet, 1906) | Aulo plu | 2 | 0.01 | gat |
| *Bothrioneurum vejdovskyanum* Štolc, 1886 | Both vej | 2604 | 9.03 | gat |
| *Limnodrilus claparedeanus* Ratzel, 1868 | Limn cla | 3 | 0.01 | gat |
| *Limnodrilus hoffmeisteri* Claparede, 1862 | Limn hof | 167 | 0.60 | gat |
| *Limnodrilus udekemianus* Claparede, 1862 | Limn ude | 6 | 0.02 | gat |
| *Limnodrilus* g. sp. div. juv. | Limn sp | 2292 | 7.95 | gat |
| *Moraviodrilus pygmaeus* Hrabě, 1935 | Mora pyg | 386 | 1.34 | gat |
| *Potamothrix hammoniensis* (Michaelsen, 1901) | Pota ham | 7 | 0.02 | gat |
| *Psammoryctides barbatus* (Grube, 1861) | Psam bar | 5482 | 19.01 | gat |
| *Psammoryctides moravicus* (Hrabě, 1934) | Psam mor | 5 | 0.02 | gat |
| *Rhyacodrilus coccineus* (Vejdovský, 1875) | Rhya coc | 41 | 0.14 | gat |
| *Rhyacodrilus falciformis* Bretscher, 1901 | Rhya fal | 1395 | 4.84 | gat |
| Rhyacodrilinae juv. with hair chaetae | Rhya juv | 1634 | 5.67 | gat |
| *Spirosperma ferox* Eisen, 1879 | Spir fer | 1 | 0.003 | gat |
| *Tubifex tubifex* (O. F. Műller, 1774) | Tubi tub | 50 | 0.17 | gat |
| Tubificidae juv. with hair chaetae | Tubi juv | 957 | 3.32 | gat |
| Naididae | | | | |
| *Chaetogaster diaphanus* (Gruithuisen, 1828) | Chae dip | 21 | 0.07 | pre |
| *Chaetogaster diastrophus* (Gruithuisen, 1828) | Chae dis | 7 | 0.02 | pre, gat, gat |
| *Nais alpina* Sperber, 1948 | Nais alp | 461 | 1.60 | gra, gat |
| *Nais barbata* O. F. Műller, 1773 | Nais bar | 108 | 0.37 | gra, gat |
| *Nais bretscheri* Michaelsen, 1899 | Nais bre | 857 | 2.97 | gra, gat |
| *Nais communis* Piguet, 1906 | Nais com | 144 | 0.50 | gra, gat |
| *Nais elinguis* O. F. Műller, 1773 | Nais eli | 5580 | 19.34 | gra, gat |
| *Nais pardalis* Piguet, 1906 | Nais par | 6 | 0.02 | gra, gat |
| *Nais pseudobtusa* Piguet, 1906 | Nais pse | 48 | 0.17 | gra, gat |
| *Nais simplex* Piguet, 1906 | Nais sim | 20 | 0.07 | gra, gat |
| *Nais variabilis* Piguet 1906 | Nais var | 1 | 0.003 | gra, gat |
| *Ophidonais serpentina* O. F. Műller, 1773 | Ophi ser | 4 | 0.01 | gra, gat |
| *Pristina aequiseta* Bourne, 1981 | Pris aeq | 56 | 0.19 | gra, gat |
| *Pristina bilobata* (Bretscher, 1903) | Pris bil | 19 | 0.07 | gra, gat |
| *Pristina rosea* (Piguet, 1906) | Pris ros | 323 | 1.12 | gra, gat |
| *Stylaria lacustris* (Linnaeus, 1767) | Styl lac | 2 | 0.01 | gra, gat |
| *Vejdovskyella intermedia* (Bretscher, 1896) | Vejd int | 22 | 0.08 | gra, gat |

*Continued on p. 122*

*Table 2.* (Continued)

| Taxon | Abbreviation | No. of individuals | % | Trophic group |
|---|---|---|---|---|
| Lumbriculidae | | | | |
| *Bythonomus lemani* (Grube, 1879) | Byth lem | 1401 | 4.86 | gat |
| *Lumbriculus variegatus* O. F. Műller, 1774 | Lumb var | 14 | 0.05 | gat |
| *Stylodrilus parvus* (Hrabě & Černosvitov, 1927) | Styl par | 813 | 2.82 | gat |
| *Stylodrilus heringianus* Claparede, 1862 | Styl her | 49 | 0.17 | gat |
| *Stylodrilus* g. sp. div. juv. | Styl juv | 3647 | 12.64 | gat |
| **Total** | | 28 842 | 100.00 | |

## *Habitat preferences*

Canonical correspondence analysis (Fig. 3) was used to extract species–environment relationships; 44 taxa and 13 environmental variables and 8 nominal variables in 84 samples were evaluated. Four nominal variables represented selected habitats – A streamline, B littoral, B gravel bar, and B streamline, and four nominal variables represented seasons – spring (March–April), spring/summer (May–June), summer (July–August), and autumn/winter (September–February). These sampling periods correspond more closely with temporal changes in oligochaete community composition than the four 'standard' seasons (e.g., winter = 21 December through 20 March). The statistical significance of environmental variables was assessed by Monte Carlo permutation test. Ranked from the most to the least: BOD, temperature, habitat A streamline, $NO_3^-$, current velocity at 40%, habitat B streamline, summer sampling and conductivity were significant.

The eigenvalues for the four axes are 0.325, 0.171, 0.084, and 0.075, respectively. According to the weighted correlation matrix, the first axis (40.8% variance of the species–environment relation) is related to BOD (0.592) and $NO_3^-$ (0.350) and thus represents the food source ($NO_3^-$ as indicator of a nutrient source for algae and diatoms). The second axis (21.6% of variance) is correlated with current velocity at 40% of the depth (0.400) and presence on habitat A streamline (−0.408), explained spatial variability. Both axes are negatively correlated with temperature (−0.334, −0.396 respectively), which reflects the summer sampling along the first axis (−0.518) and the spring/summer sampling (−0.421) along the second axis respectively (Fig. 3). The third ordination axis (10.6% of variance) is positively correlated with the presence of oligochaetes on habitat B streamline (0.391), BOD (0.376), temperature (0.360) and negatively with $NO_3^-$ (−0.401).

In the ordination diagram representatives of the family Naididae were in right upper corner showing development at lower temperatures, in spring. Most of Tubificidae representatives were negatively correlated with current velocity in 40% of the depth and occurred on habitat A streamline.

## Discussion

Most oligochaete species found in the Rokytná River occurred within a wide range of environmental conditions, demonstrating a large temporal and spatial variability. Oligochaete composition on four selected habitats did not differ significantly when comparing proportion of two feeding groups. We suggest that density changes of grazers, which were negatively correlated with the temperature and positively with discharge, were directly caused by periphyton growth. The grazers represented by members of the family Naididae have maximal development in winter and spring (Pfannkuche, 1981) when sufficient light enables the development of periphyton, also Šporka (1996) have found that the abundance of the family Naididae is dependent on the quantity of periphyton. The positive correlation of gathering collectors and the temperature again correspond with availability of food sources. They reach maximal densities in summer (Lazim & Learner, 1986), when decaying organic matter is available. Most significant relationships were associated with a very stable habitat A; the fact that gravel bars never form there enhance oligochaete growth.

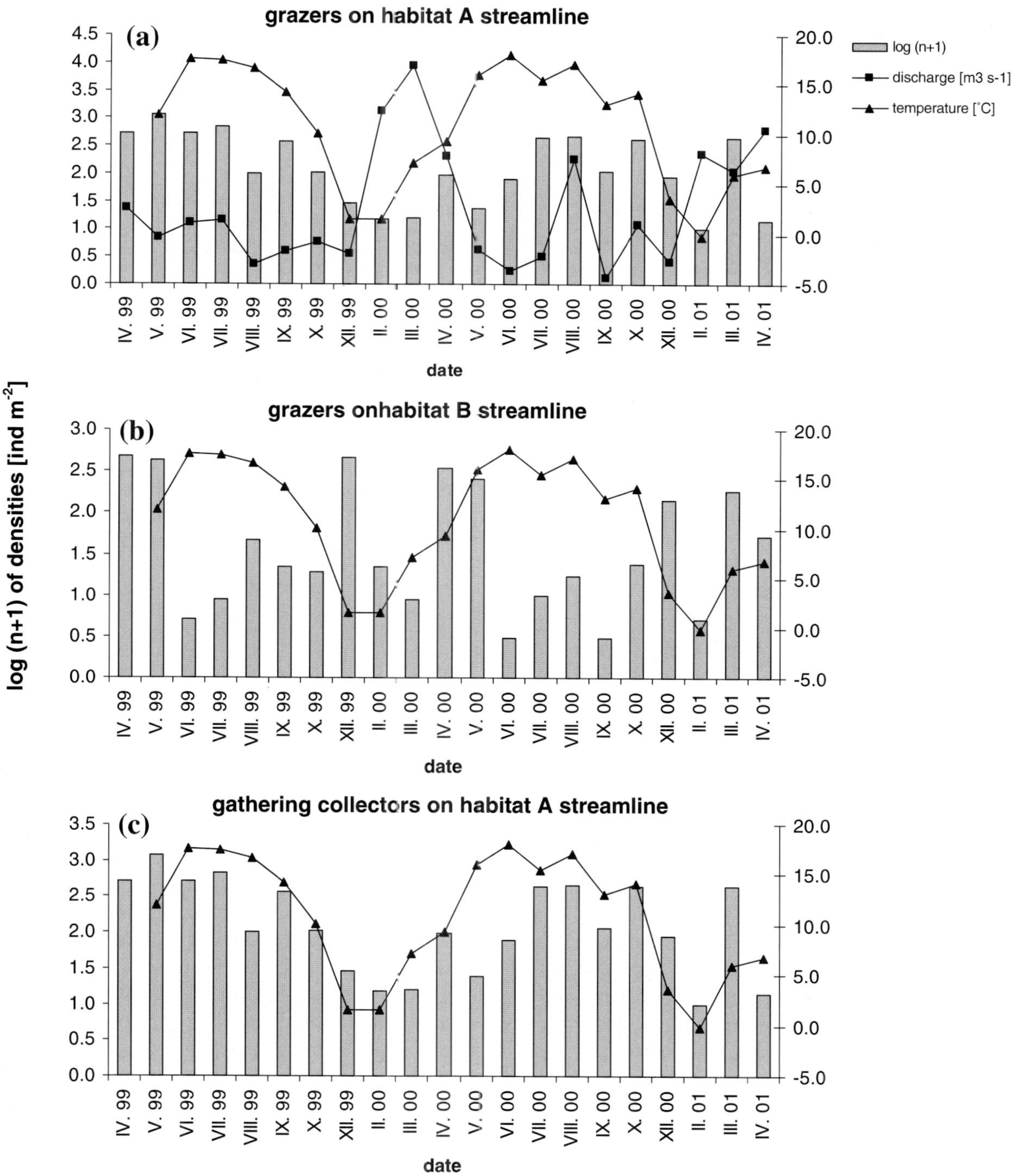


*Figure 2.* The temporal variability of grazers on habitat A streamline, positively correlated with the discharge and negatively with temperature (a) The temporal variability of grazers on habitat B streamline, negatively correlated with temperature (b) The temporal variability of gathering collectors on habitat A streamline, positively correlated with temperature (c) (Spearman rank correlation coefficient, $p < 0.05$).

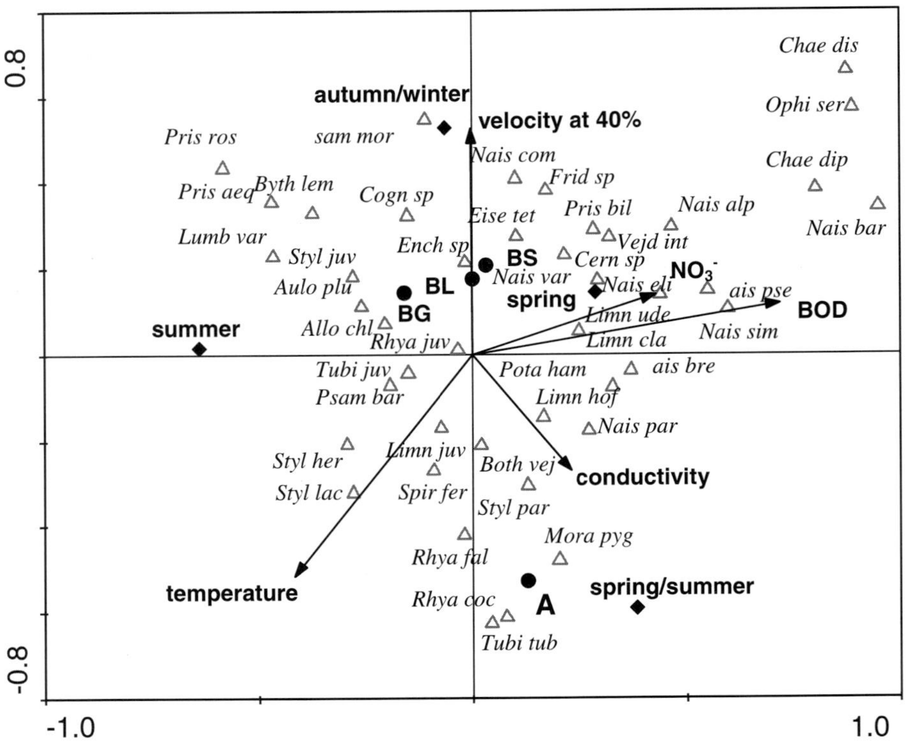


*Figure 3.* CCA diagram of the axis 1 and 2 of oligochaete taxa (for abbreviations see Table 2) in relation to environmental variables (arrows), seasons (diamonds) and sampled habitats (points): A = A streamline, BL = B littoral, BG = B gravel bar, BS = B streamline – recorded in the Rokytná River.

Variability of current velocities in different habitats of cross-section B resulted in either the movement of worms between habitats or their downward burrowing into hyporheic layers.

Habitat preferences of various species were evaluated using CANOCO. Oligochaete community composition was dependent upon organic matter input (BOD) and sources of inorganic ions $NO_3^-$ (aerobic degradation product and secondary periphyton nutrient source) along the first axis. The second axis was correlated with current velocity and presence on habitat A streamline. The most important variable determining oligochaete community composition in habitats assessed during this study was habitat A streamline in which little if any shifting of substrate occurred – it offered stable conditions for oligochaete occurrence, especially permanent availability of organic matter in substratum. Both axes, however, were also correlated with temperature and seasons of sampling, thus representing temporal variability. The values for chemical variables monitored during this study indicate that the Rokytná River varies from slightly to strongly polluted stream within a year. Detecting the habitat preferences of species in polluted waters is problematic. Some authors (Dumnicka & Pasternak, 1978; Culp et al., 1983; Prenda & Gallardo, 1992) showed that certain oligochaete species were not habitat selective. In CCA analysis, two of the four selected habitats (B littoral and B gravel bar) showed only small reaction with the oligochaete distribution. Although both were differing in environmental conditions with respect to velocity and depth, the oligochaete composition was quite similar there. The position of these environmental variables is similar in the diagram (Fig. 3), although they are not correlated. CCA analysis showed that certain habitats are vital for oligochaete occurrence (habitat A streamline was the third most important environmental variable); other variables were less important. Results were significantly influenced by temporal variability – oligochaete community composition differed both in time and in space.

We conclude that, at least in the lower reaches of a highland stream, habitats do not play as an important role as has been demonstrated in high order streams (Korn, 1963; Montanholi-Martins & Takeda, 1999, 2001). Although some authors (Montanholi-Martins & Takeda, 1999) applied a very similar sampling scheme, they found spatial distribution to prevail over the temporal one. Our conclusions are similar to those of Verdonschot (2001), who studied the influence of different substrates in lowland rivers. There were no significant relationships with the habitats if the most of the study streams included them. According to Verdonshot (2000) the disturbance caused by agricultural activities in the catchment area can cause that only ubiquists without habitat preferences settle the bottom. Most of the oligochaete species show wide ecological valences and habitat preferences are not significant when comparing relatively close sampling points in streams particular in agricultural areas. Relatively natural character of the stream is why we found two (stable) habitats important for oligochaete occurrence. On the other hand, the impact of agriculture causes extreme discharges and this promotes ubiquist species whose community composition and abundance, therefore, varies with time rather than with habitat.

## Acknowledgements

The authors would like to thank Prof. J. Gaisler (Institute of Botany and Zoology, Masaryk University) and Mr. M. J. Wetzel (Illinois Natural History Survey, Champaign, U.S.A.) for their critique of this manuscript and assistance in improving the English. We are also grateful to our team of co-workers in field and laboratory for their helpful assistance. Financial support was kindly granted by Masaryk University (No. 206/01/0902) and Ministry of Education of the Czech Republic (No. 0429).

## References

Brinkhurst, R. O., Jamieson & B. G. M. , 1971. Aquatic Oligochaeta of the World. Oliver & Boyd, Edinburgh, 860 pp.

Cranston, P. S., 1982. A key to the larvae of the British Orthocladiinae (Chironomidae). Scientific Publication of the Freshwater Biological Association 45: 1–152.

Culp, J. M., S. J. Walde & R. W. Davies, 1983. Relative importance of substrate particle size and detritus to stream benthic macroinvertebrate microdistribution. Canadian Journal of Fisheries and Aquatic Sciences 40: 1568–1574.

Cummins, K. W., 1962. An evaluation of some techniques for the collection and analysis of benthic samples with spetial emphasis on lotic waters. American Midland Naturalist 67: 477–504.

Dumnicka, E. & K. Pasternak, 1978. The influence of physicochemical properties of water and bottom sediments in the River Nida on the distribution and numbers of Oligochaeta. Acta Hydrobiologica 20(3): 215–232.

Helan, J., F. Kubíček, B. Losos, E. Sedlák & M. Zelinka, 1973. Production conditions in the trout brooks of the Beskydy Mountains. Folia Biologia. Univ J E Purkyně Brno : 105.

Helešic, J., Leichtfried, M., Wagner, F. & M. Omesová, 2005. Investigations on gravel bars and hyporheic zone in an Alpine and two Hercynian streams in Central Europe. Internationale Vereinigung für Theoretische und Angewandte Limnologie Verhandlungen 29. (In print).

Hrabě, S. (ed.), 1954. Klíč k určování zvířeny ČSR [key to the Czechoslovak fauna]. Vol. 1. ČSAV, Praha, 540 pp. (In Czech).

Hrabě, S., 1981. Vodní máloštětinatci (Oligochaeta) Československa [aquatic Oligochaeta of Czechoslovakia]. Acta Universitatis Carolinae, Biologia 1979: 1–168. (In Czech, English summ.).

Hynes, H. B. N., 1970. The Ecology of Running Waters. Liverpool University Press, Liverpool, 555 pp.

Jírovec, O., 1958. Zoologická technika [zoological technique]. Státní pedagogické nakladatelství, Praha, 342 pp (In Czech).

Kathman, R. D. & R. O. Brinkhurst, 1999. Guide to the freshwater Oligochaetes of North America. Aquatic Resources Center, College Grove, Tennessee, USA, 264 pp.

Korn, H., 1963. Studien zur Őkologie der Oligochaeten in der oberen Donau unter Berűcksichtigung der Abwassereinwirkungen. Archiv für Hydrobiologie 27: 131–182.

Lazim, M. N. & M. A. Learner, 1986. The life-cycle and production of Limnodrilus hoffmeisteri and L. udekemianus (Tubifidae; Oligochaeta) in the organically enriched Moan-Feeder Stream, Cardiff, South Wales. Archiv für Hydrobiologie 74(2): 200–225.

Martínez-Ansemil, E. & R. Collado, 1996. Distribution patterns of aquatic oligochaetes inhabiting watercourses in the Northwestern Iberian Peninsula. Hydrobiologia 334: 73–83.

Marvan, P., 1998. Řasová flóra stojatých vod, mokřadů a toků širší oblasti vlivu energetické soustavy Dukovany-Dalešice [algal flora in the region of the Dukovany-Dalešice energy complex]. Přírodověd. Sbor. Západomorav. Muzea Třebíč (Acta Scientiarum Naturalium Musei Moraviae Occidentalis Třebíč 34: 1–136. (In Czech, English summ.).

Moog, O. (ed.), 1995. Fauna aquatica Austriaca. Katalog zur Autőkologischen Einslufung Aquatischer Organismen Ősterreichs, Teil IIIA Arteninventar fűr Ősterreich. Bundesministerium fűr Land- und Forstwirtschaft, Wien, 494 pp.

Montanholi-Martins, M. C. & A. M. Takeda, 1999. Communities of benthic oligochaetes in relation to sediment structure in the upper Paraná River, Brazil. Studies on Neotropical Fauna & Environment 34: 52–58.

Montanholi-Martins M. C. & A. M. Takeda, 2001. Spatial and temporal variations of oligochaetes of the Ivinhema River and Patos Lake in the Upper Paraná River Basin, Brazil. In: Rodriguez, P. & P. F. M. Verdonschot, (eds), Aquatic Oligochaete Biology VIII. Hydrobiologia 463:197–205.

Pfannkuche, O., 1981. Distribution, abundance and life cycles of aquatic Oligochaeta (Annelida) in a freshwater tidal flat of the Elbe Estuary. Archiv für Hydrobiologie Suppl 43(4): 506–524.

Prenda, J. & A. Gallardo, 1992. The influence of environmental factors and microhabitat availability on the distribution of an aquatic oligochaete assemblage in Mediterranean River Basin. Internationale Revue der Gesamten Hydrobiologie 77(3): 421–434.

Sperber, C., 1950. A guide for the determination of European Naididae. Zoologiska Bidrag Från Uppsala 29: 45–81.

Strahler, A. N., 1957. Quantitative analysis of watershed geomorphology. American Geophysical Union Trans 38: 913–920.

Šporka, F., 1996. Macrozoobenthos – permanent fauna. In: Krno I (eds), Limnology of the Turiec River Basin (West Carpthians, Slovakia). Biologia, Bratislava 51(2):23–27.

Timm, T., 1999. A Guide to the Estonian Annelida. Estonian Academy Publishers, Tartu-Tallinn, 208 pp.

Ter Braak, C. J. F. & P. Schmilauer, 2002. Canoco Reference Manual and Cano Draw for Windows User's Guide. Software For Cononical Community Ordination (Version 4.5). Centre for Biometry, Wageningen, 500 pp.

Verdonschot, P. F. M., 2000. Soft-bottomed lowland streams: a dynamic desert. Internationale Vereinigung für Theoretische und Angewandte Limnologie Verhandlungen 27: 2577–2581.

Verdonschot P. F. M., 2001. Hydrology and substrates: determinants of oligochaete distribution in lowland streams. In: Rodriguez P. & P. F. M. Verdonschot (eds), Aquatic Oligochaete Biology VIII. Hydrobiologia 463:249–262.

Williams, D. D., 1980. Temporal patterns in recolonization of stream benthos. Archiv für Hydrobiologie 90(1): 56–74.

Zar J. H.., 1984. Biostatistical Analysis (2nd edn.). Prentice Hall, London, 428 pp.

Hydrobiologia (2006) 564:127–142
P.F.M. Verdonschot, H. Wang, A. Pinder & R. Nijboer (eds), Aquatic Oligochaete Biology IX
DOI 10.1007/s10750-005-1714-z



# Beyond masses and blooms: the indicative value of oligochaetes

Piet F.M. Verdonschot

*Alterra, Green World Research, P.O. Box 47, 6700 AA, Wageningen, The Netherlands (Fax: +31-317-424-988; E-mail: piet.verdonschot@wur.nl)*

*Key words:* Oligochaeta, assessment, streams, weighted averaging, ordination, ecological quality

**Abstract**

The European Water Framework Directive (WFD) defines a framework for assessing water bodies in Europe in the future. The conditions in the Directive impose a strong demand for "new" assessment systems. The AQEM project developed an assessment system for European streams using macroinvertebrates. Almost 900 samples were taken in about 400 streams covering 29 stream types distributed over eight countries. The role of the Oligochaeta within this European database was analysed. Almost half a million specimens of oligochaetes were collected in 772 samples. Eight families, 41 genera and 69 species were recorded, although identification emphasised the families Tubificidae and Naididae. Three countries identified oligochaetes to species level, most others restricted their identifications to easy identifiable taxa. Numbers of specimens, species, genera and families differed strongly between the countries due to method, although standardised, and taxonomic knowledge. About 50% of all collected oligochaete taxa had assigned biological and ecological indicator values for metric calculation in the AQEM assessment system. A further refinement of this indication list as well as increased coverage of oligochaete taxa was advised. Weighted averaging was used to evaluate the relation between oligochaete distribution and ecological quality class. It was concluded that when higher taxonomic levels are used in assessment, the quality evaluation results become biased. Furthermore, oligochaetes can tell us much more about the ecological status of streams than is commonly assumed. Differences in ecological optima among *Limnodrilus udekemianus*, *Ilyodrilus templetoni*, *Aulodrilus pluriseta*, *Nais communis*, and *Spirosperma ferox* are shown.

## Introduction

Masses of tubifids cover mud bottoms, explosions of naidids hover in algal blooms; no wonder many people think of oligochaetes as indicators of pollution. But are those ambassadors representative of the whole Class Oligochaeta? In the European Water Framework Directive (WFD: European Commission, 2000), a European need and approach for surface water assessment is outlined. A main starting point in the assessment is the reference condition of a water type, which is a concept new to many European countries. The reference condition is proposed as the anchor point in assessment (European Commission, 2000). The difference between a current state of a test site and its reference is scaled in five classes and taken as quality measure. The WFD generated a strong demand for either "new" assessment systems fulfilling the WFD criteria or adaptation of existing systems to meet these criteria.

The project "The development and testing of an integrated assessment system for the ecological quality of streams and rivers throughout Europe using benthic macroinvertebrates (AQEM)" (AQEM consortium, 2002; Hering et al., 2004) developed a European stream assessment system, using macroinvertebrates and environmental variables. Macroinvertebrates were used because many well described and easily identifiable species are useful for detecting changes in the environment on the regional scale desired (Resh & Unzicker,

1975; Persoone, 1979). This is a consequence of their diverse systematic origins, the duration of their life cycles (ranging from a few weeks to a few years) and the diversity of habitats they occupy (Rosenberg & Resh, 1984).

Assessing the ecological quality of European streams requires a sound and well-established methodology. In the AQEM project a large number of biological measures or metrics were tested independently for 28 of the 29 sampled European stream types. This process resulted in up to 18 core metrics, which were combined into a different multimetric index in each country.

A metric approach is most often based on a list of taxonomic entities and their biological and ecological characteristics. Typically, a list includes attributes such as functional feeding groups or habitat indications for each species. Relationships between species and their biological and environmental requirements or interactions are commonly used in applied ecology but are generally not quantified. Furthermore, the relationship between a species and an environmental condition is often unimodal (Shelford, 1913; Thienemann, 1955), i.e. a species performs best at a particular value (optimum) and performs sub-optimally or cannot survive if the value becomes either too low or too high (Odum, 1971). Assessment metrics use this knowledge to assign an indicator value, in most cases the optimum for each species, and calculate a final score for a whole species list.

Several techniques have been developed to quantify the species optimum curve, like non-least-squares regression (Gauch et al., 1974), generalised linear models (McCullagh & Nelder, 1989), or weighted averaging (Gauch, 1982). Ter Braak & Looman (1986) demonstrated the advantages of using weighted averaging despite the fact that the absence of species is disregarded.

One can question whether these techniques are useful for the application of oligochaetes in assessment. Oligochaetes have played an important role in assessment, particularly in their mass occurrences under disturbed conditions (Kolkwitz & Marsson, 1909; Hamm, 1969; Aston, 1973; Verdonschot, 1989). Mass occurrences are not only restricted to sites with severe organic pollution, but are also reported under conditions of eutrophication (e.g. Milbrink, 1973; Verdonschot, 1996) and drought episodes (Verdonschot, 1995). Also it has been argued that relationships with other environmental factors should not be neglected, such as substrate (Wachs, 1967), depth (Newrkla & Wijegoonawardana, 1987), chloride (Verdonschot, 1984), acidity (Healey & Bolger, 1984) and heavy metals (Chapman et al., 1982). These kinds of environmental relationships are often not accounted for by the current metrics. Metrics very often are restricted to higher taxonomic units like Oligochaeta or Tubificidae. Two problems occur when using the current metrics:

Higher taxonomic levels neglect differences between species.

Metrics more often do not include indicator values of oligochaetes.

To test whether these two assumptions are true the following questions were addressed:

1. What ecological indicator values did oligochaetes have at the outset in the metrics used in the project AQEM?
2. How are oligochaete taxa distributed over ecological quality classes and, more specifically, major environmental variables?
3. How should water quality assessment use oligochaetes in the future and at which taxonomic levels?

## Material and methods

### *Data collection*

The AQEM assessment system was based on a stream typology using the criteria defined by the EU Water Framework Directive under "System A". "System A" descriptors are defined by ecoregions (according to Illies, 1978), the catchment area (small 10–100, medium 100–1000, large 1000–10,000, very large > 10,000 km$^2$), geology of the catchment (siliceous, calcareous, organic), and altitude (high > 800, mid-altitude 200–800, lowland < 200 m). Using these WFD descriptors in the eight participating countries (Germany, Austria, Czech Republic, Greece, Italy, Portugal, Sweden and the Netherlands) resulted in the selection of 28 stream types.

According to the guidelines on site selection, sampling strategy and processing as described in the AQEM manual (AQEM consortium, 2002; Hering et al., 2004), macroinvertebrate samples were taken during 2000 and 2001 in approximately 400 streams in eight countries (Fig. 1). Twenty-nine stream types of different quality classes were represented. Identification was made to species level when and where possible (AQEM consortium, 2002). Due to lack on taxonomic knowledge in certain areas or for certain groups, identification was sometimes limited to higher taxonomic levels.

All 889 macroinvertebrate and environmental samples were combined into one European database, from which the oligochaetes were extracted (Table 1). Oligochaete taxa lists were adjusted to minimise differences in taxonomic resolution between countries, following the ideas for taxonomic adjustment given in Nijboer & Verdonschot

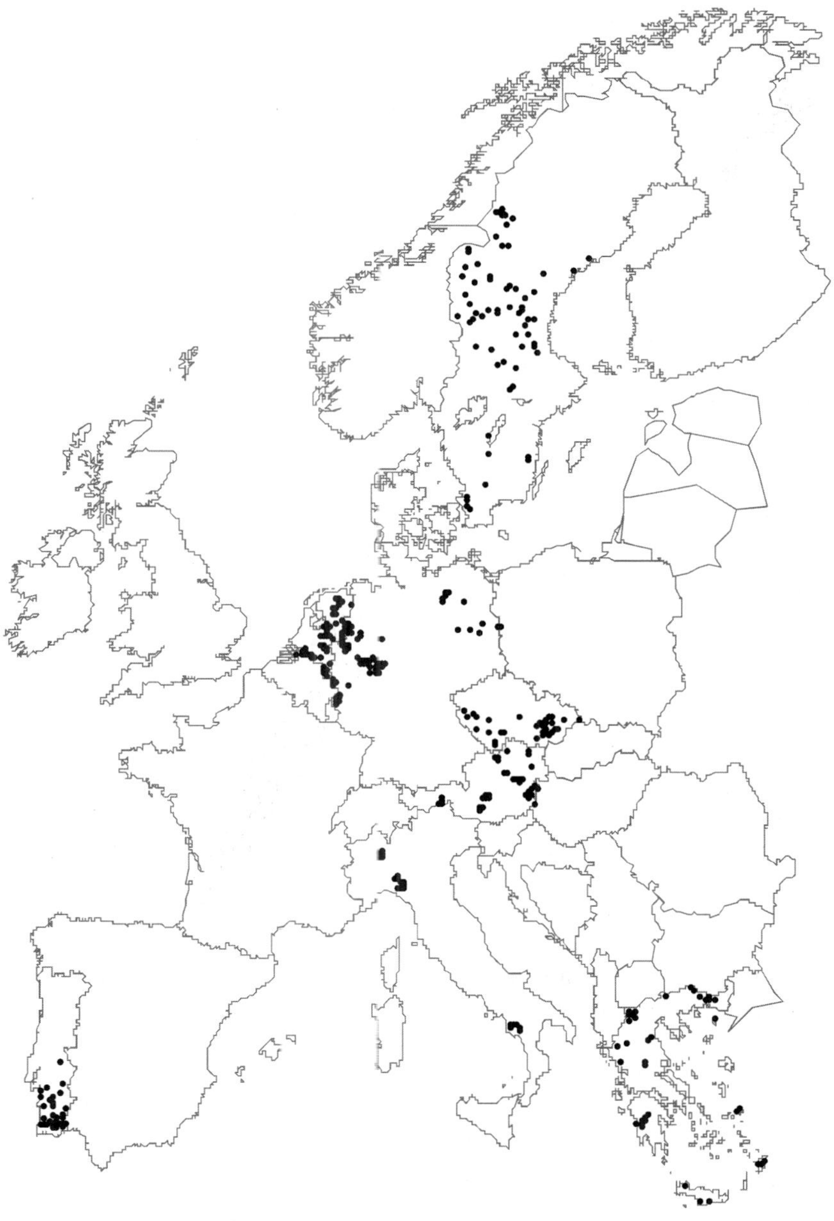

*Figure 1.* Distribution of the sampling sites over Europe.

*Table 1.* The composition of the oligochaetes present in the European AQEM database

| | Samples | Oligochaete taxa | Specimens | Families | Genera | Species |
|---|---|---|---|---|---|---|
| Number | 772 | 101 | 438,832 | 8 | 41 | 69 |

(2000), Schmidt-Kloiber & Nijboer (2004) and Vlek et al. (2004).

At each sampling site a large number of environmental parameters were recorded using a standardised "site protocol". The site protocol included 222 parameters (AQEM consortium, 2002). Environmental parameters were standardised and those variables that were measured at less than 2.5% of the sites were removed. Finally, 189 variables were included in the analysis. Two sites had too few variables measured to be included. The stream types were defined according to the WFD "System A" descriptors. The multimetric AQEM assessment system (Hering et al., 2004) was used to classify each site into an ecological quality class (EQC) ranging from 5 (high quality) to 1 (bad quality). Ultimately, the stream types and ecological quality classification were used to test the oligochaete distributions.

*Autecological information*

A large number of metrics (Table 2) were studied in the AQEM project. A number of these metrics, such as feeding types and microhabitat or current velocity preferences, require scores for the taxa. Therefore, an autecological database was generated to include this information. In order of priority, the autecological information was taken from Moog (1995), Schmedtje & Colling (1996), Verdonschot (1990), Van der Hoek & Verdonschot (1994), and additional information collected by the AQEM consortium. A distinction is made between those metrics based on taxonomic information only and those using autecological information for the individual taxa. The current indicator values of these individual taxa of oligochaetes of metrics were analysed.

Saprobity was classified using three different metrics: the old German saprobic index, the new German saprobic index and the saprobic index according to Zelinka & Marvan (1961). These three metrics represent the development of saprobic metrics over time.

Concerning the feeding type, the majority of oligochaetes were classified as gatherer-collecters; this included most tubificids. Smaller groups included grazer-scrapers (most naidids), and one predator (*Chaetogaster*).

*Indicative value of oligochaetes*

The AQEM assessment system (Hering et al., 2004) results, after a country specific multimetric calculation, in a final EQC score. This score runs from 1, low quality, up to 5, high quality. Gaussian response curves were estimated using weighted averaging (program C2 (Juggins, 2003)) and $N_2$ weighted standard deviations (Ter Braak & van Dam, 1989) to establish the distribution range of oligochaetes over ecological quality classes. Weighted averaging summarises a unimodal response curve for each species. Two parameters are of interest: $u$ = the optimum or mode for each species along the environmental gradient under study, and $t$ = the tolerance of the species to the environmental variable. The calculations were done at family, genus and species level.

Optima and tolerances were also calculated for selected environmental variables. Therefore, first a Detrended Canonical Correspondence Analysis (DCCA) (Ter Braak, 1987), as part of the program CANOCO for Windows version 4.0 (Ter Braak & Šmilauer, 1998), was executed. DCCA is a direct ordination technique, which means that the environmental variables are directly related to the species composition at the sites. The ordination axes in DCCA were calculated as linear combinations of the environmental variables. The option down-weighting of rare species was used to emphasise more commonly distributed species.

DCCA was performed for two different data matrices: one with eight families (including both the total Oligochaeta as well as total Tubificidae and Tubificidae with and without hair setae) and one with only species included. As the program calculates optima and tolerances independently for each taxon, this overlap in input did not affect the results.

*Table 2.* AQEM metrics including oligochaetes

| Ecological preference type | Metric | Taxonomic level |
|---|---|---|
| *Ecological preference based metrics* | | |
| Saprobity | Saprobic index (Zelinka & Marvan in five classes) | Species |
| | German saprobic index (old version in four classes) | Species |
| | German saprobic index (new version in four classes) | Species |
| | Dutch saprobic index | Species |
| | Czech saprobic index | Species |
| | Biological monitoring working party (BMWP) | Class |
| Acidity | Acid class (according to Braukmann in four classes) | Species |
| | Acid index (Hendrikson & Medin) | Species |
| Zonation | Zonation (percentage of community preferring a certain zone in 10 classes with an additional index) | Species |
| Current | Current preference (percentage of community preferring a certain current velocity in seven classes) | Species |
| Microhabitat | Microhabitat preference (percentage of community preferring a certain microhabitat in eight classes) | Species |
| Feeding | Feeding types (percentage of community in nine classes) | Species |
| | RETI (Rhithron feeding type index) | Species |
| Locomotion | Locomotion type (percentage of community in six classes) | Species |
| *Taxonomy based metrics* | | |
| | Abundance [ind m$^{-2}$] | Species |
| | Number of taxa | Species |
| | Average score per taxon (ASPT) | Class |
| | BMWP (Spanish version) | Class |
| | Danish strema fauna index (DSFI in two classes) | Family |
| | Belgian biotic index (BBI) | Family |
| | Indice biotico esteso (IBE in two classes) | Family |
| | Altered indice biotico esteso (IBE AQEM) | Family |
| | Diversity (Simpson index) | Species |
| | Diversity (Shannon–Wiener index) | Species |
| | Diversity (Margalef index) | Species |
| | Evenness | Species |
| | German fauna index (in five classes) | Species |
| | Portuguese index | Class |
| | Taxonomic group [%] (calculated for 30 groups) | Class |
| | Taxonomic group (number of taxa) (calculated for 32 groups) | Class |
| | Taxonomic group (abundance) (calculated for 32 groups) | Class |
| | Number of families | Family |
| | Number of genera | Genus |

The taxonomic level in the respective measures is indicated (for reference see Hering et al., 2004).

## Results

### *Data composition*

Number of samples, number of oligochaete taxa, and number of individuals differed strongly between countries (Table 3). Germany sampled the most sites and Greece the least. The highest average number of oligochaete taxa was found in Austria, the lowest average number in Sweden. The highest average number of oligochaete specimens was collected in Austria, and the lowest number in Greece.

*Table 3.* Numbers of samples, average number of oligochaetes as well as of individuals per country per sample

| Country | Number of samples | Average number of taxa | Average number of individuals |
|---|---|---|---|
| Austria | 94 | 6.1 | 2394 |
| Czech Rep. | 70 | 5.6 | 591 |
| Germany | 144 | 3.0 | 27 |
| Greece | 30 | 1.6 | 3 |
| Italy | 130 | 2.8 | 316 |
| Netherlands | 149 | 5.7 | 426 |
| Portugal | 38 | 3.9 | 1590 |
| Sweden | 117 | 1.0 | 30 |
| Total/average | 772 | 3.7 | 672 |

*Table 4.* The identification level of the Oligochaeta in each country

| Country | Number of oligochaetes | | | |
|---|---|---|---|---|
| | Oligochaete taxa | Families | Genera | Species |
| Austria | 56 | 8 | 26 | 43 |
| Czech Repub. | 43 | 7 | 24 | 32 |
| Germany | 12 | 7 | 5 | 3 |
| Greece | 12 | 5 | 10 | 0 |
| Italy | 29 | 7 | 15 | 14 |
| Netherlands | 52 | 6 | 21 | 39 |
| Portugal | 13 | 5 | 8 | 6 |
| Sweden | 1 | 1 | 0 | 0 |

From looking at the identification level of the Oligochaeta in each country, it becomes clear that differences are large (Table 4). Austria, the Czech Republic and the Netherlands identified a high number of taxa, in contrast to Sweden, which only identified to the Class Oligochaeta. Germany, Greece and Portugal mainly restricted themselves to family and genus level.

The total number of individuals for the Oligochaeta, as well as for each of the oligochaete families, was extremely low in Greece, and relatively low numbers were collected in Sweden and Germany (Table 5). The numbers in Austria were extremely high.

Species level identification was done only in the Czech Republic, Austria and the Netherlands (Table 6). This concerned mainly species belonging to the families Lumbiculidae, Naididae and Tubificidae. These different features of oligochaete taxa and numbers in the European AQEM database surely affected the analysis results.

### *Autecological information*

The AQEM autecological database contained 103 oligochaete taxa (Table 7). The available biological and ecological information was analysed for each metric. Metric values were available for about 50% of the taxa, except in the case of acidity with only 2% available, and feeding type with 90% available. The feeding type showed a very restricted distribution (Fig. 2). Though 90% of the oligochaete taxa are classified by feeding type, most appear to belong to the functional feeding

*Table 5.* Number of specimens per class and family of Oligochaeta in each country

| Class/family | Austria | Czech Republic | Germany | Greece | Italy | Netherlands | Portugal | Sweden | Total number |
|---|---|---|---|---|---|---|---|---|---|
| Oligochaeta total | 969 | | 607 | | 6 | 845 | | 3529 | 5956 |
| Enchytraeidae | 4533 | 137 | 110 | | 371 | 617 | | | 5768 |
| Glossoscolecidae | | | | | | | 53 | | 53 |
| Haplotaxidae | 39 | 100 | 31 | 20 | 27 | | | | 217 |
| Lumbricidae | 164 | 179 | 251 | 2 | 1122 | 278 | 474 | | 2470 |
| Lumbriculidae | 9751 | 5131 | 1671 | 33 | 2052 | 7745 | 29 | | 26,412 |
| Naididae | 70,102 | 4275 | 431 | 23 | 1297 | 2890 | 51,645 | | 130,663 |
| Propappidae | 17,466 | 191 | | | | | | | 17,657 |
| Tubificidae | 121,968 | 31,382 | 748 | 5 | 36,243 | 51,073 | 8217 | | 249,636 |
| Total number | 224,992 | 41,395 | 3849 | 83 | 41,118 | 63,448 | 60,418 | 3529 | 438,832 |

*Table 6.* Number of genera and species per oligochaete family in each country

| Country | Number of | Haplotaxidae | Enchytraeidae | Propappidae | Naididae | Tubificidae | Lumbriculidae | Lumbricidae |
|---|---|---|---|---|---|---|---|---|
| Austria | Genera | 1 | 5 | 1 | 7 | 7 | 4 | 1 |
| | Species | 1 | 3 | 1 | 19 | 14 | 4 | 1 |
| Czech Republic | Genera | 1 | 5 | 1 | 5 | 8 | 3 | 1 |
| | Species | 1 | 3 | 1 | 10 | 12 | 4 | 1 |
| Germany | Genera | | | | | 1 | 3 | 1 |
| | Species | | | | | | 2 | 1 |
| Greece | Genera | 1 | | | 3 | 1 | 5 | |
| | Species | | | | | | | |
| Italy | Genera | 1 | 7 | | 2 | 3 | 1 | 1 |
| | Species | 1 | 4 | | 5 | 2 | 1 | 1 |
| Netherlands | Genera | | 2 | | 7 | 8 | 3 | 1 |
| | Species | | 1 | | 17 | 17 | 3 | 1 |
| Portugal | Genera | | | | 5 | 1 | 1 | 1 |
| | Species | | | | 3 | 1 | 1 | 1 |
| Sweden | Genera | | | | | | | |
| | Species | | | | | | | |

*Table 7.* Information content for 103 oligochaete taxa present in the AQEM autecological database

| Ecological preference type | Number of classes | Number of indicators | % of oligochaetes with scores | Average score | Standard deviation |
|---|---|---|---|---|---|
| Zonation class | 10 | 50 | 49 | | |
| Microhabitat type | 8 | 65 | 63 | | |
| Saprobic index | | | | | |
| Old German index | 1 | 4 | 4 | 3.28 | 0.21 |
| New German index | 1 | 18 | 17 | 3.04 | 0.62 |
| Zelinka & Marvan | 7 | 58 | 56 | 2.25 | 0.59 |
| Current velocity class | 7 | 53 | 51 | | |
| Acidity class | 1 | 2 | 2 | 3.5 | 0.71 |
| Functional feeding group | 10 | 90 | 87 | | |
| Locomotion type | 6 | 57 | 55 | | |

category "gatherer–collector". This may be true for a number of tubificids, but it is doubtful for a number of naidids. Furthermore, most oligochaetes according to this approach belong to the locomotion categories burrower or sprawler (Table 8).

The oligochaetes in general showed a saprobic improvement when comparing the average score for taxa present using the old German saprobic index with that in the new one. This is because the old German saprobic index included mainly a few polysaprobic taxa, while the new one included a much higher number of taxa, including some with a meso- to oligosaprobic indication. Looking at the Zeklinka & Marvan scores, the number of taxa with saprobic valence scores in the oligosaprobic plus beta-mesosaprobic classes appeared to be higher than the sum of those in the alpha-mesosaprobic and polysaprobic classes (Fig. 3). The average score shows that some taxa had a much lower saprobic valence in the old indices.

The number of oligochaete taxa preferring stagnant or slow flowing waters is much higher then those restricted to fast flowing ones (Fig. 4). The taxa were more evenly distributed over the microhabitat types (Table 8) as well as over the different river zones (Table 8).

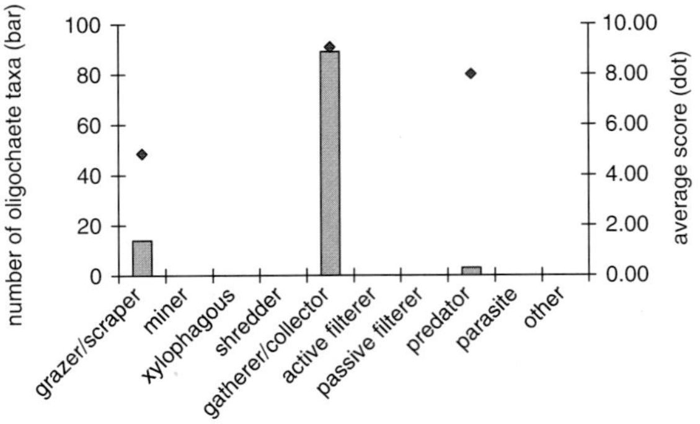


*Figure 2.* The number of oligochaete taxa in each functional feeding group and the average functional feeding score as present in the AQEM database.

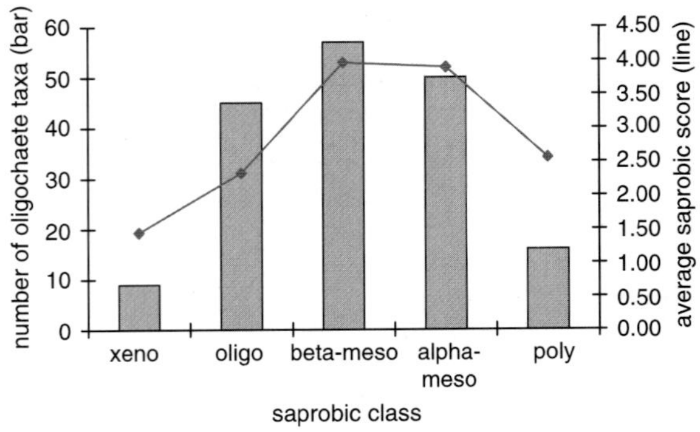


*Figure 3.* The number of oligochaete taxa in each saprobic class and the average saprobic index score per saprobic class, both according to Zelinka & Marvan (1961) as present in the AQEM database.

*Table 8.* Number of oligochaete taxa present in the AQEM autecological database with indications for locomotion type, zonation class and microhabitat type

| | Number of taxa | Average score | Standard deviation |
|---|---|---|---|
| *Locomotion type* | | | |
| Swimming/diving | 17 | – | – |
| Burrowing/boring | 31 | – | – |
| Sprawling/walking | 37 | – | – |
| (Semi)sessil | 8 | – | – |
| Other | 2 | – | – |
| *Zonation class* | | | |
| Eukrenal | 7 | 1.29 | 0.76 |
| Hypocrenal | 10 | 1.40 | 0.84 |
| Epirhithral | 23 | 1.22 | 0.52 |
| Metarhithral | 31 | 1.32 | 0.48 |
| Hyporhithral | 40 | 1.48 | 0.55 |
| Epipotamal | 47 | 1.87 | 0.71 |
| Metapotamal | 45 | 1.69 | 0.67 |
| Hypopotamal | 29 | 1.21 | 0.49 |
| Litoral | 45 | 2.00 | 0.85 |
| Profundal | 31 | 1.94 | 0.89 |
| *Microhabitat type* | | | |
| Pelal | 47 | 3.09 | 1.56 |
| Argyllal | 9 | 2.56 | 0.53 |
| Psammal | 43 | 3.19 | 1.40 |
| Akal | 21 | 2.19 | 1.03 |
| Lithal | 27 | 2.04 | 1.51 |
| Phytal | 35 | 3.71 | 2.58 |
| POM | 25 | 2.04 | 2.95 |
| Other | 5 | 1.80 | 1.10 |

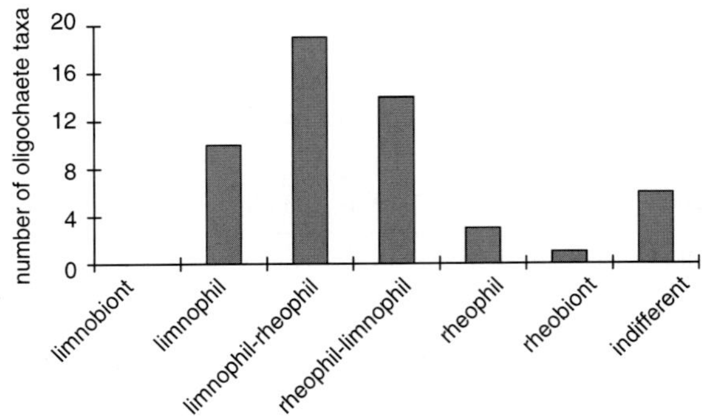


*Figure 4.* The number of oligochaete taxa in each current velocity class as present in the AQEM database.

All metrics based on ecological preferences included information on oligochaetes. The taxonomy based metrics mostly included oligochaetes, either as class or as family, except for one that included genus and one that included species level.

### *Indicative value of oligochaetes*

In total 772 samples from the AQEM database were used, to establish the relation between oligochaetes and the EQC according to the AQEM assessment system. Plotting the number of oligochaete taxa found at each site against its final EQC, there appeared to be no relationship at all (Fig. 5). Up to 12 oligochaete taxa per sample can be found in each EQC.

However, by taking the abundances into account a clear trend becomes visible (Fig. 6). The better the ecological quality becomes, the less abundant the oligochaetes get.

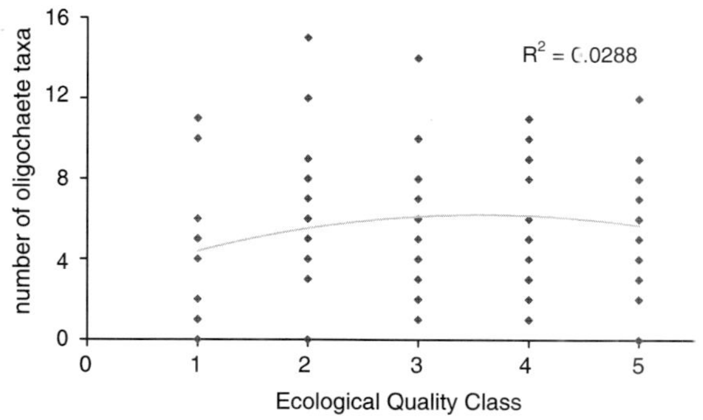


*Figure 5.* The number of oligochaete taxa per sample plotted against the ecological quality class (ranging from 1 = bad to 5 = high) according to the AQEM assessment system.

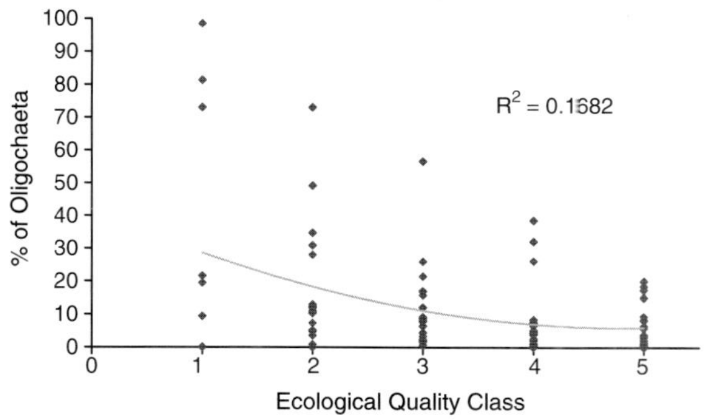


*Figure 6.* The percentages of oligochaete specimens per sample plotted against the ecological quality class (ranging from 1 = bad to 5 = high) according to the AQEM assessment system.

The average number of individuals of five selected oligochaete species (namely, *Limnodrilus udekemianus*, *Ilyodrilus templetoni*, *Aulodrilus pluriseta*, *Nais communis*, and *Spirosperma ferox*) each representing one EQC, plotted against the EQC, illustrates a clear distribution pattern (Fig. 7). This gradient in distribution along EQC's is a good example of the wide distribution of oligochaetes over the ecological quality gradient.

Table 9 includes the optima and tolerances for family, genus and species level of all oligochaetes. The table shows the differences between oligochaete taxa with respect to ecological quality. In general, the Oligochaeta prefer EQC two, but at the family level there is a difference between the Tubificidae, preferring a quality lower than two, up to the Propappidae with a class four preference. The same tendency is true for the individual genera and species (Table 9). Thus, in general, identifying oligochaetes at a higher taxonomic level obscures

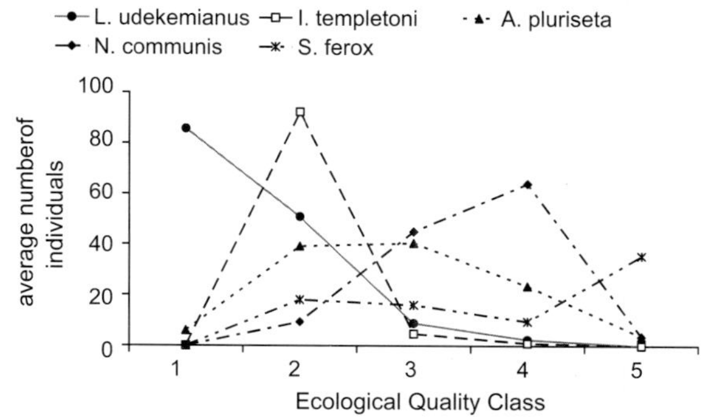


*Figure 7.* The average number of individuals of five selected oligochaete species, *Limnodrilus udekemianus* (20 samples), *Ilyodrilus templetoni* (11 samples), *Aulodrilus pluriseta* (43 samples), *Nais communis* (22 samples), and *Spirosperma ferox* (14 samples), plotted against the ecological quality class according to the AQEM assessment system.

real ecological differences at the species level. The distinction between Tubificidae with and without hair setae did not show a clear difference.

At family level the DCCA ordination showed a gradient running from families occurring in fast running streams with steep slopes (Haplotaxidae, Enchytraeidae, Propappidae) down to slow flowing, eutrophic streams with low slopes (Tubificidae) (Fig. 8). The same gradient was seen for the ordination at genus level, and is therefore not shown. Also, the species level shows a comparable pattern (Fig. 9). From the DCCA ordination, analysis two major explanatory variables were selected: altitude, representing gradients in current, slope and temperature, and total phosphorus concentration, representing organic pollution and eutrophication gradients. The optima and tolerances for family, genus and species levels with respect to altitude and total phosphorus illustrate that oligochaetes vary individually along each gradient; thus each taxon shows its own characteristic response (Table 9).

## Discussion

The AQEM project was based on a standard procedure of sampling, sorting and identification. Despite the use of these standardised field protocols and guidelines, the number of taxa, the number of specimens and the taxonomic resolution for the Class Oligochaeta differed among

*Table 9.* The optima and tolerances of the ecological quality class (EQC), altitude (m) and total phosphate [t-P μg/l] for families, genera and species

| Taxonname | EQC | | Altitude | | t-P | |
|---|---|---|---|---|---|---|
| | Optimum | Tolerance | Optimum | Tolerance | Optimum | Tolerance |
| Class/family | | | | | | |
| Oligochaeta | 2.24 | 1.16 | 232 | 182 | 440 | 746 |
| Tubificidae | 1.79 | 0.97 | 175 | 104 | 505 | 583 |
| Tubificidae with hair setae | 1.94 | 1.00 | 152 | 117 | 425 | 641 |
| Tubificidae without hair setae | 1.78 | 0.94 | 166 | 108 | 604 | 736 |
| Naididae | 2.49 | 0.99 | 280 | 207 | 426 | 1093 |
| Enchytraeidae | 3.30 | 1.51 | 583 | 388 | 128 | 255 |
| Lumbriculidae | 3.38 | 1.07 | 306 | 231 | 247 | 418 |
| Lumbricidae | 3.56 | 1.20 | 249 | 221 | 268 | 570 |
| Glossoscolecidae | 3.62 | 1.12 | 149 | 86 | 229 | 570 |
| Haplotaxidae | 3.71 | 1.37 | 402 | 248 | 498 | 612 |
| Propappidae | 4.11 | 0.93 | 473 | 219 | 59 | 70 |
| Genus | | | | | | |
| *Aulodrilus* | 2.62 | 0.69 | 158 | 213 | 183 | 503 |
| *Bothrioneurum* | 2.81 | 0.72 | 335 | 117 | 1353 | 130 |
| *Branchiura* | 3.00 | 0.94 | 88 | 25 | 25 | 347 |
| *Bythonomus* | 3.66 | 0.81 | 250 | 37 | 141 | 62 |
| *Cernosvitoviella* | 3.90 | 1.07 | 989 | 233 | 18 | 16 |
| *Chaetogaster* | 3.41 | 0.73 | 286 | 156 | 110 | 156 |
| *Cognettia* | 4.13 | 1.04 | 566 | 255 | 42 | 174 |
| *Dero* | 2.37 | 0.93 | 44 | 60 | 557 | 1252 |
| *Eiseniella* | 3.27 | 1.07 | 267 | 251 | 417 | 776 |
| *Enchytraeus* | 4.62 | 1.39 | 451 | 118 | 380 | 526 |
| *Fridericia* | 3.25 | 1.31 | 853 | 334 | 61 | 163 |
| *Haplotaxis* | 4.16 | 0.85 | 435 | 255 | 532 | 642 |
| *Henlea* | 3.01 | 1.08 | 901 | 180 | 22 | 24 |
| *Ilyodrilus* | 2.02 | 0.21 | 12 | 6 | 112 | 64 |
| *Lamprodrilus* | 5.00 | 0.94 | 1380 | 141 | 44 | 347 |
| *Limnodrilus* | 1.60 | 0.92 | 194 | 92 | 677 | 791 |
| *Lumbricillus* | 3.53 | 1.09 | 590 | 298 | 611 | 714 |
| *Lumbriculus* | 2.84 | 0.99 | 73 | 92 | 119 | 207 |
| *Marionina* | 2.88 | 1.31 | 865 | 163 | 24 | 20 |
| *Mesenchytraeus* | 3.90 | 1.47 | 1056 | 234 | 47 | 185 |
| *Nais* | 2.62 | 1.07 | 340 | 205 | 526 | 1178 |
| *Ophidonais* | 2.07 | 0.61 | 115 | 72 | 174 | 869 |
| *Pachydrilus* | 2.00 | 0.94 | 240 | 141 | 34 | 347 |
| *Piguetiella* | 4.00 | 0.94 | 420 | 141 | 1120 | 347 |
| *Potamothrix* | 1.32 | 0.80 | 201 | 70 | 583 | 661 |
| *Pristina* | 3.42 | 0.95 | 252 | 179 | 205 | 182 |
| *Pristinella* | 3.11 | 0.47 | 252 | 114 | 149 | 78 |
| *Propappus* | 4.11 | 0.93 | 473 | 219 | 59 | 70 |
| *Psammoryctides* | 2.19 | 1.12 | 202 | 121 | 792 | 997 |

*Continued on p. 137*

*Table 9.* (Continued)

| Taxonname | EQC | | Altitude | | t-P | |
|---|---|---|---|---|---|---|
| | Optimum | Tolerance | Optimum | Tolerance | Optimum | Tolerance |
| *Quistodrilus* | 1.83 | 0.71 | 3 | 2 | 170 | 82 |
| *Rhyacodrilus* | 2.70 | 1.09 | 203 | 235 | 803 | 656 |
| *Rhynchelmis* | 3.69 | 0.98 | 478 | 132 | 20 | 62 |
| *Slavina* | 3.27 | 0.84 | 128 | 134 | 142 | 70 |
| *Specaria* | 2.15 | 0.63 | 330 | 11 | 38 | 19 |
| *Spirosperma* | 4.47 | 1.11 | 376 | 70 | 485 | 824 |
| *Stylaria* | 1.94 | 0.94 | 26 | 50 | 142 | 433 |
| *Stylodrilus* | 3.58 | 1.07 | 370 | 222 | 270 | 452 |
| *Trichodrilus* | 3.22 | 1.36 | 290 | 273 | 467 | 439 |
| *Tubifex* | 1.42 | 1.00 | 172 | 89 | 421 | 455 |
| *Uncinais* | 2.00 | 0.94 | 19 | 141 | 260 | 347 |
| Species | | | | | | |
| *Aulodrilus japonicus* | 2.63 | 0.68 | 162 | 223 | 103 | 64 |
| *Aulodrilus limnobius* | 3.00 | 0.94 | 30 | 141 | 160 | 347 |
| *Aulodrilus pluriseta* | 2.57 | 0.82 | 135 | 177 | 659 | 1175 |
| *Bothrioneurum vejdovskyanum* | 2.81 | 0.72 | 335 | 117 | 1353 | 130 |
| *Branchiura sowerbyi* | 3.00 | 0.94 | 88 | 25 | 25 | 347 |
| *Bythonomus lemani* | 3.66 | 0.81 | 250 | 37 | 141 | 62 |
| *Cernosvitoviella atrata* | 3.09 | 1.01 | 833 | 36 | 20 | 13 |
| *Chaetogaster cristallinus* | 3.00 | 0.94 | 215 | 141 | 165 | 347 |
| *Chaetogaster diaphanus* | 3.80 | 0.71 | 339 | 226 | 44 | 22 |
| *Chaetogaster diastrophus* | 1.67 | 1.41 | 410 | 445 | 719 | 604 |
| *Cognettia sphagnetorum* | 4.22 | 1.02 | 510 | 215 | 28 | 12 |
| *Dero digitata* | 2.67 | 0.92 | 17 | 14 | 166 | 72 |
| *Dero obtuse* | 1.71 | 0.89 | 24 | 14 | 133 | 66 |
| *Eiseniella tetraedra* | 3.27 | 1.07 | 267 | 251 | 417 | 776 |
| *Enchytraeus albidus* | 1.00 | 0.94 | 230 | 141 | 1560 | 347 |
| *Fridericia striata* | 4.00 | 0.94 | 225 | 141 | 330 | 347 |
| *Haplotaxis gordioides* | 4.29 | 0.75 | 455 | 253 | 559 | 674 |
| *Ilyodrilus templetoni* | 2.02 | 0.21 | 12 | 6 | 112 | 64 |
| *Limnodrilus claparedeanus* | 2.18 | 0.82 | 161 | 85 | 389 | 413 |
| *Limnodrilus hoffmeisteri* | 1.57 | 0.96 | 190 | 100 | 607 | 864 |
| *Limnodrilus profundicola* | 1.08 | 0.93 | 185 | 96 | 334 | 55 |
| *Limnodrilus udekemianus* | 1.71 | 0.81 | 183 | 71 | 386 | 230 |
| *Lumbricillus lineatus* | 2.00 | 0.94 | 8 | 141 | 70 | 347 |
| *Lumbricillus pagenstecheri* | 4.08 | 1.26 | 361 | 39 | 1184 | 358 |
| *Lumbriculus variegatus* | 2.83 | 0.98 | 71 | 84 | 119 | 208 |
| *Marionina argentea* | 3.00 | 0.94 | 835 | 7 | 43 | 30 |
| *Marionina riparia* | 3.27 | 1.10 | 885 | 271 | 17 | 5 |
| *Mesenchytraeus armatus* | 3.87 | 1.49 | 1076 | 198 | 19 | 21 |
| *Nais alpina* | 3.44 | 0.90 | 549 | 185 | 23 | 47 |
| *Nais barbata* | 1.12 | 0.75 | 234 | 37 | 642 | 141 |
| *Nais behningi* | 2.37 | 0.71 | 6 | 1 | 79 | 18 |
| *Nais bretscheri* | 2.70 | 1.31 | 278 | 197 | 221 | 248 |

*Continued on p. 138*

*Table 9.* (Continued)

| Taxonname | EQC | | Altitude | | t-P | |
|---|---|---|---|---|---|---|
| | Optimum | Tolerance | Optimum | Tolerance | Optimum | Tolerance |
| *Nais communis* | 2.90 | 0.77 | 301 | 352 | 75 | 100 |
| *Nais elinguis* | 1.85 | 0.99 | 209 | 61 | 433 | 493 |
| *Nais pardalis* | 2.32 | 0.78 | 21 | 8 | 250 | 120 |
| *Nais pseudobtusa* | 1.00 | 0.45 | 193 | 18 | 346 | 41 |
| *Nais simplex* | 2.05 | 0.85 | 167 | 105 | 1141 | 487 |
| *Nais stolci* | 2.82 | 1.08 | 487 | 162 | 13 | 12 |
| *Nais variabilis* | 2.10 | 0.99 | 242 | 219 | 42 | 46 |
| *Ophidonais serpentina* | 2.07 | 0.61 | 115 | 68 | 174 | 869 |
| *Pachydrilus subterranus* | 2.00 | 0.94 | 240 | 141 | 34 | 347 |
| *Piguetiella blanci* | 4.00 | 0.94 | 420 | 141 | 1120 | 347 |
| *Potamothrix bavaricus* | 1.00 | 0.05 | 192 | 1 | 614 | 228 |
| *Potamothrix hammoniensis* | 1.66 | 1.13 | 219 | 118 | 716 | 1119 |
| *Potamothrix moldaviensis* | 1.32 | 1.08 | 191 | 31 | 356 | 110 |
| *Pristina aequiseta* | 3.00 | 0.94 | 215 | 141 | 214 | 10 |
| *Pristina amphibiotica* | 2.38 | 0.65 | 67 | 74 | 96 | 74 |
| *Pristina bilobata* | 3.16 | 0.71 | 266 | 231 | 198 | 57 |
| *Pristina foreli* | 3.61 | 1.28 | 295 | 47 | 73 | 48 |
| *Pristina longiseta* | 3.94 | 1.41 | 334 | 84 | 119 | 188 |
| *Pristina menoni* | 4.00 | 0.94 | 260 | 141 | 129 | 347 |
| *Pristina rosea* | 4.00 | 0.94 | 215 | 141 | 108 | 347 |
| *Propappus volki* | 4.11 | 0.93 | 473 | 219 | 59 | 70 |
| *Psammoryctides barbatus* | 2.19 | 1.12 | 202 | 121 | 792 | 997 |
| *Quistodrilus multisetosus* | 1.83 | 0.71 | 3 | 2 | 170 | 82 |
| *Rhyacodrilus coccineus* | 3.30 | 1.00 | 153 | 145 | 297 | 343 |
| *Rhyacodrilus falciformis* | 2.16 | 1.22 | 248 | 515 | 1255 | 355 |
| *Rhyacodrilus subterraneus* | 2.00 | 0.94 | 31 | 141 | 189 | 347 |
| *Rhynchelmis limosella* | 4.25 | 0.71 | 138 | 304 | 315 | 559 |
| *Slavina appendiculata* | 3.27 | 0.84 | 128 | 134 | 142 | 70 |
| *Specaria josinae* | 2.15 | 0.63 | 330 | 11 | 38 | 19 |
| *Spirosperma ferox* | 4.47 | 1.11 | 376 | 70 | 485 | 824 |
| *Stylaria lacustris* | 1.94 | 0.94 | 26 | 50 | 142 | 433 |
| *Stylodrilus brachystylus* | 4.54 | 0.69 | 449 | 127 | 34 | 27 |
| *Stylodrilus heringianus* | 3.27 | 1.01 | 340 | 244 | 281 | 491 |
| *Stylodrilus parvus* | 4.43 | 0.86 | 338 | 100 | 695 | 315 |
| *Tubifex ignotus* | 2.01 | 0.32 | 194 | 88 | 176 | 124 |
| *Tubifex tubifex* | 1.42 | 1.00 | 172 | 89 | 423 | 456 |
| *Uncinais uncinata* | 2.00 | 0.94 | 19 | 141 | 260 | 347 |

countries. Knowledge of Oligochaeta appeared to be sparsely distributed and depended on the presence of specialists in the respective countries. Therefore, the results presented were biased by the samples from Austria, the Czech Republic and the Netherlands, in which oligochaetes were carefully processed and identified to the species level. This, for example, explains the importance of altitude and total phosphorus as explanatory variables.

That oligochaetes play an important role in the stream ecosystems becomes clear from the number of individuals of Oligochaeta collected throughout

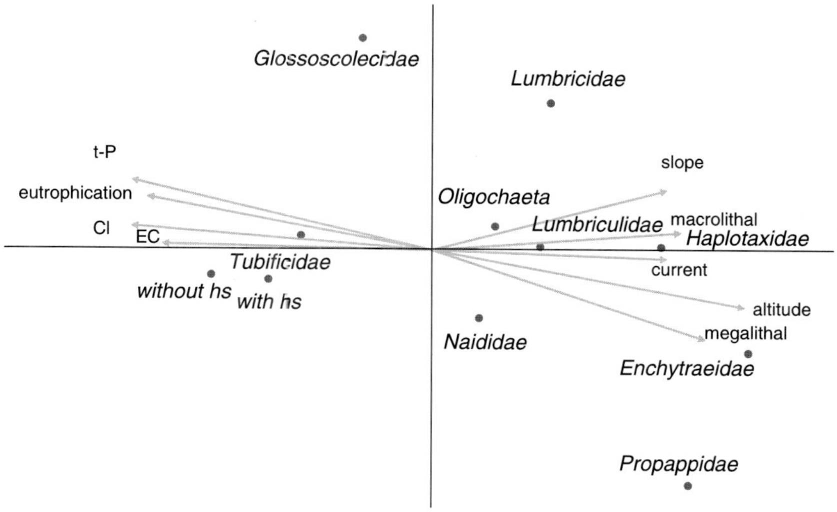


*Figure 8.* DCCA ordination diagram of the axis 1 and 2 of the families of oligochaetes (hs = hair setae).

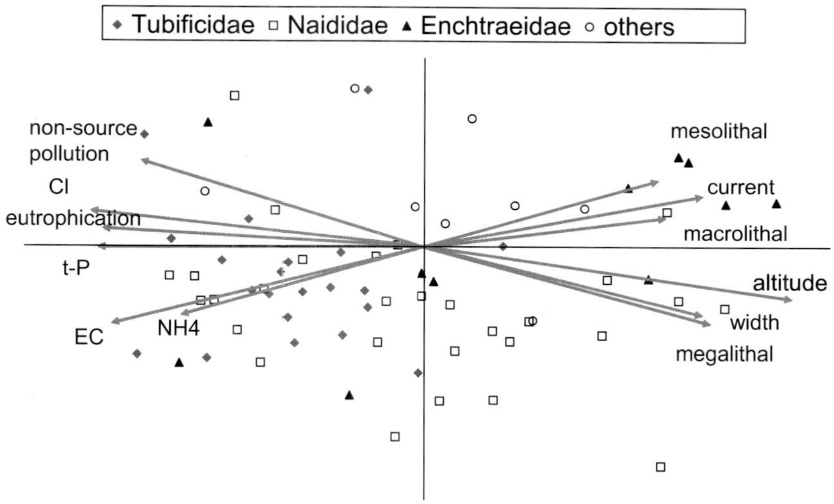


*Figure 9.* DCCA ordination diagram of the axis 1 and 2 of the species, grouped and indicated by family label, of oligochaetes.

the project, about half a million. Oligochaetes are known to occur in all kinds of aquatic habitats all over Europe (Brinkhurst & Jamieson, 1971; Timm, 1999), which argues for taking this organism group much more seriously in assessments.

The composition of a stream community is the result of the interaction of environmental and biological factors. The explanation of the distribution of oligochaetes is often based on environmental data, without regard to biological interactions. Nevertheless, the use of biological and ecological information from an autecological database would improve the usability of oligochaetes in assessment. The AQEM autecological database is an example of such. Two biological traits were explored: feeding type and locomotion type. The classification of functional feeding groups according to Merrit & Cummins (1984) was used in the AQEM autecology database. This is equivalent to the classifications by Juget & Lafont (1994) and Moog (1995). But, although most oligochaetes have comparable feeding mechanisms, their food and food preferences show subtle differences (Brinkhurst et al., 1972;

McMurtry et al., 1983; Rodriguez et al., 2001). The ecological meaning and applicability of such differences need further attention.

The locomotion type of oligochaetes appeared to be less distinctive, and values assigned to the species show little variation. Nevertheless, Sperber (1950) previously classified the naidids into 10 locomotion types based on tube and free living species that either do not swim or move, or swim with different movements. It is still a question what the ecological meaning of these categories is, but again more knowledge is needed and should be included in the autecological database. Laboratory observations and experiments could reveal the ecological significance of these groups.

The ecological information strongly focused on the indication of organic pollution. From a historical point of view this is not surprising. It is already long known that oligochaetes are primarily related to organic pollution because they use readily degradable organic matter as food (e.g. Uzunov, 1979; Milbrink, 1980). The differences between the old saprobic index, the new saprobic index and the saprobic valence reflect the development of knowledge in this field. Although Kolkwitz & Marsson (1909) indicated that different species not only preferred polysaprobic, but also $\beta$-, $\alpha$-, and oligosaprobic conditions, it took almost a century for this information to be adopted in assessment. The assumption of Lafont (1985) that the water quality responses of Tubificidae with and without hair setae would differ is not supported by these findings, because optima of both groups were in the same EQC.

The ecological indications for current velocity are in line with the expectations. Most oligochaetes generally prefer standing or slow flowing conditions (Korn, 1963). The equal class distribution of oligochaetes across microhabitats and river zones is also confirmed by other authors. Lazim & Learner (1987) showed that the distribution of tubificids was determined by the physical nature of the substratum, and that species preferences differ. The phytal preferences of naidids were shown by Verdonschot (1987, 1989) and Kairesalo & Koskimies (1987). In general, the sampling techniques used in the foregoing studies are an important question. If mixtures of habitats were sampled, which is often the case, the results can be biased. It is therefore of great importance to study the microhabitat needs of oligochaetes.

The ecological indications included in the AQEM autecological database could be extended for most oligochaete taxa, and possibly in some cases a refinement of categories can be taken in consideration, e.g. locomotion types. Furthermore, attention is needed for the acidity preference of most oligochaetes, as such information is lacking.

The altitude – eutrophication gradient shown in the analysis is strongly influenced by the countries where the samples were taken. The Netherlands mainly consist of lowland streams that are more or less eutrophied (Verdonschot, 1999), in strong contrast to the Czech Republic and Austria, which are represented by mountain streams with much lower nutrient concentrations.

## Outlook

Mass developments and blooms of oligochaetes have long been used as indicators either for organic pollution or eutrophication. That oligochaetes can tell us more about the condition of a stream is illustrated by the distribution of different taxa over the ecological quality classes. The European WFD has asked for a diagnostic tool to evaluate water quality. Diagnosis is necessary not only to assess quality, but also to indicate potential measures for recovery or restoration. Ecologists should therefore develop techniques that can tell a manager more than just a limited five class system of assessment valuation. Ecologists should extract detailed information from the species they collect in order to precisely determine the condition of a site. These results clearly show the need of using species rather than higher taxonomic levels in assessment. This information should be the basis for management advice. To be able to extract detailed information from a species list, knowledge on the niche requirements of each of the respective species is necessary. Oligochaetes are an important part of the aquatic community, they can be found everywhere, comprise a large part of the secondary productivity and play a vital role in the food chain. Thus, oligochaetes should become an important diagnostic tool for aquatic ecologists.

## Acknowledgements

The oligochaetes were identified by Ferdinand Sporka (Austria), Tjeerd-Harm van den Hoek (Netherlands), Chirstian Feld (Germany), Jana Schenkova and Prof. Frantisek Kubicek (Czech Republic). The material was sampled, sorted and processed by BOKU – University of Natural Resources and Applied Life Sciences, Vienna; the team Water of the Laboratory of the Centre of Applied Ecology of the University of Évora, Évora; National Centre for Marine Research – Institute of Inland Waters, Masaryk University and Water Research Institute TGM, Department of Environmental Assessment, Swedish University of Agricultural Sciences, Uppsala; Alterra, Wageningen. This project was completed with thanks to financial support of the European Union projects AQEM, STAR and EUROLIMPACS.

## References

AQEM consortium, 2002. Manual for the application of the AQEM method. A comprehensive method to assess European streams using benthic macroinvertebrates, developed for the purpose of the Water Framework Directive. Version 1.0, February 2002.

Aston, R. J., 1973. Tubificids and water quality: a review. Environmental Pollution 5: 1–10.

Brinkhurst, R. O. & B. G. M. Jamieson, 1971. The Aquatic Oligochaeta of the World. Oliver & Boyd, Edinburg 860 pp.

Brinkhurst, R. O., K. E. Chua & N. K. Kaushik, 1972. Interspecific interactions and selective feeding by tubificid oligochaetes. Limnology and Oceanography 17: 122–133.

Chapman, P. M., G. A.Vigers, M. A. Farrell, R. N. Dexter, E. A.Quinlan, R. M. Kocan & M. L. Landolt, 1982. Survey of biological effects of toxicants upon Puget Sound biota. I. Broadscale toxicity surveys. U.S. Dept. of Commerce, NOAA Tech. Memo, OMPA-25; 98.

European Commission, 2000. Directive 2000/60/EC. Establishing a Framework for Community Action in the Field of Water Policy. European Commission PE-CONS 3639/1/100 Rev 1, Luxemburg.

Gauch, H.G., 1982. Multivariate Analysis in Community Ecology. Cambridge U.P, Cambridge, 298 pp.

Gauch, H. G., G. B. Chase & R. H. Whittaker, 1974. Ordinations of vegetation samples by Gaussian species distributions. Ecology 55: 1382–1390.

Hamm, A., 1969. Erläuterungen zur Gütekartierung bayerischer Voralpenseen.- Der Wassergüteatlas. Münchner Beiträge zur Abwasser-, Fischerei- und Flussbiologie 15: 49–66.

Healy, B. & T. Bolger, 1984. The occurrence of species of semiaquatic Enchytraeidae (Oligochaeta) in Ireland. Hydrobiologia 115: 159–170.

Hering, D., O. Moog, L. Sandin & P. F. M. Verdonschot, 2004. Overview and application of the AQEM assessment system. Hydrobiologia 175: 1–20.

Illies, J., 1978. Limnofauna Europaea. Gustav Fischer Verlag, Stuttgart, 532 pp.

Juget, J. & M. Lafont, 1994. Theoretical habitat templetes, species traits, and species richness: aquatic oligochaetes in the Upper Rhone River and its floodplain. Freshwater Biology 31(3): 327–340.

Juggins, S., 2003. C2. A Program for Analyzing and Visualizing Palaeoenvironmental Data. User Guide. University of Newcastle, Newcastle upon Tyne, U.K. 49 pp.

Kairesalo, T. & I. Koskimies, 1987. Grazing of oligochaetes and snails on epiphytes. Freshwater Biology 17: 317–324.

Kolkwitz, R. & M. Marsson, 1909. Ökologie der tierischen Saprobien. Internationale Revue der Hydrobiologie 2: 126–519.

Korn, H., 1963. Studien zur Ökologie der Oligochaeten in der oberen Donau unter Berücksichtigung der Abwassereinwirkungen. Archiv für Hydrobiologie, Supplement 27: 131–182.

Lafont, M., 1985. Oligochaetes And Pollution In the Sediments of Rivers: An Improved Index of Biological Quality. CEMAGREF, Lyon, 10 pp.

Lazim, M. N. & M. A. Learner, 1987. The influence of sediment composition and leaf litter on the distribution of tubificid worms (Oligochaeta). A field and laboratory study. Oecologia 72: 131–136.

McCullagh, P. & J. A. Nelder, 1989. Generalized Linear Models (2nd ed.). Chapman & Hall, London, 511 pp.

McMurtry, M. J., D. J. Rapport & K. E. Chua, 1983. Substrate selection by tubificid oligochaetes. Canadian Journal of Fisheries and Aquatic Sciences 40: 1639–1646.

Merrit, R. W. & K. W. Cummins, 1984. An Introduction to the Aquatic Insects of North America. Kendall/Hunt Publ. Comp, Dubuque, 441 pp.

Milbrink, G., 1973. Oligochaete communities of Tubificidae and some Lumbriculidae in the assessment of water pollution in Swedish lakes. ZOON 1: 125–139.

Milbrink, G., 1980. Oligochaete communities in pollution biology. The European situation with special reference to lakes in Scandinavia. In Brinkhurst, R. O. & D. G. Cook (eds), Aquatic Oligochaete Biology. Plenum Publ. Corp, N.Y: 433–456.

Moog O. (ed.), 1995. Fauna Aquatica Austriaca—A Comprehensive Species Inventory of Austrian Aquatic Organisms with Ecological Data. (1st edn.), Wasserwirtschaftskataster, Bundesministerium für Land- und Forstwirtschaft, Wien.

Newrkla, P. & N. Wijegoonawardana, 1987. Vertical distribution and abundance of benthic invertebrates in profundal sediments of Mondsee, with special references to oligochaetes. Hydrobiologia 155: 227–234.

Nijboer, R. C. & P. F. M. Verdonschot, 2000. Taxonomic adjustment affects data analysis: an often forgotten error. Verhandlungen Internationale Vereinigung für Limnologie 27: 1–4.

Odum, E. P., 1971. Fundamentals of Ecology. Saunders Company, Philladelphia, 574 pp.

Persoone G., 1979. Proposal for a biotypological classification of watercourses in the European communities. In James A.

& L. Evison, (eds), Biological Indicators of Water Quality. 7.31–7.32:17–32.

Resh, V. H. & J. D. Unzicker, 1975. Water quality monitoring and aquatic organisms: the importance of species identification. Journal of Water Pollution Control Federation 47: 9–19.

Rodriguez, P., M. Martinez-Madrid, J. A. Arrate & E. Navarro, 2001. Selective feeding by the oligochaete *Tubifex tubifex* (Tubificidae, Clitellata). Hydrobiologia 158: 133–140.

Rosenberg, D. M. & V. H. Resh, 1984. Freshwater Biomonitoring and Benthic Macroinvertebrates. Chapman & Hall, New York, 488 pp.

Schmedtje U. & M. Colling, 1996. Ökologische Typisierung der Aquatischen Makrofauna. Informationsberichte des Bayerischen Landesamtes für Wasserwirtschaft 4/96; 543.

Schmidt-Kloiber, A. & R. C. Nijboer, 2004. The effect of taxonomic resolution on the assessment of ecological water quality classes. Hydrobiologia 175: 269–283.

Shelford, V. E., 1913. Animal communities in temperate America as illustrated in the Chicago region. A study in animal ecology. Geographical Society of Chicago Bulletin 5: 1362.

Sperber, C., 1950. A guide for the determination of European Naididae. Zoologiska Bidrag Från Uppsala 29: 45–78.

Ter Braak C. J. F., 1987. CANOCO A FORTRAN Program for Canonical Community Ordination by [partial] [detrended] [canonical] Correspondence Analysis, Principal Component Analysis and Redundancy Analysis (version 2.1). TNO Institute of Applied Computer Science: Wageningen, The Netherlands, 95 pp.

Ter Braak, C. J. F. & C. W. N. Looman, 1986. Weighted averaging, logistic regression and the Gaussian response model. Vegetatio 65: 3–11.

Ter Braak C. J. F. & P. Ŝmilauer, 1998. CANOCO Reference Manual and Users Guide to Canoco for Windows. Software for Canonical Community Ordination (version 4). Centre for Biometry Wageningen: Wageningen, the Netherlands, 351 pp.

Ter Braak, C. J. F. & H. Van Dam, 1989. Inferring pH from diatoms: a comparison of old and new calibration methods. Hydrobiologia 178: 209–223.

Thienemann, A., 1955. Die Binnengewässer in Natur und Kultur. Springer Verlag, Berlin, 156 pp.

Timm T., 1999. Eesti rõngusside (Annelida) määraja. A Guide to the Estonian Annelida. Tartu: Eesti Loodusuurijate Selts. ISBN 9985–231–1, 208 pp.

Uzunov, Y., 1979. Aquatic Oligochaeta: a supplement to the list of limnosaprobic bioindicators. Proceedings of the Bulgarian Academy of Sciences 32: 1101–1103.

Van der Hoek W. F. & P. F. M. Verdonschot, 1994. Functionele Karakterisering van Aquatische Ecotooptypen. IBN rapp. 072, 81 pp.

Verdonschot P. F. M., 1990. Ecological characterization of surface waters in the province of Overijssel (The Netherlands). Thesis, Wageningen, pp 255.

Verdonschot, P. F. M., 1984. The distribution of aquatic oligochaetes in the fenland area of N.W. Overijssel (The Netherlands). Hydrobiologia 115: 215–222.

Verdonschot, P. F. M., 1987. Aquatic Oligochaeta in ditches. Hydrobiologia 155: 283–292.

Verdonschot, P. F. M., 1989. The role of oligochaetes in the management of waters. Hydrobiologia 180: 213–227.

Verdonschot, P. F. M., 1995. Typology of macrofaunal assemblages: a tool for the management of running waters. Hydrobiologia 297: 99–122.

Verdonschot, P. F. M., 1996. Oligochaetes and eutrophication; an experiment over four years in outdoor mesocosms. Hydrobiologia 334: 169–183.

Verdonschot, P. F. M., 1999. Micro-distribution of oligochaetes in a soft-bottomed lowland stream (Elsbeek; The Netherlands). Hydrobiologia 139: 149–163.

Vlek, H. E., P. F. M. Verdonschot & R. C. Nijboer, 2004. Towards a multimetric index for the assessment of dutch streams using benthic macroinvertebrates. Hydrobiologia 175: 173–189.

Wachs, B., 1967. Die Oligochaetenfauna der Fliessgewässer unter besonderer Berücksichtigung der Beziehungen zwischen der Tubificiden-besiedlung und dem Substrat. Archiv für Hydrobiologie 63: 310–386.

Zelinka, M. & P. Marvan, 1961. Zur präzisierung der biologischen klassification er reinheit fliessender gewässer. Archiv für Hydrobiologie 57: 389–407.

Hydrobiologia (2006) 564:143–155
P.F.M. Verdonschot, H. Wang, A. Pinder & R. Nijboer (eds), Aquatic Oligochaete Biology IX
DOI 10.1007/s10750-005-1715-y


# Differential response of aquatic oligochaete species to increased nutrient availability – a comparative study between Estonian and Dutch raised bogs

G.A. van Duinen[1,2,*], T. Timm[3], A.J.P. Smolders[4], A.M.T. Brock[1], W.C.E.P. Verberk[1,5] & H. Esselink[1,5]

[1]*Bargerveen Foundation, Radboud University Nijmegen, P.O. Box 9010, 6500 GL Nijmegen, The Netherlands*
[2]*Department of Environmental Science, Institute for Water and Wetland Research, Radboud University Nijmegen, P.O. Box 9010, 6500 GL Nijmegen, The Netherlands*
[3]*Estonian University of Life Sciences, Centre for Limnology, 61101 Rannu, Tartumaa, Estonia*
[4]*Department of Animal Ecology and Ecophysiology, Institute for Water and Wetland Research, Radboud University Nijmegen, P.O. Box 9010, 6500 GL Nijmegen, The Netherlands*
[5]*Department of Aquatic Ecology and Environmental Biology, Institute for Water and Wetland Research, Radboud University Nijmegen, P.O. Box 9010, 6500 GL Nijmegen, The Netherlands*
(**Author for correspondence: Tel.: +31-24-3653275; Fax: +31-24-3653030; E-mail: G.vanDuinen@science.ru.nl*)

*Key words:* oligochaeta, nitrogen, phosphorus, sulphur, deposition, algae, detritus, organic matter, peat, decomposition

## Abstract

To assess the effects of increased nutrient availability on aquatic oligochaetes in raised bogs, species assemblages were compared within and between fairly pristine raised bogs in Estonia and raised bog remnants in The Netherlands. Within the pristine bog landscape a distinct pattern in the species assemblage is present. In the most nutrient-poor water bodies, in the ombrotrophic raised bog, only the fragmenting, almost never mature, acid-tolerant species *Cognettia sphagnetorum* is present. In pristine Estonian raised bogs *Nais variabilis*, *Lumbriculus variegatus* and sexually reproducing species are limited to more minerotrophic water bodies, which have a higher decomposition rate of dead organic matter and, consequently, higher nutrient availability. With ten species the lagg zone is the most species-rich part of a pristine raised bog landscape. Most of these lagg zone species are not present in Dutch bog remnants as this part of the bog landscape has long been cultivated. *Nais variabilis* occurs in the Dutch bog remnants much more frequently than in Estonian bogs, whereas the frequency of *C. sphagnetorum* and *L. variegatus* is similar between both countries. These three species respond differently to the increased nutrient availability in The Netherlands, which could be linked to differences in their diets. In contrast to pristine bog pools, *N. variabilis* in Dutch raised bog remnants is present in water bodies not influenced by minerotrophic water. In Dutch raised bog remnants the occurrence of oligochaetes is not limited anymore by nutrient availability, due to the higher atmospheric nitrogen and sulphur loads in The Netherlands. Overall, it can be concluded that the degradation of Dutch raised bogs has resulted in the loss of both the nutrient-poor parts of the landscape and the special lagg conditions.

## Introduction

Increase of nutrient availability causes shifts in aquatic and terrestrial plant species composition and productivity, and, consequently, may alter faunal species assemblages. Nutrient-poor ecosystems, like raised bogs, are especially sensitive to increased nutrient availability (e.g. Bobbink et al., 1998; Risager, 1998). Most raised bogs remaining in north-western Europe are not only affected by drainage and peat cutting (Schouwenaars, 1993), but also by increased deposition of atmospheric nitrogen (Berendse et al., 2001) and sulphur compounds (Roelofs, 1986; Leuven, 1988). Studies in lakes, rivers and ditches have shown that increased nutrient availability affects the aquatic invertebrate assemblages (e.g. Lang & Reymond, 1995; Verdonschot, 1996; Lang, 1999). Compared to the extremely nutrient poor raised bogs, those aquatic systems have a relatively high nutrient availability under unpolluted conditions. Effects of increased nutrient availability on aquatic invertebrates in raised bogs, however, are barely studied.

Comparison of the species assemblages of coleopterans, heteropterans, trichopterans, and odonates between water bodies in the fairly pristine raised bog system Nigula (Estonia) showed that variation in the species assemblage is related to natural variation in the availability of minerals and nutrients (Smits et al., 2002), among other habitat characteristics. The species assemblage of the most nutrient poor, ombrotrophic pristine bog pools in Nigula is not found in raised bog remnants in The Netherlands (Van Duinen et al., 2002). The species assemblage of Dutch bog remnants was more similar to those of the Estonian water bodies with a naturally higher nutrient availability, such as transitional mires and bog brooklets. This pattern may well be attributed to the relatively high nutrient load in The Netherlands, lifting bottlenecks for the occurrence of species in ombrotrophic water bodies. However, the causal relations giving rise to these changes in species assemblages are not yet elucidated. The species studied by Van Duinen et al. (2002) included many carnivores, whose occurrence is probably related to increased nutrient availability via the nutrient content and density of animals from a lower trophic level. The occurrence of herbivores and detritivores is expected to be more directly related to nutrient availability. This paper focuses on the aquatic oligochaetes of which many species feed either directly on dead organic matter, or on algae, fungi, and bacteria growing on organic matter (Learner et al., 1978).

In raised bogs the quality and decomposition rate of dead organic matter and the microbial community is altered by increased nitrogen and sulphur deposition. Increased N deposition results in a higher N availability in the surface water and interstitial water and in a higher N content of *Sphagnum* mosses (Lamers et al., 2000; Limpens et al., 2003a). Increased N availability also increases the cover of vascular plants such as *Molinia caerulea* and *Betula* spp. (Hogg et al., 1995; Lamers et al., 2000; Limpens et al., 2003a; Tomassen et al., 2003, 2004b). Thus, increased N deposition results in an increased input of dead organic vascular plant material to the water bodies. As a consequence, the decomposition rate of dead organic matter might increase (Lamers et al., 2000), as nutrient availability and litter quality usually limit the activity of decomposing bacteria and fungi in pristine raised bogs (Belyea, 1996; Smolders et al., 2002). Next to litter quality, a higher pH, buffering capacity, and $SO_4$ availability stimulate organic matter decomposition and N and P availability (Kok & Van de Laar, 1991; Lamers et al., 1998, 1999; Smolders et al., 2002; Tomassen et al., 2004a). $SO_4$ serves as a terminal electron acceptor for bacteria, thus increasing anaerobic oxidation of organic matter. This results in an increased decomposition rate of organic matter and an increased release of $NH_4$ and $PO_4$ in bog water bodies (Lamers et al., 1998). Growth of algae is increased by increased availability of N and P (Gulati & DeMott, 1997; Limpens et al., 2003b).

Aquatic oligochaetes feeding on algae are expected to increase with increased N and P availability in the surface water and consequent increases in algal growth, whereas detritivorous oligochaetes – feeding on dead organic matter, fungi, and bacteria – are expected to increase with a higher nutrient content and decomposition rate of dead organic matter. Here we test the hypothesis that the aquatic oligochaete species whose occurrence in pristine Estonian bogs is limited to water bodies with a higher nutrient availability are in Dutch raised bog remnants not limited anymore by nutrient availability, as nutrient availability is high overall. To test this hypothesis, the occurrence of oligochaete species as well as nutrient availability in dead organic matter, interstitial water, and surface water were compared between (1) water bodies naturally differing in nutrient availability within Estonian pristine raised bog landscapes and (2) a variety of water bodies in Dutch raised bog remnants affected by increased N and S deposition.

In this paper we will answer the following questions:

1. What oligochaete species are present in pristine Estonian raised bogs and does their presence differ between the ombrotrophic raised bog, transitional mire and lagg zone?
2. Which of these oligochaete species are present in Dutch raised bog remnants?
3. Do the oligochaete species in raised bogs respond differently to nutrient availability in surface water or nutrient content and decomposition rate of dead organic matter?

## Materials and methods

### *Study areas*

In Estonia 27 water bodies were sampled in Nigula Nature Reserve: 20 in the ombrotophic raised bog, six in the transitional mire, spring and connected pools, and one in the lagg zone. In addition, two samples were taken in the transitional mire of Valgeraba, Soomaa National Park, and two samples in the lagg of Punaraba, Endla Nature Reserve (Fig. 1). In The Netherlands samples were taken at 46 sites in seven raised bog remnants. Sampling sites included most of the various types of water body present in both pristine raised bogs (see Smits et al., 2002) and Dutch raised bog remnants (see Van Duinen et al., 2003). The water bodies sampled differed in size, water and organic matter quality, vegetation composition and structure. Despite the recent decrease in atmospheric N and S deposition in The Netherlands (RIVM, 2004), deposition is still among the highest in western Europe (Jonson et al., 1998; Alcamo et al., 2002) and N deposition is considerably higher than at the study areas in Estonia (cf. Pajuste et al., 2002; Tomassen et al., 2004b).

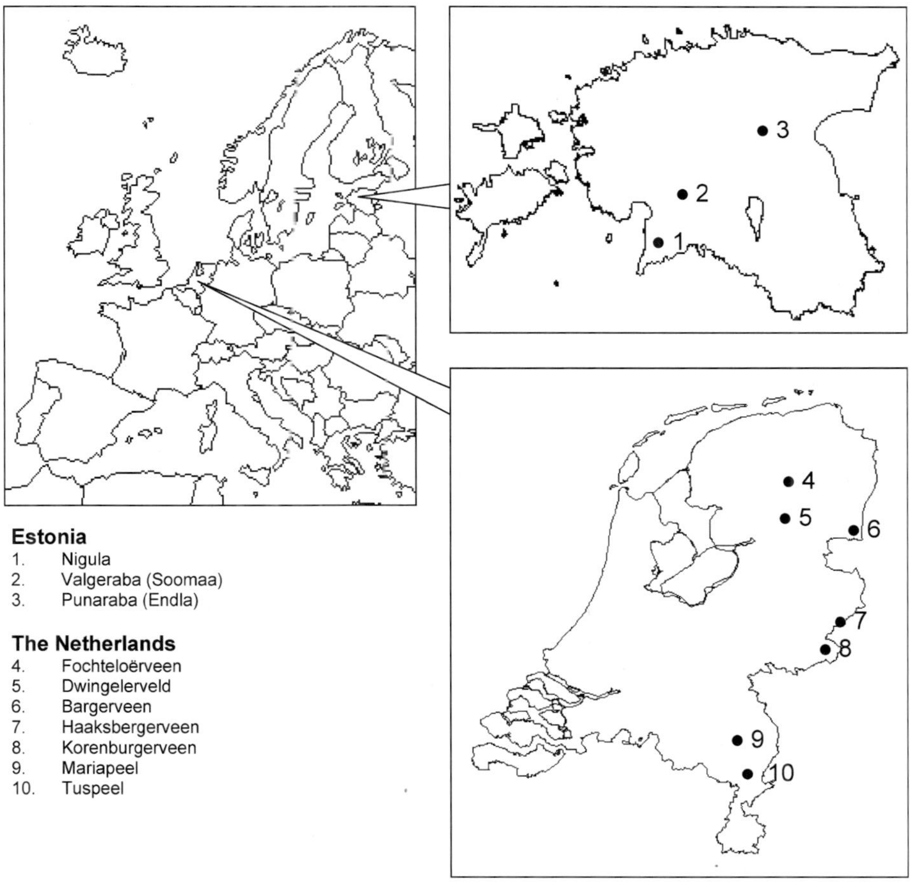


*Figure 1.* Location of study areas in Estonia and The Netherlands.

*Oligochaete sampling*

To sample most microhabitats present in the selected water bodies, a pond net (20×30 cm with 0.5 mm mesh size) was swept from the substrate and more or less open water into more dense vegetation near the shore over a distance of at least one metre. If dense vegetation was lacking, one or more longer sweeps were taken to collect at least 100 macroinvertebrates in a sample. In very dense *Sphagnum* vegetation, where invertebrate density is generally high, short sweeps (20–50 cm) were made to avoid clogging of the pond net. The Dutch sites were sampled in spring 1999. The Estonian sites were sampled in spring 2001. All samples were washed over three sieves with 2, 1, and 0.5 mm mesh sizes respectively and sorted in white trays. Oligochaetes were found in 38 of the Dutch sampling sites and in 22 of the Estonian sites. Although some oligochaetes might have been missed by the sampling method used in this study, the method did not selectively influence the species data of any category of sampling sites and did therefore not affect the findings of this study. Oligochaetes were stored in 4% formalin, identified to species level if possible using the key of Timm (1999), and deposited in the corresponding authors collection.

*Environmental variables*

In Estonia dead organic matter (top layer of accumulated peat and sedimented detritus), interstitial water, and surface water were collected in May and June 2001. At the Dutch sampling sites surface water and interstitial water were sampled one to four times between October 1998 and October 1999. Dead organic matter was collected once in August 1999. In The Netherlands interstitial water was sampled using ceramic cups pressed into the top layer of the organic matter bottom. In Estonia Rhizon soil moisture samplers were pressed into the samples of the top layer of the organic matter. These samplers (both types produced by Eijkelkamp Agrisearch Equipment, Giesbeek, The Netherlands) were connected to vacuumed PVC syringes (50 ml).

The pH of surface water was measured in the field. The pH of interstitial water samples was measured in the laboratory within 24 h after collection. After 1 mg citric acid per 5 ml of water had been added, water samples were stored at −20 °C until analysis. Organic matter was dried (48 h at 70 °C) and ground up in liquid nitrogen. Nitrogen and carbon concentrations were measured in dried samples with a CNS analyzer (type NA 1500; Carlo Erba Instruments, Milan, Italy). Digestates of ground-dried organic matter were prepared with the aid of a Milestone microwave (type mls 1200 Mega, Serisole, Italy), using nitric acid and hydrogen peroxide. Digestates were analyzed for total- P, Ca, and Mg. The analysis of the water samples included determination of the concentrations of $NO_3$, $NH_4$, $o$-$PO_4$ and S (for methods used see Van Duinen et al., 2003). The Ca + Mg content of organic matter was corrected for ash content.

*Data analysis*

Statistical significance of differences in environmental variables between the sampling sites in the ombrotrophic and minerotrophic parts of the Estonian raised bogs and between Estonian and Dutch sites was tested with an independent samples $t$-test. To assess the species' responses to water and organic matter variables linear regression analysis of chemical data and log-transformed species abundance data from Estonian and Dutch sampling sites was performed using SPSS version 11.0 for Windows.

## Results

*Estonian raised bogs*

In the Estonian raised bogs 13 oligochaete species were found (Table 1). The fragmenting enchytraeid *Cognettia sphagnetorum* was most frequently encountered and was the only species found in the water bodies of the ombrotrophic part of the raised bog landscape. This species was also present at various other sampling sites, always accompanied by other oligochaete species. In the transitional mires, spring, and connected pools assemblages of three to five oligochaete species were found. The highest number of species (11) was found in the lagg zone of Nigula bog. *Lumbriculus variegatus* and *Nais variabilis* were the second and third most frequently occurring species. These three species most frequently reproduce asexually. Sexually reproducing species were only present in the lagg zones, apart from *Tatriella slovenica*, which was also found in the spring of the bog brooklet in Nigula bog and the transitional mire of Valgeraba. In this spring, very small and fragmenting (often exhibiting regeneration either on posterior or anterior ends) tubificid juveniles were abundant. *Lumbriculus variegatus* was the largest oligochaete species found and accounted for the majority of the oligochaete biomass when present (data not shown).

The ombrotrophic water bodies in the pristine central bog, where only *C. sphagnetorum* was found, had a significantly lower pH and Ca + Mg concentration in the surface and interstitial water, compared to the more minerotrophic parts of the raised bog landscape (Table 2). The Ca + Mg content of dead organic matter was significantly lower, and the C:P quotient was significantly higher in the ombrotrophic bog, compared to the transitional mires, spring, connected pools, and lagg. No statistically significant differences were found for concentrations of $NO_3 + NH_4$ and $o$-$PO_4$ in surface and interstitial water and for the C:N quotient of organic matter among the parts of the raised bog landscape.

*Table 1*. Number of sites in which oligochaete species were encountered in the distinguished parts of the raised bog landscapes in Estonia and in the raised bog remnants in The Netherlands

| Species | Repr. | Ombrotrophic Ng $n = 20$ | Trans. mire/spring Ng/Vr $n = 8$ | Lagg Ng/Pr $n = 3$ | EST total $n = 31$ | NL total $n = 46$ |
|---|---|---|---|---|---|---|
| Naididae | | | | | | |
| *Nais variabilis* Piguet, 1906 | P | | 4 | 1 | 5 | 20 |
| *Pristina aequiseta* Bourne, 1891 | P | | | | | 1 |
| *Dero digitata* (Müller, 1774) | P | | | | | 3 |
| Tubificidae | | | | | | |
| *Rhyacodrilus coccineus* (Vejdovský, 1875) | S | | | 2 | 2 | |
| *Limnodrilus udekemianus* Claparède, 1862 | S | | | 1 | 1 | |
| *Spirosperma ferox* Eisen, 1879 | S | | | 1 | 1 | |
| *Tubifex ignotus* (Štolc, 1886) | S | | | 1 | 1 | |
| Tubificidae gen. sp. juv. (with hair chaetae) | | | 3 | | 3 | 2 |
| Enchytraeidae | | | | | | |
| *Cernosvitoviella* sp. | S | | | 1 | 1 | |
| *Cognettia sphagnetorum* (Vejdovský, 1878) | A | 11 | 4 | 1 | 16 | 18 |
| *Cognettia glandulosa* (Michaelsen, 1888) | A | | 2 | 2 | 4 | |
| Lumbriculidae | | | | | | |
| *Lumbriculus variegates* (Müller, 1774) | A | | 7 | 3 | 10 | 13 |
| *Stylodrilus brachystylus* Hrabě, 1929 | S | | | 1 | 1 | |
| *Stylodrilus heringianus* Claparède, 1862 | S | | | | | 1 |
| *Tatriella slovenica* Hrabě, 1939 | S | | 2 | 1 | 3 | |
| *Rhynchelmis tetratheca* Michaelsen, 1920 | S | | | 2 | 2 | |
| Total number of species | | 1 | 6 | 12 | 13 | 7 |

$n$ = total number of sampling sites. Ng = Nigula; Vr = Valgeraba; Pr = Punaraba. Dominant mode of reproduction according to Timm (1970): A = architomy ( = fragmentation); P = paratomy ( = budding); S = sexual; Repr. = Reproduction; Trans. = Transitional; EST = Estonia; NL = The Netherlands.

*Table 2.* Chemical data (means ± standard error) of the sampling sites in the distinguished parts of the raised bog landscapes in Estonia

| | Ombrotrophic | Transitional mire/spring | Lagg | Ombrotr. vs. other parts | |
|---|---|---|---|---|---|
| | Ng | Ng/Vr | Ng/Pr | Excl. lagg | Incl. lagg |
| | $n=20$ | $n=8$ | $n=3$ | $p$ | $p$ |
| Surface water | | | | | |
| pH | 3.9 ± 0.04 | 4.5 ± 0.22 | 5.5 ± 0.50 | * | ** |
| $o$-$PO_4$ ($\mu mol\ l^{-1}$) | 0.4 ± 0.03 | 0.8 ± 0.50 | 0.4 ± 0.15 | ns | ns |
| $NO_3 + NH_4$ ($\mu mol\ l^{-1}$) | 6.0 ± 0.74 | 13 ± 6.1 | 3.7 ± 0.71 | ns | ns |
| Ca + Mg ($\mu mol\ l^{-1}$) | 21 ± 2.5 | 68 ± 15 | 221 ± 30 | ** | ** |
| Interstitial water | | | | | |
| pH | 4.5 ± 0.07 | 5.1 ± 0.23 | 5.8 ± 0.30 | * | ** |
| $o$-$PO_4$ ($\mu mol\ l^{-1}$) | 0.2 ± 0.05 | 0.2 ± 0.04 | 4.3 ± 2.7 | ns | ns |
| $NO_3 + NH_4$ ($\mu mol\ l^{-1}$) | 9.6 ± 3.4 | 8.2 ± 3.8 | 2.5 ± 0.2 | ns | ns |
| Ca + Mg ($\mu mol\ l^{-1}$) | 25 ± 4.0 | 110 ± 28 | 312 ± 95 | ** | ** |
| Organic matter | | | | | |
| Ca + Mg ($\mu mol\ g^{-1}$ DW) | 81 ± 4.9 | 201 ± 41 | 561 ± 230 | * | * |
| C:N quotient ($g\ g^{-1}$) | 30 ± 2.2 | 31 ± 4.8 | 18 ± 1.0 | ns | ns |
| C:P quotient ($g\ g^{-1}$) | 1220 ± 105 | 698 ± 85 | 311 ± 84 | *** | *** |

Ng = Nigula; Vr = Valgeraba; Pr = Punaraba.Significance of the differences between the means of ombrotrophic pools and the other parts of the pristine bogs is indicated using the $p$-values of the $t$-test. ns = $p > 0.05$, * = $p < 0.05$, ** = $p < 0.01$, and *** = $p < 0.001$.

### *Dutch raised bog remnants*

In the Dutch raised bog remnants 7 oligochaete species were found (Table 1). Similar to the Estonian bogs, *C. sphagnetorum*, *N. variabilis*, and *L. variegatus* were the most frequent species. However, in The Netherlands *N. variabilis* was found at more sites than *C. sphagnetorum* and *L. variegatus*, whereas in Estonia *N. variabilis* was found only and exclusively at 5 of the 10 sampling sites inhabited by *L. variegatus*. The sexually reproducing species present in the Estonian lagg zones were not found in the Dutch raised bog remnants. Instead, the naidids *Pristina aequiseta* and *Dero digitata*, reproducing most frequently by paratomy (= budding), and the sexually reproducing species *Stylodrilus heringianus* were found at few sites.

The concentrations of S, $NO_3 + NH_4$ and $o$-$PO_4$ in the surface and interstitial water of the Dutch sampling sites were significantly higher than in the Estonian sites, whereas there was no difference in the pH (Table 3). The Ca + Mg content and the C:N and C:P quotients of dead organic matter were not different between Estonian and Dutch sites.

### *Species' response to water and organic matter quality*

The abundance of *L. variegatus* showed a significant relationship with the pH of surface water and interstitial water and with the C:P quotient and Ca + Mg content of dead organic matter (Table 4). The abundance of *N. variabilis* showed a significant relationship with the C:P and C:N quotient of dead organic matter. No significant relationship was found for *C. sphagnetorum*. In both Estonia and The Netherlands *C. sphagnetorum* was found at sites with the C:P and C:N quotient ranging from high to low (Fig. 2). In Estonia, *N. variabilis* and *L. variegatus* were only found if the C:P quotient was lower than 1000. This was also true for *L. variegatus* in The Netherlands, except one site where it was present in low abundance (0.75 ind./m). *N. variabilis* was encountered in four Dutch water bodies with a C:P quotient higher than 1000, but it was most abundant if the C:P and C:N quotients were low (lower than 1000 and 30, respectively). Also in the lower range of these quotients *N. variabilis* was in The Netherlands more frequent than in Estonia.

*Table 3.* Chemical data (means ± standard error) of the sampling sites in Estonia and The Netherlands

| | Estonia $n=31$ | The Netherlands $n=46$ | $p$ |
|---|---|---|---|
| Surface water | | | |
| pH | $4.2 \pm 0.1$ | $4.3 \pm 0.1$ | ns |
| $o$-$PO_4$ ($\mu mol\ l^{-1}$) | $0.5 \pm 0.1$ | $8.0 \pm 3.8$ | * |
| $NO_3 + NH_4$ ($\mu mol\ l^{-1}$) | $7.6 \pm 1.7$ | $43 \pm 6.5$ | *** |
| S ($\mu mol\ l^{-1}$) | $18 \pm 1.0$ | $56 \pm 3.2$ | *** |
| Interstitial water | | | |
| pH | $4.7 \pm 0.1$ | $4.8 \pm 0.1$ | ns |
| $o$-$PO_4$ ($\mu mol\ l^{-1}$) | $0.6 \pm 0.3$ | $26 \pm 7.4$ | *** |
| $O_3 + NH_4$ ($\mu mol\ l^{-1}$) | $8.5 \pm 2.4$ | $160 \pm 22$ | *** |
| S ($\mu mol\ l^{-1}$) | $21 \pm 1.5$ | $84 \pm 6.8$ | *** |
| Organic matter | | | |
| Ca + Mg ($\mu mol\ g^{-1}$ DW) | $159 \pm 33$ | $132 \pm 16$ | ns |
| C:N quotient ($g\ g^{-1}$) | $29 \pm 1.9$ | $28 \pm 1.6$ | ns |
| C:P quotient ($g\ g^{-1}$) | $997 \pm 91$ | $946 \pm 89$ | ns |

Significance of the differences between the means is indicated using the $p$-values of the $t$-test. ns = $p > 0.05$, * = $p < 0.05$, and *** = $p < 0.001$.

## Discussion

### *Species occurrence in pristine bogs*

The fauna of pristine raised bogs is known to be poor in aquatic oligochaete species (Harnisch, 1925, 1929; Peus, 1932). The fragmenting, almost never mature, and acid-tolerant enchytraeid *Cognettia sphagnetorum* is the single oligochaete species able to live in the acidic and extremely nutrient-poor water bodies in the central ombrotrophic raised bog (Table 1). This species is also a typical and abundant inhabitant of peatland soils, such as blanket bogs (Standen & Latter, 1977) and Fennoscandian coniferous forests (Huhta et al., 1986). Other oligochaete species are limited to those parts of the bog landscape influenced by minerotrophic alkaline water, like the lagg zones, transitional mires, and bog brooklets.

With over ten species, the lagg zone, influenced by calcareous ground water and characterised by a

*Table 4.* Results of linear regression of chemical data and log-transformed species abundance data

| | *Cognettia sphagnetorum* | | *Nais variabilis* | | *Lumbriculus variegatus* | |
|---|---|---|---|---|---|---|
| | $R^2$ | $p$ | $R^2$ | $p$ | $R^2$ | $p$ |
| Surface water | | | | | | |
| pH | 0.039 | ns | 0.027 | ns | 0.355 | $p<0.05$ |
| $o$-$PO_4$ ($\mu mol\ l^{-1}$) | 0.012 | ns | 0.021 | ns | 0.048 | ns |
| $NO_3 + NH_4$ ($\mu mol\ l^{-1}$) | 0.009 | ns | 0.003 | ns | 0.001 | ns |
| Interstitial water | | | | | | |
| pH | 0.000 | ns | 0.012 | ns | 0.109 | $p<0.05$ |
| $o$-$PO_4$ ($\mu mol\ l^{-1}$) | 0.019 | ns | 0.001 | ns | 0.000 | ns |
| $NH_4 + NO_3$ ($\mu mol\ l^{-1}$) | 0.031 | ns | 0.004 | ns | 0.058 | ns |
| Organic matter | | | | | | |
| Mg + Ca ($\mu mol\ g^{-1}$ DW) | 0.017 | ns | 0.004 | ns | 0.158 | $p<0.05$ |
| C:N quotient ($g\ g^{-1}$) | 0.019 | ns | 0.100 | $p<0.05$ | 0.080 | ns |
| C:P quotient (g g–1) | 0.008 | ns | 0.099 | $p<0.05$ | 0.194 | $p<0.05$ |

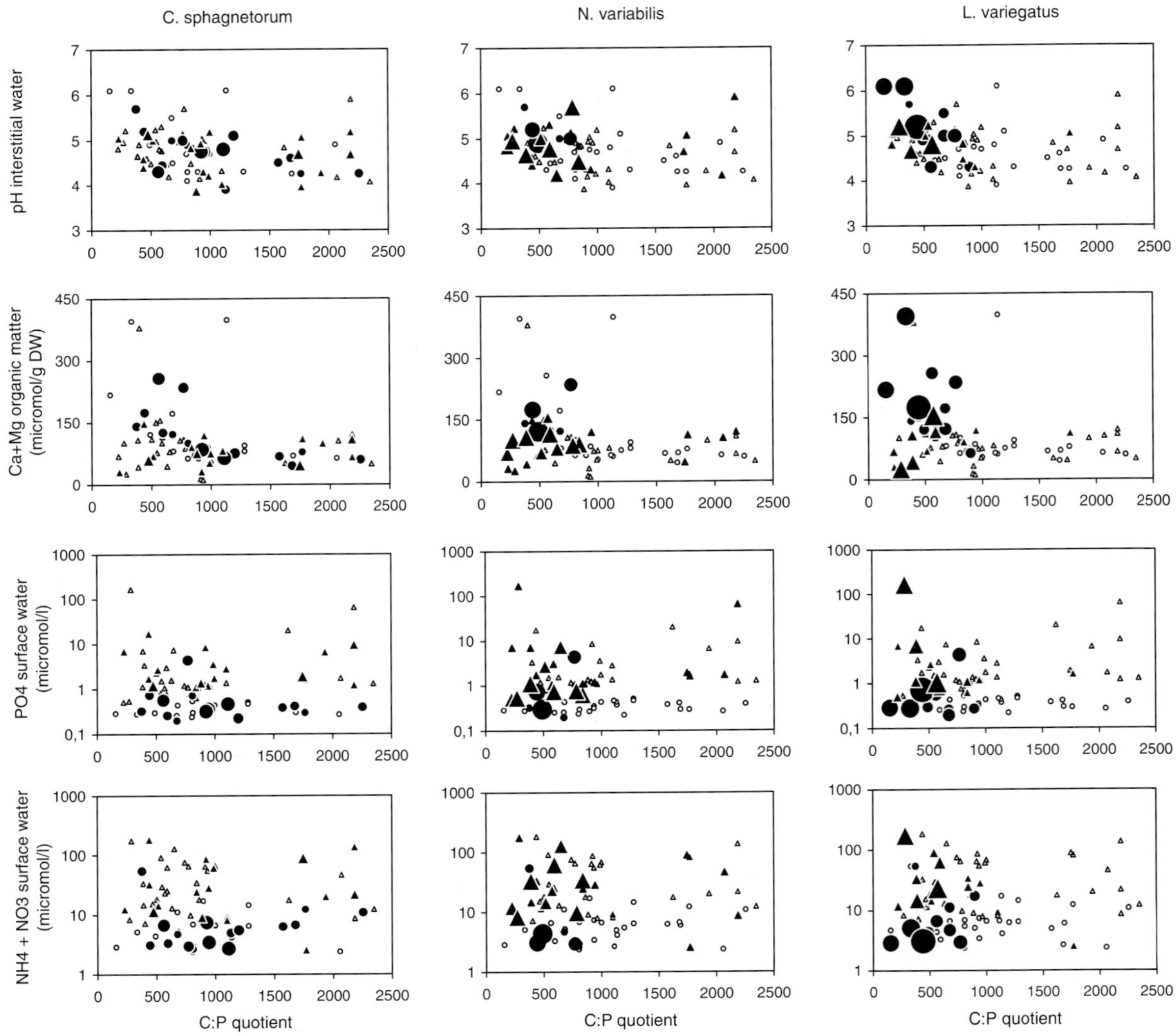


*Figure 2.* The pH in interstitial water, the Ca + Mg content in dead organic matter, and the concentration of $NH_4 + NO_3$ and $o$-$PO_4$ in surface water plotted against the C:P quotient of dead organic matter for Estonian (○) and Dutch (Δ) sampling sites. Filled symbols represent sites in which *C. sphagnetorum*, *N. variabilis* and *L. variegatus* were encountered and the larger the size of the symbol, the higher the abundance of the species. The largest size for >100 individuals per 1 m sweep with the pond net and the smallest size for <2 individuals per 1 m sweep.

relatively high abundance of vascular plants and trees, is the most species rich part of the pristine raised bog landscape. Its species assemblage (including *Rhyacodrilus coccineus*, *Tubifex ignotus*, *Tatriella slovenica*, and *Stylodrilus brachystylus*) can be regarded as a relict of the littoral zone fauna of the former lake that has been terrestrialised by the bog formation and has never been affected by eutrophication. *R. coccineus, Limnodrilus udekemianus, T. ignotus*, and *Spirosperma ferox* occur in many Estonian lakes and rivers, but the lagg community distinctly excludes the ubiquitous tubificids *Limnodrilus hoffmeisteri* and *Tubifex tubifex*, as well as the calciphilous *Psammoryctides barbatus*, also common in most Estonian lakes and rivers. *Tatriella slovenica* is a rare species in Estonia, found mostly in lakes situated between a bog and sandy hills (Timm, 1970, 1999).

Species that reproduce predominantly by sexual modes are absent from the acidic, nutrient poor part of the raised bog (Table 1). Piguet (1906) noted the absence of sexual reproduction in the species *Nais communis* and *Pristina longiseta* in Swiss bog pools. Harnisch (1925) attributed the occurrence of only asexual reproduction to the acidity and extremely low food availability in bogs. The nutrient content and abundance of food is an important factor in the growth rate of oligochaetes (Pasteris et al., 1994). Likewise, detritivorous chironomid and mayfly larvae grow more rapidly and may produce more generations per year if the available detritus contains more bacteria or has a higher P content (Ward & Cummins, 1979; Frost & Elser, 2002; Vos et al., 2002). Body growth is constrained by the nutrient content of ingested food, as more food has to be ingested to build up the body biomass when the C:P and C:N quotients are higher (Elser et al., 2000) and mineral content is lower. Streit (1978) found a higher ingestion rate and lower biomass incorporation rate for algae with a low nutritional value. The low nutrient availability may be disadvantageous for sexually reproducing oligochaetes and the acid conditions or absence of specific food components may prevent successful development of eggs and juveniles (Vos et al., 2000). Thus, within the pristine raised bog landscapes in Estonia the pattern in the occurrence and mode of reproduction of aquatic oligochaete species can be linked to the variation in pH and nutrient availability.

*Nutrient availability in pristine and polluted raised bogs*

In the more minerotrophic parts of the pristine raised bog landscape, pH and buffering capacity (with Ca and Mg as main cations) are significantly higher and the C:P quotient of the dead organic matter is significantly lower than in the ombrotrophic part (Table 2). Therefore, the decomposition rate of the dead organic matter is higher in the minerotrophic parts (Leuven & Wolfs, 1988; Kok & Van der Velde, 1994; Lamers et al., 1999; Smolders et al., 2002) and hence the nutrient availability is higher. The concentrations of $NO_3$, $NH_4$, and *o*-$PO_4$ measured in the surface and interstitial water did not differ significantly between the ombrotrophic and more minerotrophic parts of the bog landscape (Table 2) and no significant relationship was found between the species' abundance and concentrations of N and P in surface and interstitial water (Table 4). This result is not necessarily contradictory to the conclusion that the nutrient availability is higher in the more minerotrophic parts as in pristine (unpolluted) bogs the minerals released by decomposition and mineralisation of the organic matter will be taken up and incorporated in algae, bacteria, mosses, and higher plants (Lamers et al., 2000; Limpens et al., 2003a).

A higher availability of nutrients in the more minerotrophic parts of the raised bog landscape can result in a higher density and a higher nutritional value of algae for algivores (Gulati & DeMott, 1997; Elser et al., 2001). Also microbes and vascular plants, like *Menyanthes trifoliata*, *Carex* spp., and trees, can be more abundant. The organic matter formed by these plant species may have a higher nutrient content than the organic matter in the ombrotrophic parts of raised bogs and therefore be more easily decomposed by microbes (Aerts et al., 2001; Scheffer et al., 2001; Smolders et al., 2002; Limpens & Berendse, 2003; Tomassen et al., 2003, 2004a). Next, *Sphagnum* contains a number of structural and non-structural organochemical compounds (such as phenolics) which inhibit microbial activity and further restrict the decomposition of its remnants (Verhoeven & Liefveld, 1997; Smolders et al., 2002). Bacteria, algae and many higher plants are characterised by an organic matter quality which is much less resistant to decomposition processes than the *Sphagnum* litter. This implies a better decomposability of the organic matter in transitional mires and lagg zones, resulting in an increased availability of N and P and hence again in an increased growth of bacteria, algae, and leaf-decaying fungi which may stimulate rates of leaf decomposition (Grattan & Suberkropp, 2001; Gulis & Suberkropp, 2003). A higher abundance and nutrient content of algae, fungi, or bacteria in the minerotrophic parts of the pristine raised bog landscape provide food for more oligochaete species, of which *Lumbriculus variegatus* and *Nais variabilis* are the most abundant.

Due to the prolonged high deposition rates of N and S in The Netherlands, a minerotrophic influence is not necessary anymore for a higher abundance and nutrient content of algae, fungi, or bacteria. Dutch raised bog remnants contain water bodies with both a high availability of N and P in the surface water and a high C:P quotient and low Ca + Mg content of dead organic matter (Fig. 2). This combination of conditions is absent in pristine Estonian raised bogs and has been 'created' by the prolonged high deposition rates of N and S in The Netherlands. At the same time, the most nutrient-poor conditions of pristine raised bogs are not present anymore in Dutch bog remnants. This situation is schematically represented in Figure 3.

*Differential response of oligochaete species to increased nutrient availability*

*Cognettia sphagnetorum* occurred in all parts of Estonian raised bog landscapes (Table 1). No relationship was found between the abundance of this species and nutrient availability in Estonian and Dutch bogs (Table 4; Fig. 2). *C. sphagnetorum* is thus not limited by nutrient availability – and is able to live in all situations represented in Figure 3. *N. variabilis* and *L. variegatus* in Estonia were, however, found to be limited to the more minerotrophic parts of the raised bog landscape with a higher nutrient availability. Whereas the frequency of *C. sphagnetorum* and *L. variegatus* was fairly similar in both countries, *N. variabilis* was much more frequent in the Dutch bog remnants than in Estonian bogs (Table 1). These three species differed in their response to water and organic matter variables (Table 4). Probably they use different food items. *C. sphagnetorum* unselectively ingests available organic matter and uses algae and fungi as a food-source (Standen & Latter, 1977; Augustsson & Rundgren, 1998). Possibly, this species is a primary decomposer (Latter & Howson, 1978) and uses also soluble nutrients as food (Springett & Latter, 1977). According to Learner et al. (1978) and Harper et al. (1981) *N. variabilis* feeds on bacteria and diatoms, whereas *L. variegatus* feeds on detritus and algae (Moore, 1978).

In both Estonia and The Netherlands *L. variegatus* was most abundant in water bodies with a lower C:P quotient and a higher Ca + Mg content of dead organic matter and a higher pH (Table 4; Fig. 2) and, thus, with a higher decomposition rate of the organic matter (left part in Fig. 3). Increased N and P availability in the surface water in the

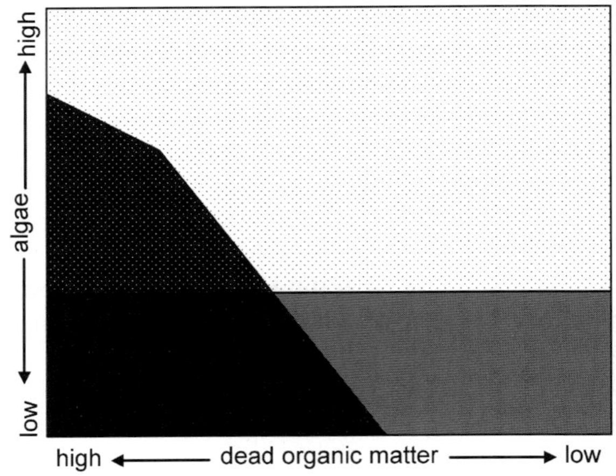


*Figure 3.* Schematic representation of the combination of dead organic matter quality and decomposition rate vs. quality and abundance of algae in pristine raised bogs (black and grey areas) and raised bogs affected by increased N and S deposition (dotted areas). The white area with black dots represents the situation absent in pristine raised bogs, 'created' by increased N and S deposition.

Dutch bog remnants, favouring growth of algae, seems to be insufficient to increase the frequency of *L. variegatus*. Therefore, we assume that *L. variegatus* predominantly feeds on organic matter and/or on the fungi and bacteria decomposing this matter and that the type of organic matter and its decomposition process are important.

Suitable conditions for *N. variabilis* in Estonian raised bogs are only present if the decomposition rate of organic matter is high (low C:P quotient and higher pH and buffering capacity). Consequently, in Estonia *N. variabilis* is only found in a subset of the water bodies in which *L. variegatus* is present. Contrary to *L. variegatus*, *N. variabilis* in The Netherlands appears to be able to live in water bodies with a higher C:P quotient of organic matter (Fig. 2; right-upper area in Fig. 3). It may therefore be concluded that *N. variabilis* is not dependent on organic matter with a high decomposition rate and the allied microbes, but responds to the abundance and nutritional value of diatoms or other algae, which are affected by N and P availability in surface water. However, it can not be precluded that in the Dutch bog remnants the prolonged high deposition rates of N and S and the increase in higher plants and algae have led to a new type of dead organic matter that is more easily decomposed by microbes, although it still may have a relatively high C:P quotient. The organic compounds produced during decomposition of this new type of organic matter may not always possess a sufficient nutritional value for *L. variegatus*. More frequently *N. variabilis* may profit from the increased abundance of bacteria and/or (micro)algae resulting from the higher availability of nutrients (N and P) in Dutch bog remnants. Therefore, in The Netherlands *N. variabilis* can be found in water bodies that are not inhabited by *L. variegatus*.

## Conclusion

The most nutrient poor water bodies in the pristine central raised bog are only inhabited by *Cognettia sphagnetorum*. Other oligochaete species in pristine Estonian raised bogs, of which *Nais variabilis* and *Lumbriculus variegatus* are most abundant, are limited to more minerotrophic water bodies with an increased decomposition rate of organic matter and a higher nutrient availability. However, in Dutch raised bog remnants the occurrence of oligochaetes is not limited anymore by N and P availability, due to increased atmospheric N and S deposition. *C. sphagnetorum*, *N. variabilis*, and *L. variegatus* respond differently to the increased availability of N and P in The Netherlands, probably because of differences in their diets In Estonia species that reproduce sexually are limited to the minerotrophic parts of the bog landscape (springs and laggs). Most of the lagg zone species were not found in Dutch bog remnants as this part of raised bog landscapes has long been cultivated and the specific conditions of laggs are not present anymore in almost all Dutch bog remnants. Overall, it can be concluded that the degradation of Dutch raised bogs has resulted in the loss of both the nutrient-poor parts of the landscape and the special lagg conditions.

## Acknowledgements

The data from the Estonian bogs couldn't have been collected without the permission, assistance, and hospitality of the staff members of Nigula Nature Reserve, Endla Nature Reserve, and Soomaa National Park, and the hard work of Judith Bosman, Juhan Javoiš, Monique Peeters, and Michel Smits. In The Netherlands, Staatsbosbeheer, Vereniging Natuurmonumenten, and Stichting het Limburgs Landschap gave permission to enter their reserves and to take samples. Jan Kuper, Sandra Lomans, and Theo Peeters helped sorting these samples. Hilde Tomassen (Dep. Aquatic Ecology & Environmental Biology, Radboud University Nijmegen) provided water quality data of several Dutch sampling sites. Jelle Eygensteyn provided assistance during the chemical analyses and performed the C and N analysis. Hein van Kleef, Rob Leuven, Piet Nienhuis, Henk Siepel, Gerard van der Velde, and two anonymous referees gave useful comments on earlier drafts of the manuscript. Tarmo Timm was supported by the Core grant # 0362482s03 "Formation and changes of the biological diversity under the human impact in different lake types (Estonia – Denmark – Netherlands)". This research project is part of the national research programme 'Survival Plan for Woodland and Nature', funded by the Dutch Ministry of Agriculture, Nature and Food Quality.

## References

Aerts, R., B. Wallén, N. Malmer & H. De Caluwe, 2001. Nutritional constraints on *Sphagnum*. Journal of Ecology 89: 292–299.

Alcamo, J., P. Mayerhofer, R. Guardans, T. van Harmelen, J. van Minnen, J. Onigkeit, M. Posch & B. de Vries, 2002. An integrated assessment of regional air pollution and climate change in Europe: findings of the AIR-CLIM Project. Environmental Science & Policy 5: 257–272.

Augustsson, A. K. & S. Rundgren, 1998. The enchytraeid *Cognettia sphagnetorum* in risk assessment: Advantages and disadvantages. Ambio 27: 62–69.

Belyea, L. R., 1996. Separating the effects of litter quality and microenvironment on decomposition rates in a patterned peatland. Oikos 77: 529–539.

Berendse, F., N. van Breemen, H. Rydin, A. Buttler, M. M. P. D. Heijmans, M. R. Hoosbeek, J. A. Lee, A. Mitchell, T. Saarinen, H. Vasander & B. Wallén, 2001. Raised atmospheric $CO_2$ levels and increased N deposition cause shifts in plant species composition and production in *Sphagnum* bogs. Global Change Biology 7: 591–598.

Bobbink, R., M. Hornung & J. G. M. Roelofs, 1998. The effects of air-borne nitrogen pollutants on species diversity in natural and semi-natural European vegetation. Journal of Ecology 86: 717–738.

Elser, J. J., W. F. Fagan, R. F. Denno, D. R. Dobberfuhl, A. Folarin, A. Huberty, S. Interlandi, S. S. Kilham, E. McCauley, K. L. Schulz, E. H. Sieman & R. W. Sterner, 2000. Nutritional constraints in terrestrial and freshwater food webs. Nature 408: 578–580.

Elser, J. J., K. Hayakawa & J. Urabe, 2001. Nutrient limitation reduces food quality for zooplankton: *Daphnia* response to seston phosphorus enrichment. Ecology 82: 898–903.

Frost, P. C. & J. J. Elser, 2002. Growth responses of littoral mayflies to the phosphorus content of their food. Ecology Letters 5: 232–240.

Grattan, R. M. & K. Suberkropp, 2001. Effects of nutrient enrichment on yellow poplar leaf decomposition and fungal activity in streams. Journal of the North American Benthological Society 20: 33–43.

Gulis, V. & K. Suberkropp, 2003. Effect of inorganic nutrients on relative contribution of fungi and bacteria to carbon flow from submerged decomposing leaf litter. Microbial Ecology 45: 11–19.

Gulati, R. D. & W. R. DeMott, 1997. The role of food quality for zooplankton: remarks on the state-of-the-art, perspectives and priorities. Freshwater Biology 38: 753–768.

Harnisch, O., 1925. Studien zur Ökologie und Tiergeographie der Moore. Zoologische Jahrbücher, Abteilung für Systematik, Ökologie und Geographie der Tiere 51: 1–166.

Harnisch, O., 1929. Die Biologie der Moore. Die Binnengewässer VII. Stuttgart, 146 pp.

Hogg, P., P. Squires & A. H. Fitter, 1995. Acidification, nitrogen deposition and rapid vegetational change in a small valey mire in Yorkshire. Biological Conservation 71: 143–153.

Harper, R. M., J. C. Fry & M. A. Learner, 1981. A bacteriological investigation to elucidate the feeding biology of *Nais variabilis* (Oligochaeta, Naididae). Freshwater Biology 11: 227–236.

Huhta, V., R. Hyvönen, P. Kaasalainen, A. Koskenniemi, J. Muona, I. Mäkelä, M. Sulander & P. Vilkamaa, 1986. Soil fauna of Finnish coniferous forests. Annales Zoologici Fennici 23: 345–360.

Jonson, J. E., J. Bartnicki, K. Olendrzynski, H. A. Jakobsen & E. Berge, 1998. EMEP Eulerian model for atmospheric transport and deposition of nitrogen species over Europe. Environmental Pollution 102: 289–298.

Kok, C. J. & B. J. Van de Laar, 1991. Influence of pH and buffering capacity on the decomposition of *Nymphaea alba* L. detritus in laboratory experiments: a possible explanation for the inhibition of decomposition at low alkalinity. Verhandlungen der Internationalen Vereinigung für Theoretische und Angewandte Limnologie 24: 2689–2692.

Kok, C. J. & G. van der Velde, 1994. Decomposition and invertebrate colonization of aquatic and terrestrial leaf material in alkaline and acid still water. Freshwater Biology 31: 65–75.

Lamers, L. P. M., H. B. M. Tomassen & J. G. M. Roelofs, 1998. Sulfate induced eutrophication and phytotoxicity in freshwater wetlands. Environmental Science & Technology 32: 199–205.

Lamers, L. P. M., C. Farhoush, J. M. van Groenendael & J. G. M. Roelofs, 1999. Calcareous groundwater raises bogs; the concept of ombrotrophy revisited. Journal of Ecology 87: 639–648.

Lamers, L. P. M., R. Bobbink & J. G. M. Roelofs, 2000. Natural nitrogen filter fails in polluted raised bogs. Global Change Biology 6: 583–586.

Lang, C., 1999. Contrasting responses of oligochaetes (Annelida) and chironomids (Diptera) to the abatement of eutrophication in Lake Neuchâtel. Aquatic Sciences 61: 206–214.

Lang, C. & O. Reymond, 1995. Empirical relationships between oligochaete communities and phosphorus in lakes before and after the abatement of eutrophication. Archiv für Hydrobiologie 136: 237–245.

Latter, P. M. & G. Howson, 1978. Studies on microfauna of blanket bog with particular reference to Enchytraeidae II. Growth and survival of *Cognettia sphagnetorum* on various substrates. Journal of Animal Ecology 47: 425–448.

Learner, M. A., G. Lochhead & B. Hughes, 1978. A review of the biology of British Naididae (Oligochaeta) with emphasis on the lotic environment. Freshwater Biology 8: 357–375.

Leuven, R. S. E. W., 1988. Impact of acidification on aquatic ecosystems in The Netherlands. Thesis Nijmegen.

Leuven, R. S. E. W. & W. J. Wolfs, 1988. Effects of water acidification on the decomposition of *Juncus bulbosus* L. Aquatic Botany 31: 57–81.

Limpens, J. & F. Berendse, 2003. How litter quality affects mass loss and N loss from decomposing *Sphagnum*. Oikos 103: 537–547.

Limpens, J., F. Berendse & H. Klees, 2003a. N deposition affects N availability in interstitial water, growth of *Sphagnum* and invasion of vascular plants in bog vegetation. New Phytologist 157: 339–347.

Limpens, J., J. T. A. G. Raymakers, J. Baar, F. Berendse & J. D. Zijlstra, 2003b. The interaction between epiphytic algae, a parasitic fungus and *Sphagnum* as affected by N and P. Oikos 103: 59–68.

Moore, J. W., 1978. Importance of algae in the diet of the oligochaetes *Lumbriculus variegatus* (Müller) and *Rhyacodrilus sodalis* (Eisen). Oecologia 35: 357–363.

Pajuste, K., H. Iher, T. Truuts, M. Kört, M. Kivistik, M. Uri & R. Kolk, 2002. Sademete keemia. In Roose, A. (ed.) Eesti Keskkonnaseire 2001. Estonian Ministry of the Environment, Tallinn: 28–32. [In Estonian].

Pasteris, A., C. Bonacina & G. Bonomi, 1994. Observations on cohorts of *Tubifex tubifex* cultured at different food levels, using cellulose substrate. Hydrobiologia 278: 315–320.

Peus, F., 1932. Die Tierwelt der Moore. Handbuch der Moorkunde III. Bornträger verlag, Berlin. [In German].

Piguet, E., 1906. Observations sur les Naididées. Revue suisse de Zoologie 14: 185–317.

Risager, M., 1998. Impacts of nitrogen on *Sphagnum* dominated bogs with emphasis on critical load assessment. PhD Thesis, Department of Plant Ecology, Botanical Institute, University of Copenhagen, Denmark.

RIVM, 2004. Natuurbalans 2004. Sdu uitgevers, Den Haag. [In Dutch].

Roelofs, J. G. M., 1986. The effect of airborne sulphur and nitrogen deposition on aquatic and terrestrial heathland vegetation. Experientia 42: 372–377.

Scheffer, R. A., R. S. P. Van Logtestijn & J. T. A. Verhoeven, 2001. Decomposition of *Carex* and *Sphagnum* litter in two mesotrophic fens differing in dominant plant species. Oikos 92: 44–54.

Schouwenaars, J. M., 1993. Hydrological differences between bogs and bog-relicts and consequences for bog restoration. Hydrobiologia 265: 217–224.

Smits, M. J. A., G. A. van Duinen, J. G. Bosman, A. M. T. Brock, J. Javoiš, J. T. Kuper, T. M. J. Peeters & H. Esselink, 2002. Species richness in a species poor system: aquatic macroinvertebrates of Nigula raba, an intact raised bog system in Estonia. In Schmilewski, G. & L. Rochefort (eds). Proceedings of the International Peat Symposium – Peat in Horticulture – Quality and Environmental Changes, pp 283–291.

Smolders, A. J. P., H. B. M. Tomassen, L. P. M. Lamers, B. P. Lomans & J. G. M Roelofs, 2002. Peat bog restoration by floating raft formation: the effects of groundwater and peat quality. Journal of Applied Ecology 39: 391–401.

Springett, J. A. & P. M. Latter, 1977. Studies on the microfauna of blanket bog with particular reference to Enchytraeidae I. Field and laboratory tests of micro-organisms as food. Journal of Animal Ecology 46: 959–974.

Standen, V. & P. M. Latter, 1977. Distribution of a population of *Cognettia sphagnetorum* (Enchytraeidae) in relation to microhabitats in a blanket bog. Journal of Animal Ecology 46: 213–219.

Streit, B., 1978. A note on the nutrition of *Stylaria lacustris* (Oligochaeta: Naididae). Hydrobiologia 61: 273–276.

Timm, T., 1970. On the fauna of the Estonian Oligochaeta. Pedobiologia 10: 52–78.

Timm, T., 1999. A Guide to the Estonian Oligochaeta. Estonian Academy Publishers, Tartu-Tallinn.

Tomassen, H. B. M., A. J. P. Smolders, L. P. M. Lamers & J. G. M. Roelofs, 2003. Stimulated growth of *Betula pubescens* and *Molinia caerulea* on ombrotrophic bogs: role of high levels of atmospheric nitrogen deposition. Journal of Ecology 91: 357–370.

Tomassen, H. B. M., A. J. P. Smolders, L. P. M. Lamers & J. G. M. Roelofs, 2004a. Development of floating rafts after the rewetting of cut-over bogs: the importance of peat quality. Biogeochemistry 71: 69–87.

Tomassen, H. B. M., A. J. P. Smolders, J. Limpens, L. P. M. Lamers & J. G. M. Roelofs, 2004b. Expansion of invasive species on ombrotrophic bogs: desiccation or high N deposition? Journal of Applied Ecology 41: 139–150.

Van Duinen, G. A., A. M. T. Brock, J. T. Kuper, T. M. J. Peeters, M. J. A. Smits, W. C. E. P. Verberk & H. Esselink, 2002. Important keys to successful restoration of characteristic aquatic macroinvertebrate fauna of raised bogs. In Schmilewski, G. & L. Rochefort (eds), Proceedings of the International Peat Symposium – Peat in Horticulture – Quality and Environmental Challenges, pp 292–302.

Van Duinen, G. A., A. M. T. Brock, J. T. Kuper, R. S. E. W. Leuven, T. M. J. Peeters, J. G. M. Roelofs, G. van der Velde, W. C. E. P. Verberk & H. Esselink, 2003. Do restoration measures rehabilitate fauna diversity in raised bogs? A comparative study on aquatic macroinvertebrates. Wetlands Ecology and Management 11: 447–459.

Verdonschot, P. F. M., 1996. Oligochaetes and eutrophication; an experiment over four years in outdoor mesocosms. Hydrobiologia 334: 169–183.

Verhoeven, J. T. A. & W. M. Liefveld, 1997. The ecological significance of organochemical compounds in *Sphagnum*. Acta Botanica Neerlandica 46: 117–130.

Vos, J. H., M. A. G. Ooijevaar, J. F Postma & W. Admiraal, 2000. Interaction between food availability and food quality during growth of early instar chironomid larvae. Journal of the North American Benthological Society 19: 158–168.

Vos, J. H., P. J. Van den Brink, F. P. van den Ende, M. A. G. Ooijevaar, A. J. P. Oosthoek, J. F. Postma & W. Admiraal, 2002. Growth response of a benthic detritivore to organic matter composition of sediments. Journal of the North American Benthological Society 21: 443–456.

Ward, G. M. & K. W. Cummins, 1979. Effects of food quality on growth of a stream detritivore, *Paratendipes albimanus* (Meigen) (Diptera: Chironomidae). Ecology 60: 57–64.

Hydrobiologia (2006) 564:157–169
P.F.M. Verdonschot, H. Wang, A. Pinder & R. Nijboer (eds), Aquatic Oligochaete Biology IX
DOI 10.1007/s10750-005-1716-x


# Toxicity of Santander Bay sediments to the euryhaline freshwater oligochaete *Limnodrilus hoffmeisteri*

P. Rodriguez[1,*], J. Arrate[1], M. Martinez-Madrid[2], T. B. Reynoldson[3], V. Schumacher[4] & J. Viguri[5]
[1]*Dpt. Zoología y Biología Celular Animal, University of the Basque Country, Apdo.644, 48080, Bilbao, Spain*
[2]*Dpt. Genetica, Antropología Física y Fisiología Animal, University of the Basque Country, Bilbao, Spain*
[3]*Acadia Centre for Estuarine Research, Department of Biology, Acadia University, Box 115 B4P 2R6, Wolfville, Nova Scotia, Canada*
[4]*University of Aachen, Aachen, Germany*
[5]*E.T.S.I.I.T. Department of Chemical Engineering, University of Cantabria, Santander, Spain*
*(*Author for correspondence: E-mail: pilar.rodriguezr@ehu.es)*

*Key words:* toxicology, sediment bioassay, estuary, *Limnodrilus hoffmeisteri*

## Abstract

The freshwater euryhaline oligochaete *Limnodrilus hoffmeisteri* was selected for use in bioassays with polluted sediment from Santander Bay. It is easy to culture; is tolerant of low to moderate, up to 15‰, salinity; and is common in oligohaline conditions in European and North American estuaries. Worms were collected from an estuarine population and kept in unpolluted sediment for between 2 and 4 weeks, under laboratory conditions, at 7–8.5‰ salinity and 22.5 °C. Sediment from different sites in Santander Bay were sieved through 250 $\mu$m mesh and adjusted to a salinity of approx. 7‰ prior to the bioassays, either by adding sea water or dechlorinated tap water to the overlying water. High levels of ammonia in some sediment, which may confound results in ecotoxicity bioassays, were reduced by oxidation of the sediments in shallow trays. Sediment bioassays were performed with sexually mature *Limnodrilus hoffmeisteri* worms in 250-ml beakers, with a 1:3 ratio of sediment:water and 4 worms per baker. Endpoints in the 14-day bioassay were % mortality, adult final biomass, % adults that have shown resorption of the clitellum, number of cocoons, and burrowing behaviour. It was possible to rank the sites according to their toxicity using both mortality rates and sublethal effects. The control site had the following values for the endpoints: 5% mortality, ($\bar{x} \pm$ sd) 2.40 $\pm$ 1.52 cocoons per beaker and 1.271 $\pm$ 0.470 mg dw adult final biomass. The most toxic sediment resulted in 65% mortality, resorption of the clitellum in 67% of the adults, no production of cocoons and a low final biomass ($\bar{x} \pm$ sd = 0.681 $\pm$ 0.489 mg dw per adult). A second site had high mortality (60%) and no reproduction, although resorption of the clitellum did not occur in surviving animals. The remaining sites showed similar mortality (35–42%), and at only one of them was low reproduction observed (0.8 $\pm$ 0.447 cocoons per beaker). Behavioural effects, measured as length of galleries in a fixed area of the test-vessel at the end of the bioassay, were significant compared with control at only one site. Multivariate analysis showed the mortality gradient to be the strongest, with a second unassociated gradient representing clitellum resorption. The mortality gradient was associated with Cu and Zn concentration, and PAH and Pb possibly with resorption.

## Introduction

This study is part of a larger environmental assessment study of chemical pollution of sediments in the Santander Bay, in the north of Spain (Irabien et al., 2001; Viguri et al., 2002). The objective of this part of the work was to examine the use of oligochaetes in the assessment of chronic toxicity of estuarine sediments. This exploratory study includes an examination of the invertebrate communities at the study sites to establish the presence and abundance of oligochaetes. Also to test the validity of a chronic bioassay and identify appropriate endpoints for environmental risk assessment, using sediment samples taken from the upper estuaries of Santander Bay.

Aquatic oligochaetes can be abundant and, in some cases, the dominant taxon in estuarine benthic communities. The paucity of suitable test species from estuarine environments makes oligochaetes potentially useful in conducting acute and chronic tests from freshwater through estuaries and into marine waters (Chapman, 2001). In 1982, Giere and Pfannkuche published a review of the ecology of estuarine and marine oligochaetes and described the role that the salinity gradient and the sediment characteristics play in the faunistic composition of estuarine benthos. With the exception of a few euryhaline species (marine or freshwater species), most species are limited to one of the three intertidal zones (oligo-, meso- or polyhaline) whose limits are generally accepted, respectively, at 5, 18 and 25‰ salinity. The euryhaline freshwater tubificid *Limnodrilus hoffmeisteri* is a common and abundant component of the estuarine communities in both oligohaline and ß mesohaline zones (up to 10‰ salinity) of estuaries of Europe and North America (Giere & Pfannkuche, 1982). There are few true estuarine oligochaete species that could be used in sediment toxicity testing in Europe, possible species are *Heterochaeta costata*, *Monopylephorus rubroniveus* among the tubificids, and several *Lumbricillus* and *Enchytraeus* species among the enchytraeids. However, in Europe (*Lumbricillus lineatus*: Giere & Hauschidt, 1979; *Marionina subterranea*: Giere, 1980) or North America (*Monopylephorus cuticulatus:* Chapman et al., 1982a, b; Brinkhurst et al., 1983, Chapman, 1987) only limited toxicity testing has been performed with estuarine oligochaetes. These species are not always readily available for culturing in the laboratory, and their identification requires specialist examination of the population. This makes the euryhaline tubificid *Limnodrilus hoffmeisteri* a good candidate for estuarine-sediment toxicity testing. Table 1 summarises the criteria that have been considered in selecting *L. hoffmeisteri* as a test organism for estuarine-sediment toxicity assessment.

*Limnodrilus hoffmeisteri* is among the indigenous species that occur in mildly polluted sites of estuaries close to the laboratory, and occurs in high numbers. Both *Monopylephorus rubroniveus* and *Heterochaeta costata*, although previously reported from unpolluted sites (Rodriguez, 1983), were not found in this study. The fact that *L. hoffmesiteri* has a cosmopolitan distribution and is relatively easy to culture and identify make it a good potential candidate as a test species (Table 1). Among the species that inhabit the estuaries, it has the advantage that it has already been used as test-organism and its sensitivity and biology are well documented, although primarily in freshwater conditions. Chapman et al. (1982a, b) performed acute aqueous toxicity tests and demonstrated the sensitivity of *L. hoffmeisteri* to some metals and oil. Brinkhurst et al. (1983) measured respiration rates of the species at different pH, temperature and dissolved oxygen, combining these variables in different exposure

*Table 1.* Criteria used in this study for the selection of *Limnodrilus hoffmeisteri* as a test-organism for sediment toxicity assessment in estuaries

| Criteria | |
|---|---|
| 1. Data on its sensitivity to a range of chemicals | In acute aqueous and in sediment chronic freshwater toxicity tests |
| 2. Contact with the sediment | Maximum: infaunal, sediment-dweller detritivorous organism |
| 3. Culture under laboratory conditions | In freshwater. Unknown in brackish waters |
| 4. Taxonomical identification | Easy for non-specialists |
| 5. Ecological relevance | High in oligo- and mesohalines parts of estuaries |
| 6. Geographic distribution | Cosmopolitan. Common and abundant in polluted estuaries and brackish seas of Europe and North America |

conditions with toxicants. The first chronic whole-sediment toxicity tests with this species were performed by Wiederholm et al. (1987). This test included *L. hoffmeisteri* with other aquatic species, and measured survival, growth and reproduction after several months of exposure. This species has also been used in sediment toxicity tests where both reproductive impairment, egestion rates and sediment avoidance were reported (Lotufo & Fleeger, 1996), and in microcosms together with *Tubifex tubifex* to investigate toxicity of PAHs (Reible et al., 1996). *Limnodrilus hoffmeisteri* is also being considered in field and laboratory studies on toxicant bioaccumulation (Connell et al., 1988; Egeler et al., 1997).

The biology of *L. hoffmeisteri* in estuaries has also been described by several authors. The species has a strong preference for fine grain sediments (< 63 $\mu$m) (Giere & Pfannkuche, 1982), which may influence culturing and sediment toxicity tests. Hunter (1981) reviewed data on the effect of salinity on the survival of the species in both laboratory and field studies, and the reproductive cycle has been described by Pfannkuche (1981) in the Elbe Estuary (Germany). Salinity tolerance limits of *L. hoffmeisteri* have a maximum value of about 15‰ (12–13‰: Brinkhurst & Simmons, 1968 in the Bay of San Francisco (USA); 10.7‰ at 5 °C, and 14.7‰ at 20 °C: Birtwell & Arthur, 1980 in the estuary of the River Thames, England). However, acclimation can also affect tolerance limits as Chapman & Brinkhurst (1980) demonstrated with *L. hoffmeisteri* from the Fraser River (Canada); worms previously acclimated to 5‰ increased their tolerance to salinity from 1[illegible] to 14.5‰. In a study of Italian estuaries, Casellato (1994) reported *L. hoffmeisteri* at sites where water salinity reached up to 30‰ (Casellato, 1994) in the Brenta Estuary, where the species was regularly associated with *Heterochaeta costata* (Casellato & Poja, 1984). However, interstitial salinity that directly affects worm distributions do not follow daily salinity variations in the overlying water, and these data should be critically interpreted. Interestingly, Chapman & Brinkhurst (1981) noted that salinity may also affect the maturation process in the worms since no mature individuals were found over 9‰ salinity in the interstitial water of Fraser River Estuary, while immature worms were found up to 14‰ salinity.

Based on these data, it was decided to use *Limnodrilus hoffmesiteri* as the test species in chronic bioassays of estuarine sediments from Santander Bay. Individuals were used after an acclimation period in the laboratory at salinity levels that allowed both survival and reproduction. Survival, burrowing activity, growth and reproduction were evaluated as endpoints for sediment toxicity.

## Materials and methods

### *Study sites and sampling*

Figure 1 shows the site locations in Santander Bay (north of Spain). The Bay is an important tourist area, but also has industrial development (mainly metallurgical and chemical) along the shoreline, as well as a harbour and urban centre (250,000 inhabitants). Sites were selected in each of the main estuaries that form the Bay which is at the confluence of several rivers. Sampling was done during low tide in May (2000) and salinity values were measured in the overlying water at each site. As the sampling was conducted at low tide, salinity of the overlying water was substantially influenced by freshwater.

Invertebrate samples were taken from an area of 20×20 cm$^2$, to 10 cm depth using a spade, and washed through 500-$\mu$m mesh. Samples were then fixed in formalin in the field. Sediment samples for bioassays were also taken to a 10 cm depth, from a separate sample, with a pre-washed plastic spade. Two litres of sediment was wet-sieved in the field through 250-$\mu$m mesh, to eliminate the indigenous fauna and macroalgae. The sediment sample was transported in polypropylene bottles to the laboratory, within 12 h and kept in the dark, on ice.

### *The acclimation period in the laboratory*

The culture used in the sediment bioassays was sampled from a population in the Mundaka Estuary (River Oka, Vizcaya, Spain). The culture was started with animals taken from one sampling occasion. Worms were placed in unpolluted sediment from the Butrón Estuary (Vizcaya, Spain), which had been sieved through 250-$\mu$m mesh to eliminate coarse particles and indigenous fauna. Physico-chemical characteristics of water of the Butrón Estuary at the sampling sites are the

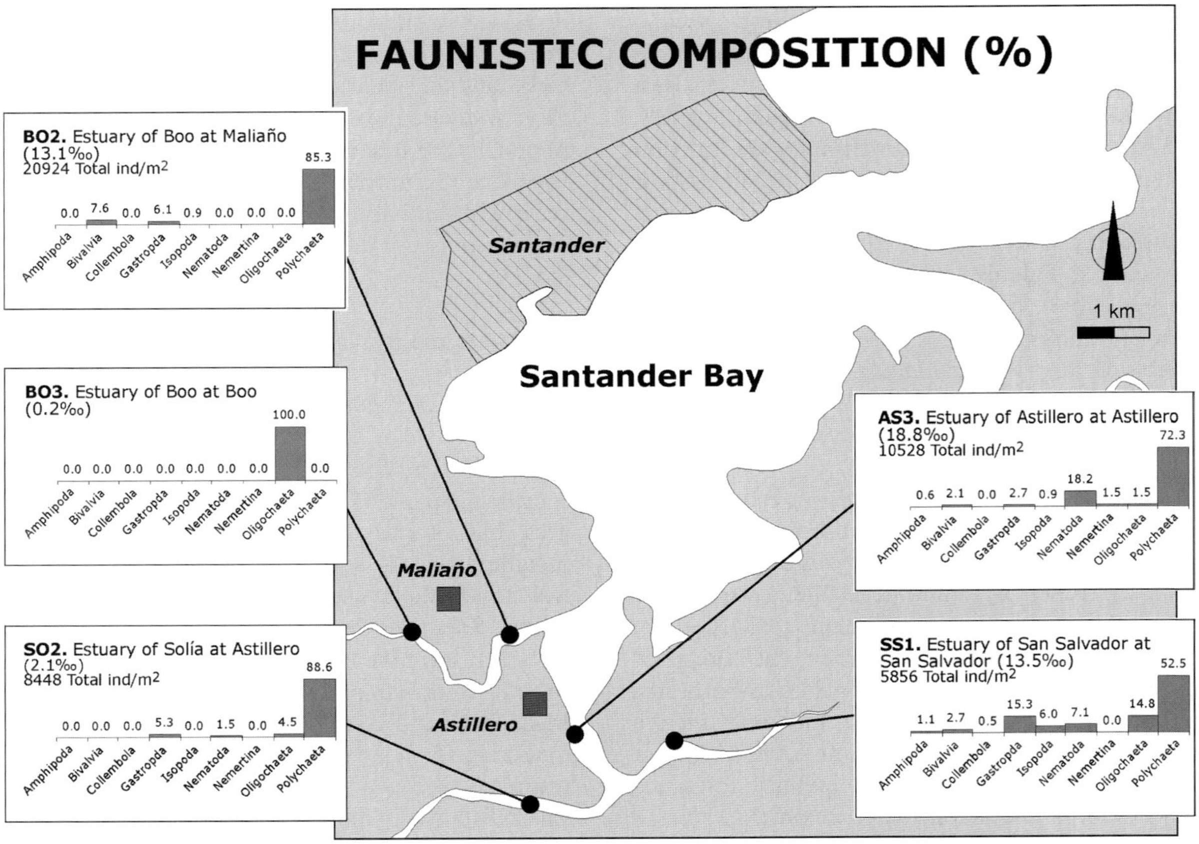


*Figure 1.* Location of study sites in the upper estuaries of Santander Bay (Northern Spain) and general macroinvertebrate composition and density of the faunistic communities. Into brackets it is indicated the water salinity at low tide.

following: DO = 7.0 mg $l^{-1}$, pH = 8.08, conductivity = 578 $\mu$S $cm^{-1}$, alkalinity = 74 mg $CO_3Ca$ $l^{-1}$, salinity = 0‰, $PO_4^{-3}$ = 0.78 mg $l^{-1}$, $NO_3^-$ = 4.84 mg $l^{-1}$, $NO_2^-$ = 1.0 mg $l^{-1}$. Sediment characteristics at this site are: pH = 7.27, N–$NH_3$ (porewater) = 1.36 mg $l^{-1}$, LOI (fraction < 63 $\mu$m) = 3.37%, TOC (fraction < 63 $\mu$m) = 1.96%, metals (in $\mu$g $g^{-1}$ dw, data provided by the Basque Government): Cr = 35.0, Cd = 0.05, Cu = 7.6, Fe = 14.250, Mn = 83.0, Ni = 14.0, Pb = 53.0, Zn = 62.0, Hg = 0.57, As = 3.6, organic compounds (in $\mu$g $kg^{-1}$ dw, data provided by the Basque Government): HCH = 0.20, DDT = 2.31, $\sum$9PAHs = 21.80, $\sum$9PCBs = 10.49.

In the acclimation vessels, a 3-cm layer of this sediment was added with 2–3 cm of overlying water. Different salinities were established by mixing dechlorinated tap water with sea water, avoiding daily increments of more than 3‰, to achieve the desired salinity. Based on literature data, we decided to maintain the salinity below 9‰ to avoid potential negative effects of salinity on the reproduction of the worms. Salinity in the cultures was maintained at 7–8‰ and normal behaviour, as well as cocoon production and growth of juveniles, was observed. Worms used in the bioassays were acclimated to this salinity for 2 to 4 weeks before the experiments were performed.

### *Sediment pretreatment*

Before testing, sediments were conditioned to salinity levels and ammonia concentrations appropriate to survival and reproduction of *L. hoffmeisteri*.

### *Salinity*

The salinity in the interstitial water of sediments was dependent on both site location in the bay and daily tide cycles at the time of sampling. High test-water salinity may result in false positives in both mortality and reproductive impairment of worms exposed to estuarine sediments. To avoid this, all sediments were mixed with either dechlorinated tap water or sea water (depending on the required salinity values) until the desired salinity was adjusted up or down to 7–8‰. Sediment was placed in a shallow tray (25×40 cm) with water and it was mixed by hand several times per day. Usually, sandy sediments required less time than silt–clay rich sediments. All the water used in this process was retained and used as overlying water in the bioassay, thus reducing losses of chemical compounds.

*Unionised ammonia*
In organically rich estuarine sediments, high levels of unionised ammonia in the test-vessel can also cause mortality or reproductive impairment (Ankley et al., 1990; Borgmann, 1994), confounding the results of the bioassay. Ammonia concentration of the porewater was reduced by strong aeration of the overlying water while sediment salinity was being adjusted. High levels of ammonia in the study sediment porewater required, in some cases, up to 4 days aeration to be reduced to below the 2 mg $l^{-1}$ target. Ammonia was measured in porewater using a Hach model DR/200 spectrophotometer (Nessler method). Porewater was extracted by centrifugation of sediment at 3500 rpm, 4 °C, during 30 min, followed by filtration of the overlying water through a 0.45-$\mu$m mesh filter. The proportion of sediment to water in the test-unit was 1:3 as an additional measure to minimise effects due to ammonia concentration.

*The influence of particle-size distribution on burrowing behaviour*

The influence of sediment particle-size characteristics on the worm burrowing behaviour was measured as the length of galleries in a fixed area of the beaker surface. This experiment was conducted with two sediments of different particle-size characteristics, from the Butrón Estuary. Salinity was adjusted by mixing dechlorinated tap water and sea water as described above and sediment was distributed to three replicates for each sediment type under the same conditions as the bioassay (see below). Three worms were added to each replicate and left in the dark, at 22 ± 1 °C in an incubator, with aeration, for 2 weeks. Photographs were taken after 1 and 2 weeks. Mean lengths of the galleries were compared using a Student's *t*-test.

*Bioassay conditions*

Once the sediment had been treated, it was distributed in 250-ml volume test-vessels (50-ml sediment and 150-ml overlying water) and food supplement (40-mg Tetramin per vessel) was added. We used glass vessels, pre-washed in acidified water and rinsed in acetone. Test-vessels filled with sediment and water were left in the incubator at 22 ± 1 °C, in the dark, with slight aeration. After settling overnight, four sexually mature worms were added to each vessel. Five replicates were used in control and test-sediments with a two-week exposure. Every few days, evaporative losses of water were replaced by adding dechlorinated tap water. Dissolved oxygen (DO) and salinity of the overlying water were measured once per week in 2 replicates, selected randomly among the test-vessels. Two additional vessels (without animals) per treatment were added for measuring DO, pH, ammonia, nitrites and nitrates, and phosphates in the overlying water, and ammonia in porewater) at the beginning and at the end of the bioassay.

At the end of the 2-week period, the following endpoints were measured: percent survival, the percentage of individuals that had resorbed the clitellum, adult final biomass, number of cocoons and the length of the galleries. On completion of the bioassay, a digital photograph of the galleries contained in a fixed area was taken through the glass surface of each test-vessel. The gallery length was measured from digitised images. The sediment was then washed through a 500 and 250-$\mu$m mesh to recover the worms and cocoons. Sieving was done with water from the pre-treatment sediment, that had been kept at 4 °C, in the dark. Sieved sediment was examined under the stereoscope microscope to count the cocoons, these are not easily recognised by eye due to the mineral particles that adhere to the cocoon. Adult worms from each beaker were separated in Petri dishes for 3 h to allow the gut to empty, and faeces were regularly removed with a pipette. Worms were weighed in washed pre-weighed glass microfiber Whatman filters (2.5 cm ø) after being dried at 80 °C in the oven, using a microbalance Sartorius Micro M3P (precision 1 $\mu$g).

Several computer programs (Flash®, Adobe-Photoshop® and Canvas®) were used to acquire and analyse the images. Statistical analyses were performed with SPSS 10.0 for MacIntosh (2000). Percentages were arcsine transformed and other variables transformed when necessary to normalise the data and achieve homogeneity of the variances. Shapiro-Wilk was used as a normality test and the *F* test and Levene's statistic to test variances prior to use either the Student's *t*-test or ANOVA, respectively. When data did not meet normality or variances were not homogeneous after transformation, the non-parametric Mann–Whitney *U*-test was used to compare data from each site to the control.

To integrate test endpoints and rank the relative toxicity at the sites, as well as examine the interactions among the bioassay endpoints, we used hybrid multidimensional scaling (Belbin, 1991). This is an ordination method that makes no assumptions about the distribution of the data. However, because the data are measured in different units they were range standardised. Euclidean distance was used as the distance measure. To examine the contribution of the various test endpoints to the bioassay ordination and to examine the relationship between sediment contaminants and chemistry with the bioassay ordination we used principal axis correlation, a method analogous to linear multiple regression. This is a method that allows either the original variables or another set of variables (in this case the chemical data) to be placed in ordination space. It also provides a correlation coefficient of the original variables with the ordination matrix. The significance of the correlation was estimated using Monte Carlo simulation (100 random arrangements of the matrix). All multivariate analyses were performed using PATN software (Belbin, 1993).

## Results

Examination of the benthic community composition revealed that aquatic oligochaetes were present in all the study sites, except at BO2. Site BO3 showed very low diversity and with the exception of a few gammarids, it had nearly 100% oligochaetes (*Lumbricillus* sp, *Paranais litoralis* and *Monopylephorus rubroniveus*). In the remaining sites, oligochaetes comprised less than 15% of the total fauna. Figure 1 shows the general composi-

*Table 2.* Physico-chemical characteristics of the study sites the day 1/day 14 of the bioassay, and granulometric composition of the sediments, expressed as percentage in % dry weight of medium and fine sand (fraction of 63–250 $\mu$m particle diameter) and silt/clay (below 63 $\mu$m diameter) of the study sediments and control (the Butrón Estuary)

| | Control | AS3 | BO2 | BO3 | SO2 | SS1 |
|---|---|---|---|---|---|---|
| **Overlying water** | | | | | | |
| $O_2$ [%] | 84/88 | 92/91 | 84/85 | 86/85 | 86/84 | 91/88 |
| $O_2$ [mg $l^{-1}$] | 7.3/7.5 | 8.1/8.2 | 7.2/7.4 | 7.5/7.4 | 7.3/7.4 | 7.9/7.8 |
| pH | 7.69/8.19 | 7.73/8.21 | 8.14/8.0 | 8.06/8.28 | 8.07/8.34 | 7.19/8.30 |
| $NO_2^-$ [mg $l^{-1}$] | < DL | < DL | < DL | < DL | < DL | < DL |
| $NO_3^-$ [mg $l^{-1}$] | 6.16/2.2 | 5.28/ < DL | 6.16/6.16 | 3.52/3.08 | 8.36/1.32 | 4.84/0.82 |
| $PO_4^{3-}$ [mg $l^{-1}$] | 1.24/0.69 | 0.75/0.39 | 1.31/– | 0.99/1.27 | 1.77/0.34 | 1.8/0.22 |
| Salinity [‰] | 6.8/8.0 | 6.2/8.0 | 6.3/7.8 | 7.1/8.0 | 6.2/7.6 | 6.8/7.8 |
| **Sediment** | | | | | | |
| pH | 7.15/7.64 | 7.57/7.86 | 7.51/8.15 | 7.78/7.96 | 7.17/7.6 | 7.32/7.97 |
| N–$NH_3$ (porewater) [mg $l^{-1}$] | 0.75/3.77 | –/0.20 | 0.56/7.84 | 1.11/0.70 | 1.09/2.61 | 0.50/1.86 |
| Sand (63–250 $\mu$m) [%] | 58.7 | 60.1 | 65.2 | 67.2 | 65.4 | 82.9 |
| Silt/clay (< 63 mm) [%] | 41.3 | 39.9 | 34.8 | 32.8 | 34.6 | 17.1 |

DL, Detection limit.

tion of the macroinvertebrate fauna at the study sites.

Throughout the bioassay, dissolved oxygen in the overlying water of the test-vessels was always >74% saturation (>6.4 mg $l^{-1}$) and salinity fluctuated between 6.2 and 8.0‰. Although the level of ammonia in the porewater was relatively low on day 1 of the bioassay (below 1.5 mg $l^{-1}$ N–$NH_3$), it increased by the end of the bioassay, mainly at BO2 (7.8 mg $l^{-1}$ N–$NH_3$) (Table 2). The unionised ammonia in the sediment porewater could act as a confounding factor in the bioassay site at BO2, although sediment pH was never over 9.

Table 3 shows the metals and PAH concentrations in the study sediments. These data show the sediments to be of very different quality based on chemical concentration. Previous authors (Irabien et al., 2001; Viguri et al., 2002) have classified these sediments to include those as Heavily-Polluted by metals and PAHs (BO3 and BO2) and Non-Polluted by PAHs but Moderately-Polluted by metals (AS3, SO2 and SS1). Sediments from the Boo Estuary (BO2 and BO3) have high concentrations of Zn, Pb, Cu and Mn and of PAHs, which have values of 1033 $\mu$g $g^{-1}$ TOC and 8404 $\mu$g $g^{-1}$ TOC, respectively. The metal and PAH values reflect the severe environmental deterioration of this estuary from industrial activity (ferroaloys and foundries) and uncontrolled discharge of industrial effluents along the river in the Boo Estuary. Samples from Sites SS1, AS3 and SO2 showed substantially elevated levels of Fe, which are more likely related to the historical input of large amounts of solid wastes derived from local mining activities. Sediment concentrations of Cr, Ni and Zn were moderate, with locally elevated levels in the Solia Estuary (SO2), probably related to the discharge of untreated effluents from local plating industries located upstream. These sediments also had lower concentrations of PAHs (from 114 to 180 $\mu$g $g^{-1}$ TOC) compared to BO2 and BO3.

*Table 3.* Metals, PAHs and %TOC (whole sediment) in the upper layer of sediments of Santander Bay at study sites measured through X-Ray Fluorescence Spectrometry and High Performance Liquid Chromatropgraphy (HPLC), respectively (from Irabien et al., 2001; Viguri et al., 2002)

| | AS3 | BO2 | BO3 | SO2 | SS1 |
|---|---|---|---|---|---|
| (% dw) | | | | | |
| $SiO_2$ | 45.75 | 40.57 | 29.43 | 37.16 | 38.52 |
| $Al_2O_3$ | 10.47 | 12.36 | 7.57 | 14.32 | 12.62 |
| $Fe_2O_3$ | 12.08 | 9.75 | 4.44 | 22.16 | 18.95 |
| MnO | 0.41 | 1.11 | 8.61 | 0.58 | 0.69 |
| MgO | 2.47 | 3.14 | 2.17 | 2.08 | 2.38 |
| CaO | 5.41 | 4.5 | 11.29 | 1.78 | 1.68 |
| $Na_2O$ | 3.15 | 2.03 | 2.14 | 1.56 | 2.93 |
| $K_2O$ | 1.93 | 2.02 | 1.37 | 2.09 | 1.94 |
| $TiO_2$ | 0.54 | 0.56 | 0.36 | 0.62 | 0.63 |
| $P_2O_5$ | 0.28 | 0.42 | 0.63 | 0.47 | 0.34 |
| $\mu$g $g^{-1}$ dw | | | | | |
| Co | 14 | <dl | 9 | 3 | <dl |
| Cr | 271 | 210 | 74 | 1035 | 270 |
| Cu | 82 | 119 | 186 | 111 | 72 |
| Ni | 61 | 43 | 40 | 203 | 70 |
| Pb | 131 | 345 | 1257 | 229 | 149 |
| V | 114 | 141 | 73 | 162 | 161 |
| Zn | 831 | 1315 | 3554 | 2295 | 900 |
| Σ16 PAHs ($\mu$g $g^{-1}$ TOC) | 180.0 | 1033.3 | 8404.1 | 130.0 | 113.6 |
| % TOC | 1.0 | 2.4 | 4.1 | 2.0 | 2.2 |

Lethal and sublethal responses in *Limnodrilus hoffmeisteri*, from exposure to chemically contaminated sediment from Santander Bay, are summarised in Table 4. At all the study sites, worms showed significantly higher mortality than the control (Mann–Whitney *U*-test, $p < 0.05$). The sediments from SO2 and BO3 had the highest mortality where 60 and 65%, respectively, of the worms died. Mean mortality in the controls was low (5%). In two replicates from Site AS3, one nereid was found with only one and no test worms, respectively, and predation was suspected.

In some of the sites, clitellum resorption was recorded at the end of the test period. In sites from the Boo Estuary (BO2 and BO3) the percentages were very high (see Table 4). It should be noted that reproductive impairment may also be a result of other factors, as at Sites SO2 and AS3, where the clitellum was observed and there was complete reproductive failure.

Cocoon production in the control batches was low (mean ± sd = 2.40 ± 1.52 cocoons per test-vessel). Among the study sites, cocoons were only found at SS1 (mean ± sd = 0.80 ± 0.45 cocoons per test-vessel), but the number of cocoons was significantly lower than in the control (*t*-test, $p = 0.028$).

The adult final biomass in the control sediment was the highest (mean ± sd = 1.271 ± 0.467 mg dw). However, there were no significant differences in the final biomass of the adult worms between the study sites and the control (ANOVA, $p = 0.340$).

Finally, the behaviour of the worms in the bioassays appeared to be normal at all the sites. The gallery length, measured after 2 weeks, using digitised images in the computer by Canvas®, was significantly lower than the control at only Site SO2 (*t*-test, $p = 0.002$) (Table 4). However, it is important to note that particle-size distribution in sediments can influence the burrowing behaviour of *L. hofmeisteri*, as demonstrated in a separate experiment with two different sediments from the Butrón Estuary. The length of the galleries built by 3 worms in a sandy sediment was significantly higher than the length in a silt-clay sediment, after 1 week (*t*-test, $p = 0.003$) and 2 weeks (*t*-test, $p = 0.014$).

The ordination (stress 0.092) of the bioassay endpoints (mean values) shows a gradient from the Control site to BO3 and SO2, with Sites SS1, BO2 and AS2 being intermediate in position (Fig. 2). The strongest gradient (axis 2) is driven largely by mortality, and the second gradient is one where clitellum resorption is negatively associated with biomass, i.e. resorption of reproductive organs results in reduced biomass. However, these data suggest (Fig. 2a) that there is no association between resorption and mortality. Site BO3 had the highest mortality and resorption (plus loss of biomass). Site SO2 although having high mortality showed no evidence of resorption. Burrowing and cocoon production did not contribute significantly ($p > 0.05$) to the observed pattern (Table 5). Only Cu and Zn (Fig. 2b) were significantly ($p < 0.05$)

*Table 4.* Mortality (%), clitelum resorption (% individuals), adult final biomass (mg dry weight), number of cocoons per test-unit, and length of the galleries (cm) measured on a fixed surface area of the test-vessel, at the end of the 2-week sediment bioassay with *Limnodrilus hoffmeisteri*

| Site | Mortality (%) $\bar{x} \pm$ sd | Clitellum resorption (%) | Final biomass (mg dw) $\bar{x} \pm$ sd | No. cocoons $\bar{x} \pm$ sd | Length of the galleries (cm) $\bar{x} \pm$ sd |
|---|---|---|---|---|---|
| Control | 5.0 ± 11.2 | 10.5 | 1.271 ± 0.467 | 2.40 ± 1.52 | 7.26 ± 0.20 |
| AS3$^{(1)}$ | 41.6 ± 14.4* | 0 | 1.160 ± 0.170 | 0.00 | 8.11 ± 2.07 |
| BO2 | 35.0 ± 22.4* | 23.1 | 0.987 ± 0.421 | 0.00 | 7.53 ± 2.76 |
| BO3 | 65.0 ± 28.5* | 57.1 | 0.681 ± 0.489 | 0.00 | 6.85 ± 1.72 |
| SO2 | 60.0 ± 28.5* | 0 | 0.993 ± 0.403 | 0.00 | 5.30 ± 0.40* |
| SS1 | 40.0 ± 28.5* | 25.0 | 0.900 ± 0.184 | 0.80 ± 0.45* | 8.65 ± 2.00 |

*Differences with control are significant at $p < 0.05$. For information on statistical tests, see text. ($n = 5$ for all endpoints except length of galleries, where $n = 3$).

$^{(1)}$In 2 replicates one nereid was found and only one and zero oligochaetes, respectively. Since it was suspected death by predation, these replicates were not considered for the calculation of endpoints.

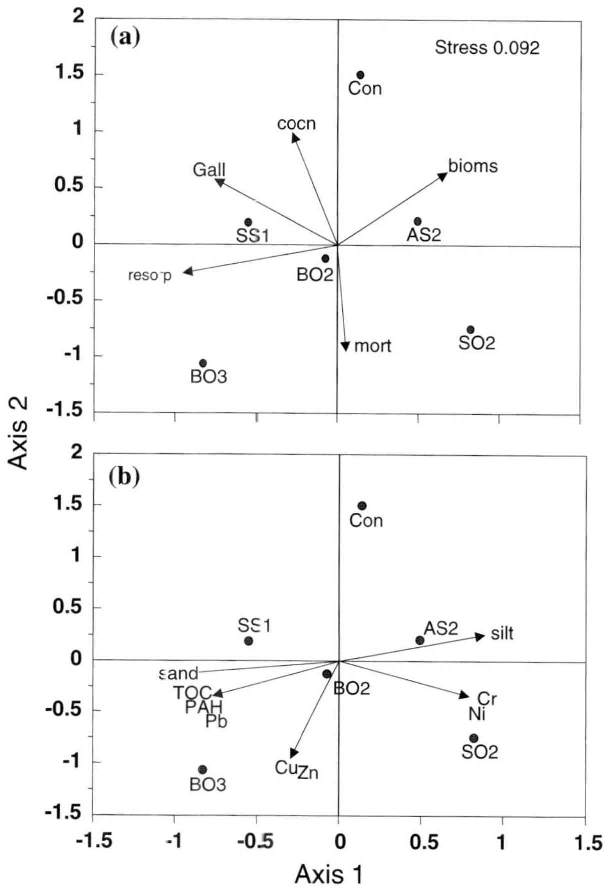


*Figure 2.* Multidimensional scaling ordination plots (stress = 0.092) for five study sites and a control together with the endpoints produced by *Limnodrilus hoffmeisteri* sediment bioassay (a) and environmental variables (b). (Abbreviations (a) Con = control; gall = length of galleries, cocn = number of cocoons, mort = mortality, bioms = worm final biomass, resorp = clitellum resorption, (b) TOC = total organic carbon, PAH = total PAHs, Cu = copper, Pb = lead, Cr = chromium, Ni = nickel, Zn = zinc).

associated with the ordination response, although most of the other contaminants had high *r* values (Table 5). It is noteworthy that Cu and Zn are strongly associated with each other (Fig. 2b) and associated with the mortality vector (Fig. 2a), whereas TOC, PAH and Pb are associated with sand, and Cr and Ni with silt (Fig. 2b). PAH and Pb may be correlated with the resorption of reproductive tissue, as they are associated with the resorption vector (Fig. 2a).

## Discussion

The importance of aquatic oligochaetes in estuarine ecosystems has been documented by numerous authors (Birtwell & Arthur, 1980; Diaz, 1980, 1989; McLusky et al., 1993; Seys et al., 1999). In Santander Bay the utility of the group has been demonstrated in the analysis of invertebrate communities. The need for sediment bioassays in a comprehensive, integrated approach to estuarine pollution has been identified by several authors (e.g. Gandrass et al., 2000; Den Besten et al., 2003). Crustaceans and/or molluscs are the most frequently used estuarine species in sediment bioassays (e.g. MacPherson & Chapman, 2000; Byrne & O'Halloran, 2001; Marsden &Wong, 2001). However, in the upper portions of estuaries, areas that have often suffered industrial and urban pollution impacts, the invertebrate community is frequently dominated by tidal freshwater oligochaete taxa (e.g. McLusky et al., 1980, 1993; Henderson, 1983), and some of these species, such as *L. hoffmeisteri*,

*Table 5.* Correlation coefficients between *Limnodrilus* bioassay endpoints and environmental variables with Ordination axes [In bold appear the variables with significant correlation]

| Variable | r |
|---|---|
| Biomass | **0.9910** |
| Mortality | **0.9731** |
| Resorption | **0.9634** |
| Cocoon | 0.8868 |
| Galleries | 0.6979 |
| Cu | **0.9675** |
| Zn | **0.9373** |
| Cr | 0.8700 |
| Pb | 0.8687 |
| TOC | 0.8653 |
| Ni | 0.8561 |
| PAH | 0.8297 |
| sand | 0.5678 |
| silt | 0.5678 |

may have potential as test-organisms for chronic sediment bioassays.

This study shows a sediment bioassay with *L. hoffmeisteri* to demonstrate that levels of contaminants were bioavailable and had an effect on the biota. The sediment at the Site BO3 situated at the head of the Boo Estuary is the most toxic since it showed highest percentages of mortality and clitellum resorption, no cocoon production and low final biomass (approximately half that of the control). The Site SO2, at the Solía Estuary, also showed high mortality and no production of cocoons, although adults did not resorb the clitellum. Sites at the estuaries of Astillero (AS3), Boo (BO2) and San Salvador (SS1) showed similar mortality (40–60%), although some reproduction occurred at SS1, which is the least toxic site.

Effects of pollutants on the burrowing activity of worms has been observed by other authors (White & Keilty, 1988; Keilty et al., 1988; Meller et al., 1998; Bettinetti & Provini, 2002). For instance, White & Keilty (1988) and Keilty et al. (1988) quantified the number of unburrowed (i.e. more than an estimated 75% of the body was above sediment surface) individuals. This is a subjective classification of burrowed vs unburrowed individuals because it is not always possible to determine what portion of the animal is below the sediment surface, and therefore it could result in an incorrect classification. Photographing galleries is a simple and quantitative method of describing burrowing activity. However, the results were not wholly satisfactory because of the influence of sediment particle-size composition on the results. A separate experiment with two sediments sampled from the Butrón Estuary (control estuary) showed significant differences in the length of galleries in sandy sediments compared with silt–clay rich sediments. It was surprising that the burrowing activity of the worms was only reduced significantly at Site SO2. Other more toxic sediments did not caused such behavioural response. The sediment at this site had a particle-size distribution comparable to AS3, BO2 and BO3, and as with other bioassay endpoints, when comparing results from different sediments, other variables may confound the value of the endpoint, such as differences in organic content. Thus, it appears that different toxicants induce two distinct responses in *Limnodrilus*, an effect on survival and an induction of clitellum resorption with loss of biomass, which are not necessarily associated.

These are only preliminary results. Difficulties in eliminating confounding factors such as ammonia from porewater can influence worm reproduction or survival in the test and compromise the results. This may have been the case at Site BO2 where the level of porewater ammonia increased during the bioassay and probably required water renewal during the test. The presence of ammonia can confound the interpretation of results in sediment bioassays and several techniques have been recommended to reduce its toxic effect in test organisms (Ankley et al., 1990). Experiments performed with oligochaete worms in aqueous tests showed high mortality at very low levels of ammonia. For instance, Schubauer-Berigan et al. (1995) studied the influence of pH on ammonia toxicity to the worm *Lumbriculus variegatus* in flow-through aqueous tests of 10-d duration. At a pH of 7.8, similar to our study conditions, in porewater, these authors calculated $NH_3$ -$LC_{50}$ of only 0.768 mg $l^{-1}$. However, Monda et al., (1995) reviewed the 96-h $LC_{50}$ values for unionised ammonia of macroinvertebrate species that ranged from 1.0 to 10.0 mg $l^{-1}$. These authors stressed the importance of source, culture and handling of organisms, as well as the presence of substrate in the tests, when interpreting differences in ammonia toxicity. We should also consider the possibility of additive, synergistic or antagonistic interactions of pollutants with ammonia that can complicate the interpretation of the results (Ankley et al., 1990).

*Nereis diversicolor* can consume oligochaetes (Giere & Pfannkuche, 1982) and sediment should be carefully sieved to avoid predation by indigenous fauna on the test organisms. Replicates where this may have occurred were not used in the data analysis. Cocoon production in the control batches was lower than expected. Lotufo & Fleeger (1996) found a production of 19.6–23.6 offspring (unhatched cocoons plus juveniles) per worm, in 4 weeks, in control sediment. Lower production in our study may be a result of a shorter duration of the bioassay; however, we cannot discount the presence of ammonia in the porewater which may have induced reproductive impairment. Borgmann (1994) found reproductive impairment in *Hyalella azteca*, in a 10-week duration test, at a concentration of 0.32 mM $NH_3$ (=4.46 mg/l N–$NH_3$, pH $= 8.43 \pm 0.34$). These data suggest that at Site BO2 ammonia may have caused reproductive impairment, regardless of possible sensitivity differences between the taxa. The pH also influences ammonia toxicity as in aqueous solution ammonium ions, ammonia and hydroxide ions are in equilibrium, and the toxicity is dependent on the amount of unionised ammonia present and it is therefore dependent on pH (Hellawell, 1986). High pH values in porewater at BO2, over pH 8 at the end of the bioassay, could have increased ammonia toxicity. At the remaining sites, it is assumed that mortality and reproductive impairment is a result of sediment toxicity associated with pollutants, since neither high pH nor high ammonia concentration were observed.

In conclusion, although these results are preliminary, they do support the use of *Limnodrilus hoffmesiteri* in ecotoxicological assessment of sediment as in sediment bioassays, worm survival, clitellum resorption, growth, reproduction and burrowing behaviour can be measured. Acceptable negative control criteria of less than 10% mortality can be achieved. Despite the fact that *Limnodrilus hoffmeisteri* is not a true estuarine species, but an euryhaline freshwater species it is a suitable species for sediment bioassays, particularly for oligo- and ß mesohaline sections of the estuaries, or freshwater-dominated estuaries. The species can be cultured satisfactorily under controlled laboratory conditions and can be easily acclimated to a constant salinity of about 7–8‰. Confounding factors of salinity and unionised ammonia need to be considered. For example, it may be necessary to adjust salinity of all sediments to a value that allows survival and reproduction of the test species. Ammonia in the sediment porewater must be controlled and kept low. Aeration followed by removal of overlying water may be necessary. The relative sensitivity of the species to ammonia concentration should be established in sediment toxicity tests to interpret results of bioassays, when this toxicant is present.

It would be desirable to have a complete suite of oligochaete test species to perform bioassays also for α-mesohaline and polyhaline sites of estuaries. The possibility of using other oligochaete species such as *Heterochaeta costata* and *Monopylephorus rubroniveus* in estuarine sediment bioassays should be explored.

## Acknowledgements

This work has been funded by a Project of the Marcelino Botín Foundation I/GC 00154.310-7197, Eduardo Fernández Miguel Section, under the 3rd Proposal of Environment and Support Development in Cantabria (The Coastal Area), in the year 2000. We are indebted to Dr M. Jesus Irabien and Dr Iñaki Yusta for providing data on sediment chemistry of Santander Bay. We gratefully acknowledge the Water Division of the Basque Government (Dirección de Aguas, Red de Vigilancia de la Calidad de Aguas litorales) for providing information on the characteristics of the sediment in the estuary of the River Butrón (year 1999).

## References

Ankley, G. T., A. Katko & J. Arthur, 1990. Identification of ammonia as an important sediment-associated toxicant in the Lower Fox River and Green Bay, Wisconsin. Environmental Toxicology and Chemistry 9: 312–322.

Belbin, L., 1991. Semi-strong hybrid scaling, a new ordination algorithm. Journal of Vegetation Science 2: 491–496.

Belbin, L., 1993. PATN, Pattern Analysis Package. Division of Wildlife and Ecology . CSIRO, Canberra, Australia.

Bettinetti, R. & A. Provini, 2002. Toxicity of 4-nonylphenol to *Tubifex tubifex* and *Chironomus riparius* in 28-day whole sediment tests. Ecotoxicology and Environmental Safety 53: 113–121.

Birtwell, J. K. & D. R. Arthur, 1980. The ecology of tubificids in the Thames Estuary with particular reference to *Tubifex costatus* (Claparède). In Brinkhurst, R. O. & D. G. Cook (eds), Aquatic Oligochaete Biology. Plenum Press, New York: 331–381.

Borgmann, U., 1994. Chronic toxicity of ammonia to the amphipod *Hyalella azteca*; importance of ammonium ion and water hardness. Environmental Pollution 86: 329–335.

Brinkhurst, R. O. & M. L. Simmons, 1968. The aquatic Oligochaeta of the San Francisco Bay system. California Fish and Game 54: 180–194.

Brinkhurst, R. O., P. M. Chapman & M. A. Farrell, 1983. A comparative study of respiration rates of some aquatic oligochaetes in relation to sublethal stress. International Revue der gesamten Hydrobiologie 68: 683–699.

Byrne, P. A. & J. O'Halloran, 2001. The role of bivalve molluscs as tools in estuarine sediment toxicity testing: a review. Hydrobiologia 465: 209–217.

Casellato, S., 1994. Oligochaete fauna of estuarine areas and lagoons on the northern Adriatic coast (Italy). Bolletino di Zoologia 61: 261–269.

Casellato, S. & R. Poja, 1984. Ecology of tubificids in the lower reaches of the rivers Adige and Brenta (N. E. Italy). Bolletino di Zoologia 51: 339–352.

Chapman, P., 1987. Oligochaete respiration as a measure of sediment toxicity in Puget Sound (Washington). In Brinkhurst, R. O. & R. J. Diaz (eds), Proc. 3rd International Symposium on Aquatic Oligochaetes. Hydrobiologia 155: 249–258.

Chapman, P., 2001. Utility and relevance of Aquatic Oligochaetes in ecological risk assessment. In Rodriguez, P. & P. F. M. Verdonschot (eds), Proc. 8th International Symposium on Aquatic Oligochaeta. Hydrobiologia 463: 149–169.

Chapman, P. M. & R. O. Brinkhurst, 1980. Salinity tolerance in some selected Aquatic Oligochaetes. International Revue der gesamten Hydrobiologie 65: 499–505.

Chapman, P. M. & R. O. Brinkhurst, 1981. Seasonal changes in interstitial salinities and seasonal movements of subtidal benthic invertebrates in the Fraser River Estuary, B.C. Estuarine, Coastal and Shelf Science 12: 49–66.

Chapman, P. M., M. A. Farrell & R. O. Brinkhurst, 1982a. Relative tolerances of selected aquatic oligochaetes to individual pollutants and environmental factors. Aquatic Toxicology 2: 47–67.

Chapman, P. M., M. A. Farrell & R. O. Brinkhurst, 1982b. Relative tolerances of selected aquatic oligochaetes to combination of pollutants and environmental factors. Aquatic Toxicology 2: 69–78.

Connell, D. W., M. Bowman & D. W. Hawker, 1988. Bioconcentration of chlorinated hydrocarbons from sediment by oligochaetes. Ecotoxicology and Environmental Safety 16: 293–302.

Diaz, R. J., 1980. Ecology of freshwater and estuarine tubificidae. In Brinkhurst, R. O. & D. G. Cook (eds), Aquatic Oligochaete Biology. Plenum Press, New York: 319–330.

Diaz, R. J., 1989. Pollution and tidal benthic communities of the James River Estuary, Virginia. In Kaster J. L. (ed.), Proc. 4th International Symposium on Aquatic Oligochaete Biology. Hydrobiologia 180: 195–212.

Den Besten, P., E. De Deckere, M. Babut, A. del Power Valls, C. Zago & A. Oen, 2003. Biological effects-based sediment quality in ecological risk assessment for European waters. Journal of Soils and Sediments 3: 144–162.

Egeler, P., J. Römbke, M. Meller, Th. Knacker, C. Franke, G. Studinger & R. Nagel, 1997. Bioaccumulation of Lindane and Hexachlorobenzene by tubificid sludgeworms (Oligochaeta) under standardized laboratory conditions. Chemosphere 35: 835–852.

Gandrass, J., W. Salomons, U. Foerstner, (eds), 2000. River sediments and related dredged material in Europe. Scientific background from the viewpoints of chemistry, ecotoxicology and regulations. Dredged Material in the Port of Rotterdam. Part II. GKSS Research Centre, Geesthacht, Germany.

Giere, O., 1980. The impact of crude oil and oil dispersants on the marine oligochaete *Marionina subterranea*. Cahiers de Biologie Marine 21: 51–60.

Giere, O. & D. Hauschidt, 1979. In Naylor, E. & R. G. Hartnoll, (eds), Cyclic phenomena in marine plants and animals. Proc. 13th European Marine Biology Symposium, Pergamon Press, Oxford: 113–122.

Giere, O. & O. Pfannkuche, 1982. Biology and ecology of marine oligochaeta, a review. In Barnes, M. (ed), Oceanography and Marine Biology, Vol. 20. Aberdeen University Press, Aberdeen: 173–309.

Hellawell, J. M., 1986. Biological Indicators of Freshwater Pollution and Environmental Management. Elsevier Applied Science, London.

Henderson, A. R., 1983. Long term monitoring of the macrobenthos of Upper Clycle Estuary. Water Science and Technology 16: 359–373.

Hunter, J., 1981. Survival strategies in tubificids in the Thames and other estuaries. In Jones, N. V. & Wolff W. J., (eds), Feeding and Survival Strategies of Estuarine Organisms. Plenum Publ. Corp., New York, Marine Sciences 15: 53–63.

Irabien, M. J., I. Yusta, A. Alvaro, J. Soto, J. Gómez, J. Viguri, J. Verde & A. Irabien, 2001. Cronología de la contaminación antropogénica por Fe, Zn, Pb y PAHs en sedimentos de la bahía de Santander. In 1er Foro Nacional sobre Gestión Integral de Zonas Costeras. 2–5 October, 2001. Santander: 140.

Keilty, T. J., D. S. White & P. F. Landrum, 1988. Short-term lethality and sediment avoidance assays with endrin-contaminated sediment and 2 oligochaetes from Lake Michigan. Archives of Environmental Contamination and Toxicology 17: 95–101.

Lotufo, G. R. & J. W. Fleeger, 1996. Toxicity of sediment associated Pyrene and Phenanthrene to *Limnodrilus hoffmeisteri* (Oligochaeta, Tubificidae). Environmental Toxicology and Chemistry 15: 1508–1516.

MacPherson, C. A. & P. Chapman, 2000. Copper effects on potential sediment test organisms: The importance of appropriate sensitivity. Marine Pollution Bulletin 40: 656–665.

McLusky, D. S., M. Teare & P. Phizacklea, 1980. Effects of domestic and industrial pollution on distribution and abundance of aquatic oligochaetes in the Forth Estuary. Helgoländer Meeresunterssuchungen 33: 384–392.

McLusky, D. S., S. C. Hull & M. Elliot, 1993. Variations in the intertidal and subtidal macrofauna and sediments along a salinity gradient in the upper Forth Estuary. Netherlands Journal of Aquatic Ecology 27: 101–109.

Marsden, I. D. & C. H. T. Wong, 2001. Effects of sediment copper on a tube-dwelling estuarine amphipod, *Paracorophium excavatum*. Marine and Freshwater Research 52: 1007–1014.

Meller, M., P. Egeler, J. Rombke, H. Schallnas, R. Nagel & B. Streit, 1998. Short term toxicity of lindane, hexachlorobenzene, and copper sulphate to tubificid sludgeworms (Oligochaeta) in artificial media. Ecotoxicology and Environmental Safety 39: 10–20.

Monda, D. P., D. L Galat, S. E. Finger & M. S. Kaiser, 1995. Acute toxicity of ammonia (NH3-N) in sewage effluent to *Chironomus riparius*: II. Using a generalized linear model. Archives of Environmental Contamination and Toxicology 28: 385–390.

Pfannkuche, O., 1981. Distribution, abundance and life cycles of aquatic Oligochaeta (Annelida) in a freshwater tidal flat of the Elbe Estuary. Archive für Hydrobiologie, Suppl., 43: 506–524.

Reible, D. D., V. Popov, K. T. Valsaraj, L. J. Thibodeaux, F. Lin, M. Dikshit, M. A. Tudaro & J. W. Fleeger, 1996. Contaminant fluxes from sediment due to tubificid oligochaete bioturbation. Water Research 30: 704–714.

Rodriguez, P., 1983. Contribution à la connaisasance de la faune d'oligochètes aquatiques du Pays Basque et zones limitrophes. Annales de Limnologie 19: 93–100.

Schubauer-Berigan, M. K., P. D. Monson, C. W. West & G. T. Ankley, 1995. Influence of pH on the toxicity of ammonia to *Chironomus tentans* and *Lumbriculus variegatus*. Environmental Toxicology and Chemistry 14: 713–717.

Seys, J., M. Vincx & P. Meire, 1999. Spatial distribution of oligochaetes (Clitellata) in the tidal freshwater and brackish parts of the Schelde Estuary (Belgium). In Healy, B., T. B. Reynoldson & K. A. Coates, (eds), Proc. 7th International Symposium on Aquatic Oligochaetes. Hydrobiologia 406: 119–132.

Viguri, J. R., J. Verde & J. A. Irabien, 2002. Environmental assessment of polycyclic aromatic hydrocarbons (PAHs) in surface sediments of the Santander Bay, Northern Spain. Chemosphere 48: 157–165.

White, D. S. & T. J. Keilty, 1988. Burrowing avoidance assays of contaminated Detroit River sediments, using the freshwater oligochaete *Stylodrilus heringianus* (Lumbriculidae). Archives of Environmental Contamination and Toxicology 17: 673–681.

Wiederholm, T., A. M. Wiederholm & G. Milbrink, 1987. Bulk sediment bioassays with five species of freshwater oligochaetes. Water Air Soil Pollution 36: 131–154.

Hydrobiologia (2006) 564:171–181
P.F.M. Verdonschot, H. Wang, A. Pinder & R. Nijboer (eds), Aquatic Oligochaete Biology IX
DOI 10.1007/s10750-005-1717-9



# Oligochaete assemblages in the hyporheic zone and coarse surface sediments: their importance for understanding of ecological functioning of watercourses

Michel Lafont[1,*] & Anne Vivier[2]
[1]*Cemagref, Biology Research Unit, 3 bis Quai Chauveau, C.P. 220, F-69336, Lyon, Cedex 09, France*
[2]*ENGEES, SHU Laboratory, 1 Quai Koch, 67000, Strasbourg, France*
*(*Author for correspondence: E-mail: lafont@lyon.cemagref.fr)*

*Key words:* running waters, coarse sediments, hyporheic zone, Oligochaeta

**Abstract**

The hyporheic zone and its interactions with coarse surface sediments is increasingly reported by aquatic ecologists because the water exchanges between surface and subsurface are important factors for the understanding of the ecosystem functioning. However, the hyproheic oligochaete assemblages have received less attention than other assemblages such as crustaceans. In addition, studies investigating the incidence of pollution in watercourses have mostly focused on the benthic zone and have neglected the hyporheic zone. Some examples are given from an unpolluted glacial river (Roseg), polluted plains rivers (Moselle, Rhône) and a protected wetland in an urbanized environment. The hyporheic zone kept the memory of past and present incidences of pollution, in particular when downwellings of polluted surface waters to the hyporheic zone predominated. The Active hydrologic Exchange Describers between surface and subsurface (AED oligochaete species) were the same in the glacial river Roseg, the rivers Rhône and Moselle and the urbanized wetland. The predominance of pollution-tolerant species like *Limnodrilus hoffmeisteri* was observed in polluted groundwater as well as in polluted surface coarse sediments. Moreover, the urbanized wetland exhibited a high species richness, suggesting that the hyporheic zone is a reservoir of species. The oligochaete communities enable biologists to simultaneously assess the pollution incidence, the permeability of coarse habitats, the water exchanges between surface and subsurface, and give an approximate measure of the metabolic activities in the sediments. Consequently, the simultaneous study of surface and hyporheic oligochaete assemblages is of great interest when considering the ecological functioning of watercourses and the incidence of pollution inputs.

## Introduction

This paper does not represent an exhaustive survey of research contributions and as such is very incomplete regarding the study of surface coarse and hyporheic sediments for the watercourse functioning evaluation. We tried to underline what should be reasonably attainable when dealing with the ecology of oligochaete assemblages living in coarse surface sediments and the hyproheic zone, and their importance for understanding of ecological functioning of watercourses.

The hyporheic zone may be defined as the saturated interstitial areas beneath the stream bed and into the stream banks that contain some proportion of channel water (Gibert et al., 1990; Vervier et al., 1992; White, 1993). Underestimated for a long time, its relations with surface waters and its importance for understanding the ecology of river corridors is now well recognized (Hynes, 1983; Danielopol, 1984, 1989; Gibert et al., 1990; Triska et al., 1990; Bretschko, 1991; Bencala, 1993; Valett et al., 1993; Brunke & Gonser, 1997; Boulton, 2000; Jones & Mulholland, 2000; Hancock, 2002; Malard et al.,

2002, 2003). Ward (1989) pointed out the four dimensional (length, width, depth, time) structure of lotic ecosystems. Fellows et al. (2001) demonstrated that 40–93% of the whole stream respiration occurred in the hyporheic zone. These authors also demonstrated that important ecological processes can occur in the shallow but hydrologically dynamic hyporheic zone.

The bottom of watercourses formed in coarse-grained sediments is related to the hyporheic system and is the prime habitat where water exchanges between surface and subsurface occur. Coarse sediments generally predominate in the majority of freshwater ecosystems in France, excluding the deep zones of lakes, reservoirs, some back waters and aquatic habitats in the wetlands, but including the littoral zone of many lakes, including Lake Geneva. For example, the deep zones of the river Rhône are composed of coarse sand and gravel at 8 m depth downstream of the urban area of Lyon, although the width of the river is about 200 m (Lafont et al., 1992). Even in the slow-flowing river Saône, at Site Auxonne (150 m width, current velocity less than 0.1 m $s^{-1}$), the fine-grained sediments are restricted to the bank areas where coarse sediments also predominate. The deep zone (−4 m) is composed of gravel and coarse sand (Lafont, 1989 and unpublished data).

The benthic habitats of running waters formed by coarse sediment and the hyporheic habitats display a high permeability. The efficiency of permeable sediments in the degradation of organic matter was for a long time recognized (Williams & Taylor, 1968; Solbe et al., 1974; Solbe, 1975; Bou, 1977) and used for purification processes in wastewater treatment plants (Bruce et al., 1983, cited by Datry et al., 2003). Their study in natural watercourses presents a challenging task for present and future researches. Research is now focusing on this domain and numerous recent papers have underlined the strong physicochemical and biological connections between surface and groundwater, and their importance for the ecological study of lotic ecosystems (Boulton, 2000; Hancock, 2002; Boulton et al., 2003). To date biological studies were mostly devoted to metabolic and microbial activities (see Jones & Mulholland, 2000; Griebler et al., 2001), or to crustacean assemblages (see Griebler et al., 2001). The oligochaete assemblages received much less attention and research on that subject progressed more slowly than the biogeochemical and hydrological studies (Stanley & Jones, 2000), with generally a disconnection between the consideration of surface and subsurface zones (Boulton, 2000).

## Interstitial oligochaete assemblages and watercourse functioning

### *Overview*

Oligochaete assemblages are well-known as common inhabitants of surface sediments and hyporheic zone, and they are also significant colonizers of groundwater, especially in deep karstic zones (Malard, 2001). The consideration of a batch of 93 species demonstrated that at least 62 species were collected both in surface and groundwater (Lafont, 1989). Moreover, the oligochaete assemblages may occur at all depths, from the surface to the deep hyporheos (Weigelhofer & Waringer, 2003). In addition, it is less mentioned that the surface coarse sediments shelter the richest assemblages of oligochaetes (Lafont, 1989). Oligochaetes firmly keep their historical status of "sludge worms", giving the impression that they are only an accessory fauna in coarse sediments, whilst the Arthropods are the main faunal elements. Moreover, the use of unsuitable filtration methods, for example 0.500 mm mesh-size which is commonly used in macrobenthos studies, leads to a significant underestimation of oligochaete densities (Nalepa & Robertson, 1981) and the loss of about 50% of the species (Lafont, 1989). The species richness of small oligochaetes is greater, and this is particularly true in the interstices of the surface coarse sediments and hyporheic system, which generally shelter the smallest species. For example, in the Family Tubificidae, *Branchiura sowerbyi*, *Tubifex tubifex* or *Limnodrilus* spp. are giant species compared to members of the Rhyacodrilinae. Even *Nais elinguis* may be a giant compared to *Rhyacodrilus falciformis*, small Phallodrilinae, or species of the genus *Cernosvitoviella*.

The hyporheic oligochaetes have been less studied than other assemblages (e.g. crustaceans), with the exception of pioneer contributions like those of Botea (1963), Ladle (1971), Ferrarese & Sambugar (1976), Bird (1982), Schwank (1982a, b), Dole (1983), Gaschignard (1984), Dumnicka (1986), Juget (1980, 1984, 1987), Juget & Dumnicka (1986), Strayer & Bannon-O'Donnell (1988).

The lack of studies involving both invertebrate communities and hydrologic measurements has been pointed out (Boulton, 2000), but this is not surprising because multidisciplinary researches are generally more difficult to organize. However, multidisciplinary studies put to the light more interesting findings than the sole consideration of biology or hydrology (Lafont et al., 1992; Valett et al., 1993).

*Examples from case studies*

*River Rhône at Grand Gravier site: definition of AED species*

A first example is given by the comparison between oligochaete species and hydrologic measurements at a suburban polluted site (the 'Grand Gravier') of the River Rhône downstream of the suburban area of Lyon (Lafont & Durbec, 1990). The hydrodynamic parameters were measured at a site about 500 m long. Within this site, two stations of 50 m each were selected and the filtration velocity of water through the sediments was assessed by piezometric measurements (2 transects per station). Station 1 (transects T1 and T2, Table 1) was characterized by a significant lower filtration velocity than Station 2 (T3 and T4). At both stations, oligochaeta were sampled in surface coarse sediments by means of a Surber net and a metallic corer driven into the substrate by a hammer (depth of the samples: 0–100 mm). A mesh-size of 0.160 mm was retained for the sediment filtration. The identification of oligochaetes was conducted to the level of species and informed us about several exciting features:

(1) although a polluted site, the species richness was relatively high (33 species);
(2) the biological quality of Station 2 (greater filtration velocity) was significantly better than that of Station 1, and this was related to a decrease of the percentages of pollution-tolerant Tubificidae;
(3) at Station 2, a significant increase in the percentages of *Gianius riparius* (*Phallodrilus riparius*), *Stylodrilus* spp., *Pristina* and *Pristinella* spp., *Marionina argentea*, *Spirosperma velutinus* and *Cernosvitoviella atrata* was observed. These species were already known as groundwater dwellers and their significant abundances in surface sediments related to upwellings of groundwater (Botea, 1963; Juget, 1980, 1984, 1987; Gashignard, 1984; Juget & Dumnicka, 1986; Lafont, 1989).

A second field investigation was performed in the same site at three different depths (−1.5, −4 and −8 metres, Table 1). Biological samples were taken using the same Surber type net operated by divers and the mean filtration velocities were available at these stations (Lafont et al., 1992). A significant increase in percentages of groundwater dwellers (*Stylodrilus* spp., *S. velutinus*, *Rhyacodrilus* sp., *Pristinella osborni*) was observed at 4 m depth, where the filtration velocities

*Table 1.* Structure of oligochaete assemblages at four transects at the Grand Gravier site (River Rhône)

| Transects | Station 1 | | Station 2 | | Deep zone | | |
|---|---|---|---|---|---|---|---|
| | T1 | T2 | T3 | T4 | −1.5 m | −4 m | −8 m |
| % Tubificidae (1) | 60 | 72 | 26.5 | 44.5 | 8 | 42.5 | 42 |
| % AED | 2.5 | 4 | 21.5 | 48 | 2 | 11 | 1.5 |
| Filtration flow velocities | | | | | | | |
| (max flow velocity, m day$^{-1}$) | 23 | | 33 | | | | |
| (mean flow velocity, m day$^{-1}$) | | | | | 2.7 | 6.3 | 2.3 |

(1): pollution-tolerant Tubificidae, excluding *Spirosperma velutinus* and *Psammoryctides barbatus*; AED: Active Exchange Describers (*Pristina* and *Pristinella* spp., *Cernosvitoviella* sp., *Stylodrilus* sp., *S. velutinus, Rhyacodrilus* sp.); from Lafont & Durbec (1990); Lafont et al. (1992).

through the sediments were greater (mean filtration velocity: 6.3 m day$^{-1}$). Conversely, the pollution-tolerant species did not significantly decline. More surprising was the increase of the percentages of *Psammoryctides barbatus* at −4 m, probably in relation to the predominance of coarse sand (Lafont et al., 1992). The results were more complex than expected, but suggested that the relations between filtration velocities and the percentages of groundwater-dwelling oligochaetes followed a linear pattern.

After these two field studies, the groundwater-dwelling oligochaetes were considered as Active Exchange Describers between surface and subsurface (AED species, Lafont et al., 1992), because (1) they were primarily groundwater dwellers and; (2) they predominated in the surface layers where filtration velocities through hyporheic sediments were greater. A list of Active Exchange Describers (AED) was proposed (Lafont et al., 1992) and is constantly improved (Table 2).

*River Roseg*

In the glacial river Roseg (Switzerland), the upwellings of groundwater were measured by chemical analyses of the $SiO_2$ contents. Malard et al. (1999) showed that silica was an effective tracer of groundwater input to the river. Benthic oligochaetes were sampled with a Hess-sampler (sampling area: 0.043 m$^2$). Hyporheic oligochaetes were collected by driving a mobile pipe to a depth of 30 cm below the streambed (Malard et al., 2001). Ten litres of interstitial water were extracted using a Bou–Rouch pump (Bou & Rouch, 1967). The benthic and hyporheic samples were filtered through a 0.100 mm mesh size. The hyporheic oligochaete assemblages exhibited significant species richness and abundances only at stations where upwellings of groundwater occurred (Lafont & Malard, 2001). Nearly all the collected species belonged to AED indicator taxa (Table 2), with the exception of *Nais communis*, a superficial species suspected to have developed morphological and physiological adaptations to subterranean life (Lafont & Malard, 2001). The benthic oligochaetes followed quite similar distribution patterns at stations where significant upwellings of groundwater occurred (Malard et al., 2001). The same predominant species were collected both in the surface and hyproheic sediments. The main difference was that the composition, abundance and distribution of oligochaete communities were more stable over time in the hyporheic system than in the benthic system.

*River Moselle*

In the river Moselle (France), the situation of surface water downwelling areas was also investigated, and we started from the hypothesis that they might be indicated by the invasion of subsurface areas by surface-dwelling oligochaete species. The oligochaete communities were collected at 5 stations both in surface and hyporheic sediments in June, August and October 1992 (Lafont et al., 1996). Benthic oligochaetes were sampled with a Surber net sampler (sampling area: 0.04 m$^2$). Hyporheic oligochaetes were collected by driving a mobile pipe to a depth of 30 cm below

*Table 2.* List of Active Exchange Describers (AED); from Juget & Lafont (1994); Lafont (unpublished data); Lafont et al. (1992 and accepted); Sambugar et al. (1999)

| Hydrologically validated AED |
|---|
| *Spirosperma velutinus*, all the Rhyacodrilinae and Phallodrilinae species, including *Rhyacodrilus coccineus*, all the species belonging to the Genera *Pristina* and *Pristinella*, *Chaetogaster parvus*, *Stylodrilus* spp., including *S. heringianus***, *Rhynchelmis* spp., *Trichodrilus* spp., *Bichaeta sanguinea* (all the Lumbriculidae, except *Lumbriculus variegatus*), *Cernosvitoviella* spp., including *C. atrata*** and *C. carpatica***, *Achaeta* spp., *Marionina argentea*** *Haplotaxis gordioides***; *Dorydrilus michaelseni***; *Propappus volki*** |
| Suspected AED (need for validation) |
| *Bothrioneurum* spp., *Aulodrilus pigueti*, *Quistadrilus multisetosus*, *Haber* spp., *Lumbricillus fennicus*, *Tubifex kessleri*, *Mesenchytraeus armatus*, *Nais communis*** (eyeless populations) |

**mentioned in the surface and hyporheic sediments of the glacial river Roseg (Malard et al., 2001).

the streambed, and 10 l of interstitial water were extracted using a Bou–Rouch pump (Bou & Rouch, 1967). The benthic and hyporheic samples were filtered through a 0.160 mm mesh size. The hydrologic measurements were restricted to the annual hydrographs of the river (1992). The use of a Factorial Correspondence Analysis (FCA) on the relative abundances of oligochaete species, collected in benthic and hyporheic samples, allowed to identify the three species groups which explained the first plane of the FCA (F1–F2 = 37% of the total inertia). Group 1 included two superficial Naididae species, *Nais barbata* and *Nais bretscheri*. Group 2 was characterized by the predominance of superficial Tubificidae taxa, which are well-known for their pollution-tolerance (*Limnodrilus hoffmeisteri* and immature worms without hair setae). Group 3 comprised the Active Exchange Describers (AED) (*P. volki*, *Pristina longiseta*, *P. aequiseta foreli* and *P. jenkinae*). The seasonal evolution of oligochaete assemblages at Station 1 (the less polluted situation, no water pumping) showed that, both in surface sediments and the hyporheic system, the assemblages changed from being a superficial Naididae population (Group 1) in June, to a population of AED species (Group 3) in August and October (Fig 1). At Station 4, the most polluted situation (paper-mill industry) with intensive groundwater pumping, surface sediments and the hyporheic system were predominantly colonized by taxa of Group 2 (pollution-tolerant genus *Limnodrilus*, Fig. 2). The dynamics of hydrological functioning, illustrated by the hydrographs, were consistent with the distribution of oligochaete assemblages at Station 1. The hyporheic system was invaded by surface species at the end of high surface water discharges (June, previous predominance of surface water downwellings). On the contrary, both surface sediments and the hyporheic system were invaded by AED species when low riverbed discharges occurred (August–October), suggesting that the river was predominantly fed by upwellings of groundwater. At Station 4, the predominance of the superficial and pollution-tolerant genus *Limnodrilus*, both in superficial sediments and the hyporheic system, was considered as the consequence of intensive pumping, the groundwater table being constantly low and downwellings of polluted surface water constantly predominating either during high or low riverbed discharges.

*River Rhône at Chasse-sur-Rhône site*

At Chasse-sur-Rhône, a polluted site of the river Rhône (downstream of the urban area of Lyon), the goal was to assess the vulnerability of a water-pumping area for drinking water supply (Durbec et al., 1992). The oligochaete samples were collected both in surface and hyporheic sediments in October 1992, under low river discharge conditions. The sampling methodology was strictly similar to the one used for studying the river

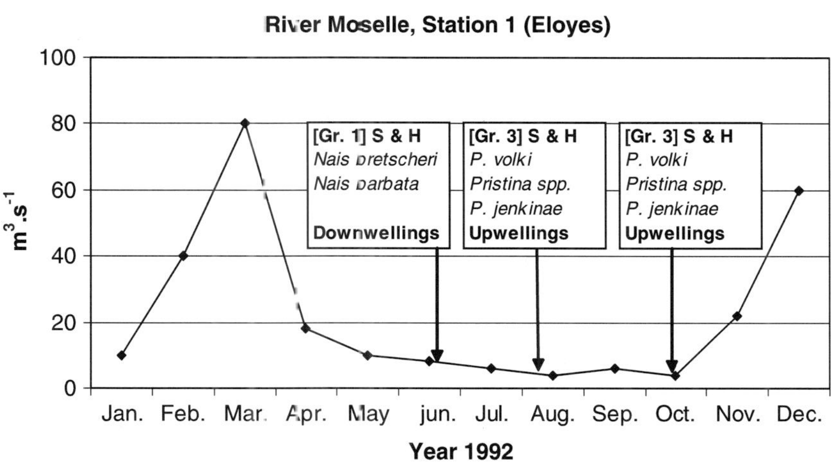


*Figure 1.* River Moselle at Station 1 (slight pollution, no water-pumping in ground water); annual evolution of the river discharges ($m^3\ s^{-1}$) and groups of oligochaete species (Gr1, Gr3) that predominated both in surface sediments (S) and the hyporheic layer (H); the vertical arrows correspond to the sampling dates of oligochaete assemblages.

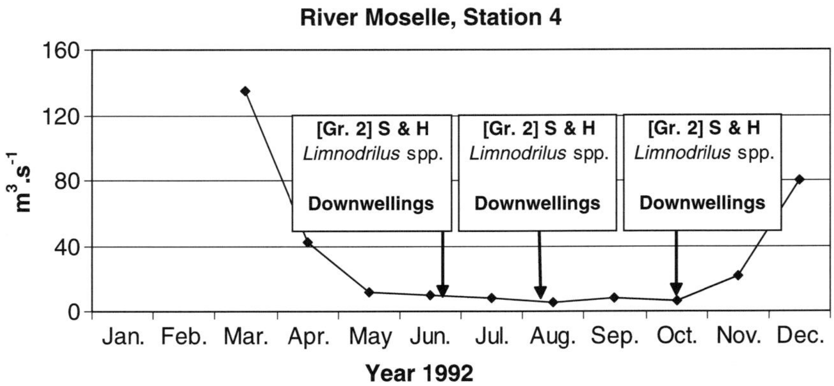


*Figure 2.* River Moselle at Station 4 (strong pollution, water-pumping in ground water); annual evolution of the river discharges ($m^3\ s^{-1}$) and group of oligochaete species (Gr2) that predominated both in surface sediments (S) and the hyporheic layer (H); the vertical arrows correspond to the sampling dates of oligochaete assemblages.

Moselle. The 5 studied stations in the river were facing five pumping wells (P1–P5) located into the banks. The percentages of pollution-tolerant species of Tubificidae were higher in the hyporheic zone than in the surface sediments at Stations P1 and P4, equal at P2 and P3 and lesser at P5 (Fig. 3). The biological features suggested: (a) the predominance of downwellings of surface water at the whole site and the vulnerability of the groundwater resource to surface pollution, with Station P3 being the most vulnerable, and; (b) the storage of pollution in the hyporheic system of P1 and P4, due to former downwellings. One sampling season was considered, however the biological results were in accordance with those of the hydrological study (Durbec et al., 1992), except for P2, where the hydrological study did not mention the vulnerability to pollution.

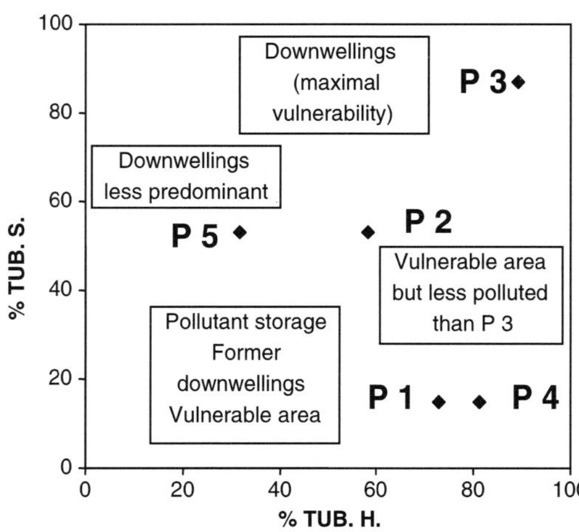


*Figure 3.* Plot of the river Rhône at Chasse-sur-Rhône site (October 1992); P1 to P5: sampling stations; percentages of pollution-tolerant species of Tubificidae in the surface (% TUB.S.) and hyporheic (% TUB.H.) layers.

*Case of an urbanized site (Rhône basin)*

At the Crépieux-Charmy site (urban area of Lyon), an alluvial zone of about 375 ha was delineated for the protection of the groundwater resource devoted to drinking water supply. The oligochaete and crustacean assemblages were sampled on three occasions (June and November 1993, July 1994) in the coarse surface and hyporheic sediments at 12 stations of the river Rhône, the Miribel and the Jonage canals. The sampling methodology was the same as the one used for studying the river Moselle and Chasse-sur-Rhône site. The surface waters were only slightly polluted as the area is upstream of the industrial activities (Camus & Lafont, 1994). The oligochaete species richness accounted for at least 59 species, with rare species like *Trichodrilus intermedius*. When considering the 13 species only found at the Grand Gravier (Lafont & Durbec, 1990; Lafont et al., 1992) and three others only found at Chasse-sur-Rhône (Durbec et al., 1992), the coarse and hyporheic sediments of these three urbanized sites (Crépieux-Charmy, Gand-Gravier, Chasse-sur-Rhône) sheltered at least 75 species. The examination of the percentages of superficial species in the hyporheic layer and of the AED species in the surface sediments enabled a map to be drawn of the hydrologic exchange zones (Fig. 4). These results were in accordance with the global view of hydrologic evaluation, with the predominance of upwelling processes during low surface water discharges (July 1994, Fig. 4). The Crépieux-Charmy site gives an illustration of how to exploit oligochaete assemblages in integrated functional studies, although further refinements are expected (e g. comparison of invertebrate assemblage patterns with hydrological modelling). This study also demonstrated the importance of surface coarse and hyporheic sediments for the protection of oligochaete biodiversity.

## Discussion, conclusion

Very similar distribution patterns of surface and hyporheic oligochaete assemblages were obtained from different watercourses – the Moselle, the Rhône and the glacial river Roseg. Nine of the 16 oligochaete taxa collected from the river Roseg (Malard et al., 2001) were found in the Crépieux-Charmy and Grand Gravier areas. The 6 Active Exchange Describers (AED) of the river Roseg were all collected in the river Rhône (Table 2). The main difference is that the number of AED species is greater in the Rhône than in the Roseg, in relation to a less harsh environment.

The pollution-tolerant species found in the hyporheic zone of the river Moselle are the same as those that dwell in polluted surface sediments (Lafont et al., 1996). The same applies to karstic areas (Malard, 2001). The great interest when considering oligochaete assemblages lies in their simultaneous values as water exchange and pollution indicators, which are related to their ability to colonize both surface and groundwater habitats. The studies of the river Moselle and river Rhône at the Chasse-surRhône site led us to the view that the hyporheic system keeps the 'memory' of past and present pollutions in stony streams and rivers (Lafont et al., 1996). This concept was strengthened by results obtained from the study of the ecological incidence of intermittent pollutions (combined sewer overflows CSOs) at four stations of a suburban stream (stream Chaudanne, Lafont et al., in press). The pollution indicators *Tubifex tubifex* and *Lumbricillus rivalis* predominated, from April 2000 to April 2001, in the hyporheic system at Station 4, located 500 m downstream of the CSOs, indicating that the hyporheic interstices became soaked with polluted sludge. This situation

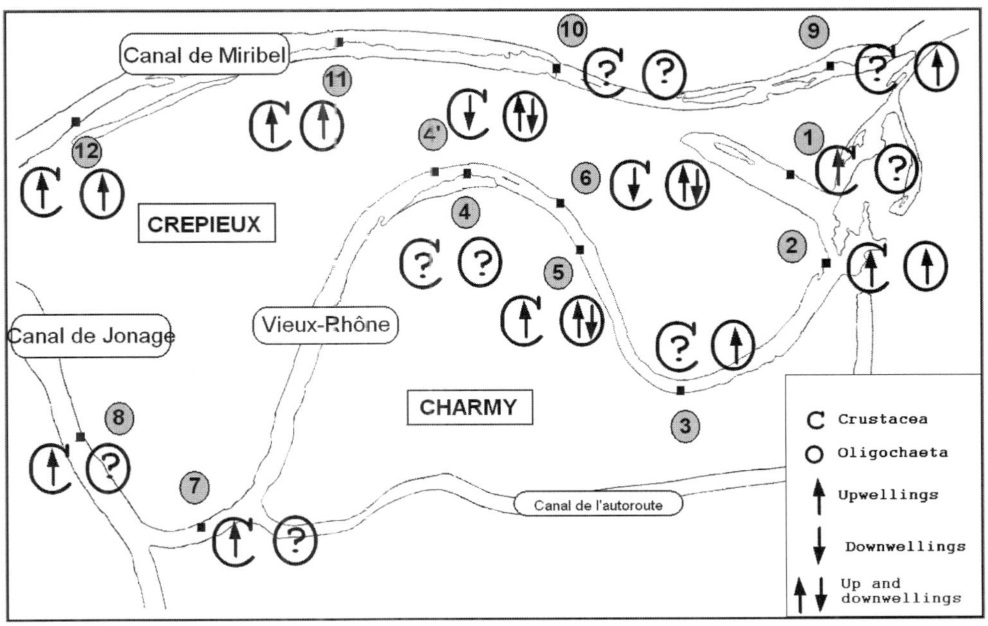


*Figure 4.* The Crépieux-Charmy site; the ways of the water exchanges between surface and subsurface were mapped in July 1994 by the examination of crustacean (C) and oligochaete (O) assemblages.

was related to high ammonium salt contents in the hyporheic waters and suggested that downwellings of polluted waters predominated. Moreover, *T. tubifex* and *L. rivalis* significantly occurred in the coarse surface sediments at Station 4 only when the CSOs were high and stream discharges very low (August to the beginning of October 2000). These results emphasized the general view that coarse surface sediments are more exposed to the variations of surface hydraulic conditions and more unstable than the hyporheic system (Giere, 1993), a situation also illustrated by the study of the river Roseg (Malard et al., 2001). In consequence, the benthic sediments give incomplete diagnoses of pollution effects.

The concept of the vulnerability of groundwater to a surface pollution was previously proposed (Lafont et al., 1992, 1996). However, this was a simplistic view as the hyporheic zone may also contaminate the surface sediments, a feature which is well-documented regarding nutrients (Holmes, 2000) -that probably exists in the rivers Moselle and Rhône. Moreover, the surface sediments are not always permanently saturated and can become exposed during low water discharges, or removed during floods. Conversely, the hyporheic system is generally permanently saturated, even when the surface stream-bed has dried up, and it remains intact during floods, particularly in the deeper layers (McElravy & Resch, 1991). Because of this, the hyporheic zone may accumulate more pollutants. In conclusion, for fundamental or applied studies, the sole consideration of the surface sediments gives us incomplete information (Ferrarese & Sambugar, 1976; Hynes, 1983; Rouch & Danielopol, 1987; Danielopol, 1989; Ward & Palmer, 1994). In the case of pollution studies, it is a serious gap which might lead to a biased view about the state of ecological degradation. Furthermore, as the self-purification capacities of a river are strongly related to the hydric exchanges between surface and subsurface (Hancock, 2002), the biological assessment of downwellings (predominance of superficial species in the hyporheic zone) and upwellings (predominance of AED species in the surface sediments) would be very useful in the functional studies of watercourses where coarse sediments predominate.

Unfortunately, most biomonitoring schemes focus on the surface habitats and neglect the hyporheic system. In France, the IBGN index (invertebrate index, Afnor, 1992), the diatom index IBD (Afnor, 2000) and the oligochaete index IOBS (Afnor, 2002) only refer to the surface habitats. The biomonitoring of hyporheic zones remains disconnected from that of the benthic system, except some valuable recent proposals (Boulton, 2000; Boulton et al., 2003). This situation is not satisfactory because of the close connections between surface and groundwater.

The value of oligochaetes as pollution indicators is well-recognized, as well as their strong relations with organic matter which is for example attested by their role in wastewater treatments (Bou, 1977; Bruce et al., 1983, cited by Datry et al., 2003). Oligochaete metabolic activity in coarse sediments and the hyporheic zone is well less known. This is an important ecological trait which has not been studied enough, but is in the course of being developed in field (Datry et al., 2003) and experimental case studies (Mermillod-Blondin et al., 2003).

The value of oligochaete assemblages as indicators of pollution, metabolic activities and water exchange is shared with other biocenoses. However the attractive characteristic of oligochaetes is that they might enable biologists to simultaneously assess the effects of pollution, dynamics of water exchanges and metabolic activities. Conversely, oligochaetes are not universal indicators and other taxocenoses must be considered to supplement and refine the diagnoses. The simultaneous use of crustacean and oligochaete communities at the Crépieux-Charmy site (Fig. 4) is a good example of an integrated ecological diagnosis. If the ecological value of oligochaetes in surface coarse sediments and the hyporheic system was until now underestimated, it is also important not to overestimate their importance.

The present and relevant classifications of groundwater organisms (stygoxene, stygophile and stygobite organisms) (Claret et al., 1999) were historically based on the examination of crustacean assemblages or aquatic insect larvae. However, oligochaete species do not easily fit into these classifications and new proposals must be made. The hyporheic zone is generally considered as an ecotone and a refuge during floods or drought (Gibert, 2001). Regarding oligochaete communities, the hyporheic zone might be above all a reservoir of species, and probably the main route for the dissemination of species (Lafont & Malard, 2001). These observations are consistent with the concepts of Bretschko (1991) and Naiman et al. (1988, cited by Bretschko 1991).

As a conclusion, it is suggested that the study of coarse surface sediments and hyporheic systems is indispensable to understand the ecological functioning of watercourses. The fine sediments must be considered in biomonitoring studies (Lafont et al., 2001), but in watercourses where the coarse sediments predominate, the interest of fine sediments for the understanding of river functions declines when their percentage cover of the river bed is reduced (Lafont, 2001). Our present research focuses now on the study of surface coarse sediments and the hyporheic system, with special reference to the comparison between the distributions of invertebrate assemblages, the modelling of hydrological exchanges and the geomorphological background.

## Acknowledgements

We thank D. Danielopol for his kind proposal of commenting on our manuscript, also F. Malard for his help before the Nine International Symposium on Oligochaete Biology. We are gratefully indebted to the two anonymous reviewers for their comments that enabled us to significantly improve the manuscript. This work was partly funded by the European Project URBEM (Urban Water Enhancement Methods).

## References

Afnor, 1992. Détermination de l'indice biologique global normalisé (IBGN). NF T 90-350.

Afnor, 2000. Détermination de l'Indice biologique Diatomées (IBD). NF T 90-354.

Afnor, 2002. Détermination de l'indice oligochètes de bioindication des sédiments (IOBS). NF T 90-390.

Bencala, K. E., 1993. A perspective on stream-catchment connections. Journal of the North American Benthological Society 12: 44–47.

Bird, G. J., 1982. Distribution, life cycle and population dynamics of the aquatic enchytraeid *Propappus volki* (Oligochaeta) in an English chalkstream. Holarctic Ecology 5: 67–75.

Botea, F. R., 1963. Contributii la studiul răspîndirii olgochetelor limicole din R.P.R. Valea Motrului IV: 401–409.

Bou, C., 1977. Conséquences écologiques de l'extraction des alluvions récentes dans le cours moyen du Tarn. Bulletin d'Ecologie 8: 435–444.

Bou, C. & R. Rouch, 1967. Un nouveau champ de recherches sur la faune aquatique souterraine. Comptes Rendus de l'Académie des Sciences, Paris 265: 369–370.

Boulton, A. J., 2000. River ecosystem health down under: assessing ecological conditions in riverine groundwater zones in Australia. Ecosystem Health 6: 108–118.

Boulton, A. J., W. F. Humphreys & S. M. Eberhard, 2003. Imperilled subsurface waters in Australia: Biodiversity, threatening processes and conservation. Aquatic Ecosystem Health and Management 6: 41–54.

Bretschko, G., 1991. Bed sediments, groundwater and stream limnology. Verhandlungen Internationale Vereiningung für Theorische und Angewandte Limnologie 24: 1957–1960.

Brunke, M. & T. Gonser, 1997. The ecological significance of exchange processes between rivers and groundwater. Freshwater Biology 37: 1–33.

Camus, J. C. & M. Lafont, 1994. Etude de la vulnérabilité à la pollution du champ captant de Crépieux-Charmy (69). Utilisation des oligochètes et des microcrustacés, Cemagref, 48 pp.

Claret, C., P. Marmonier, M. J. Dole-Olivier, M. Creuzé des Châtelliers, A. J. Boulton & E. Castella, 1999. A functional classification of interstitial invertebrates: supplementing measures of biodiversity using species traits and habitat affinities. Archiv für Hydrobiologie 145: 385–403.

Danielopol, D. L., 1984. Ecological investigations on the alluvial sediments of the Danube in the Vienna area – a phreatobiological project. Verhandlungen Internationale Vereiningung für Theorische und Angewandte Limnologie 22: 1755–1761.

Danielopol, D. L., 1989. Groundwater fauna associated with riverine aquifers. Journal of the North American Benthological Society 8: 18–35.

Datry, T., F. Hervant, F. Malard, L. Vitry & J. Gibert, 2003. Dynamics and adaptive responses of invertebrates to suboxia in contaminated sediments of a stormwater infiltration basin. Archiv für Hydrobiologie 156: 339–359.

Dole, M. J., 1983. Le domaine aquatique souterrain de la plaine alluviale du Rhône à l'est de Lyon. I: diversité hydrologique and biocénotique de trois stations représentatives de la dynamique fluviale. Vie and Milieu 33: 219–229.

Dumnicka, E., 1986. Naididae (Oligochaeta) from subterranean waters of West Indian islands. Distribution, taxonomic remarks and description of a new species. Bijdragen tot de Dierkunde 56: 267–281.

Durbec, A., M. Lafont & J. C. Camus, 1992. Etude de la vulnérabilité à la pollution du champ captant de Chasse-sur-Rhône (38). Burgéap, Cemagref, 16 pp.

Fellows, C. S., H. M. Valett & C. N. Dahm, 2001. Whole-stream metabolism in two montane streams: contribution of the hyporheic zone. Limnology and Oceanography 46: 523 531.

Ferrarese, U. & B. Sambugar, 1976. Ricerche sulla fauna interstiziale iporreica dell'Adige in relazione allo stato di inquinamento del Fiume. Rivista di Idrobiologia. 15: 47–127.

Gaschignard, O., 1984. Impact d'une crue sur les invertébrés benthiques d'un bras mort du Rhône. Verhandlungen Internationale Vereiningung für Theorische und Angewandte Limnologie 22: 1997–2001.

Gibert, J., 2001. Basic attributes of groundwater ecosystems. In Griebler, C., D. L. Danielopol, J. Gibert, H. P. Nachtnebel & J. Notenboom (eds), Groundwater Ecology. European Communities, 39–52.

Giere, O., 1993. Meiobenthology. The microscopic fauna in aquatic sediments. Springer-Verlag, Berlin Heidelberg 328 pp.

Griebler, C., D. L. Danielopol, J. Gibert, H. P. Nachtnebel & J. Notenboom, 2001. Groundwater Ecology, European Communities, 413 pp.

Gibert, J., M. J. Dole-Olivier, P. Marmonier & P. Vervier, 1990. Surface water/groundwater ecotones. In Naiman, R. J. & H. Décamps (eds), Ecology and Management of Aquatic-Terrestrial Ecotones 4. UNESCO and Parthenon Publishing Group, Paris, 199–225.

Hancock, P. J., 2002. Human impacts on the stream-groundwater exchange zone. Environmental Management 29: 763–781.

Holmes, R. M., 2000. The importance of ground water to stream ecosystem function. In Jones, J. B. & P. J. Mulholland (eds), Streams and Ground Waters. Academic Press, San Diego, 137–148.

Hynes, H. B. N., 1983. Groundwater and stream ecology. Hydrobiologia 100: 93–99.

Jones, J. B. & P. J. Mulholland, 2000. Streams and Ground Waters. Academic Press, San Diego, 425 pp.

Juget, J., 1980. Aquatic oligochaeta of the Rhône-Alpes area: Current research priorities. In Brinkhurst, R. O. & D. G. Cook (eds), Aquatic Oligochaete Biology. Plenum Press, New York and London, 241–251.

Juget, J., 1984. Oligochaeta of the epigean and underground fauna of the alluvial plain of the French upper Rhône (biotypological try). Hydrobiologia 115: 175–182.

Juget, J., 1987. Contribution to the study of Rhyacodrilinae (Tubificidae, Oligochaeta), with the description of two new stygobiont species from the alluvial plain of the French Upper Rhône, *Rhyacodrilus amphigenus*, sp. n. and *Rhizodriloides phreaticola*, g. n., sp. n. Hydrobiologia 155: 107–118.

Juget, J. & E. Dumnicka, 1986. Oligochaeta (incl. Aphanoneura) des eaux souterraines continentales. In Botosaneanu, L. (ed.), Stygofauna mundi. E. J. Brill, Leiden, 234–244.

Juget, J. & M. Lafont, 1994. Theoretical habitat templets, species traits, and species richness: aquatic Oligochaetes in the Upper Rhône River and its floodplain. Freshwater Biology 31: 327–340.

Ladle, M., 1971. Studies on the biology of oligochaetes from the phreatic water of an exposed gravel bed. International Journal of Speleology 3: 311–316 +4 Plates (100–103).

Lafont, M., 1989. Contribution à la gestion des eaux continentales: utilisation des oligochètes comme descripteurs de l'état biologique et du degré de pollution des eaux et des sédiments. Thèse de Doctorat ès Sciences, Université Lyon I, 403 pp.

Lafont, M., 2001. A conceptual approach to the biomonitoring of freshwater: the Ecological Ambience System. Journal of Limnology 60(Suppl. 1): 17–24.

Lafont, M. & A. Durbec, 1990. Essai de description biologique des interactions entre eau de surface et eau souterraine: application à l'évaluation de la vulnérabilité d'un aquifère à la pollution d'un fleuve. Annales de Limnologie 26: 119–129.

Lafont, M. & F. Malard, 2001. Oligochaete communities in the hyporheic zone of a glacial river, the Roseg River, Switzerland. Hydrobiologia 463: 75–81.

Lafont, M., A. Durbec & C. Ille, 1992. Oligochaete worms as biological describers of the interactions between surface and groundwater: a first synthesis. Regulated Rivers, Research and Management 7: 65–73.

Lafont, M., J. C. Camus & A. Rosso, 1996. Superficial and hyporheic oligochaete communities as indicators of pollution and water exchange in the River Moselle, France. Hydrobiologia 334: 147–155.

Lafont, M., J. C. Camus, A. Fournier & E. Sourp, 2001. A practical concept for the ecological assessment of aquatic ecosystems: application on the river Dore in France. Aquatic Ecology 35: 195–205.

Lafont, M., A. Vivier, S. Nogueira, P. Namour & P. Breil, 2006. Surface and hyporheic Oligochaete assemblages in a French suburban stream. Hydrobiologia 564: 183–193.

Malard, F., 2001. Groundwater contamination and ecological monitoring in a mediterranenan karst ecosystem in southern France. In Griebler C., D. L. Danielopol, J. Gibert, H. P. Nachtnebel & J. Notenboom (eds), Groundwater Ecology. European Communities, 183–194.

Malard, F., K. Tockner & J. V. Ward, 1999. Shifting dominance of subcatchment water sources and flow paths in a glacial floodplain (Val Roseg, Switzerland). Arctic, Antarctic, and Alpine Research 3: 135–150.

Malard, F., M. Lafont, P. Burgherr & J. V. Ward, 2001. A comparison of longitudinal patterns in hyporheic and benthic oligochaete assemblages in a glacial river. Arctic, Antarctic, and Alpine Research 33: 457–466.

Malard, F., K. Tockner, M. J. Dole-Olivier & J. V. Ward, 2002. A landscape perspective of surface-subsurface hydrological exchanges in river corridors. Freshwater Biology 47: 621–640.

Malard, F., D. Galassi, M. Lafont, S. Doledec & J. V. Ward, 2003. Longitudinal pattern of invertebrates in the hyporheic zone of a glacial river. Freshwater Biology 48: 1–17.

McElravy, E. P. & V. Resch, 1991. Distribution and seasonal occurrence of the hyporheic fauna in a northern California stream. Hydrobiologia 220: 233–246.

Mermillod-Blondin, F., M. Creuzé des Châtelliers & M. Gérino, 2003. Effects of the interaction between tubificid worms on the functioning of hyporheic sediments: an experimental study in sediment columns. Archiv für Hydrobiologie 156: 203–223.

Nalepa, T. F. & A. Robertson, 1981. Screen mesh size affects estimates of macro- and meio-benthos abundance and biomass in the Great Lakes. Canadian Journal of Fisheries and Aquatic Sciences 38: 1027–1034.

Rouch, R. & D. L. Danielopol, 1987. L'origine de la faune aquatique souterraine, entre le paradigme du refuge et le modèle de la colonisation active. Stygologia 3: 345–372.

Sambugar, B., N. Giani & E. Martinez-Ansemil, 1999. Groundwater oligochaetes from Southern-Europe. Tubificidae with marine phyletic affinities: new data with description of a new species, review and consideration on their origin. Mémoires de Biospéologie XXVI: 107–116.

Schwank Von, P., 1982a. Turbellarien, Oligochaeten und Archianneliden des Breitenbachs und anderer oberhessischer Mittelgebirgsbäche. III. Die Taxozönosen der Turbellarien und Oligochaeten in Fließgewässern – eine synökologische Gliederung. Archiv für Hydrobiologie/Supplementum 62: 191–253.

Schwank Von, P., 1982b. Turbellarien, Oligochaeten und Archianneliden des Breitenbachs und anderer oberhessischer Mittelgebirgsbäche. IV. Allgemeine Grundlagen der Verbreitung von Turbellarien und Oligochaeten in Fließgewässern. Archiv für Hydrobiologie/Supplementum 62: 254–290.

Solbe, J. F., 1975. Annelida. In Curds, C. R. & H. A. Hawkes (eds), Ecological Aspect of Used Water Treatment. Academic Press, London, 305–335.

Solbe, J. F., P. G. Ripley & T. G. Tomlinson, 1974. The effects of temperature on the performance of experimental percolating filters with and without mixed macro-invertebrate populations. Water Research 8: 557–573.

Stanley, E. H. & J. B. Jones, 2000. Surface-subsurface interactions: past, present, and future. In Jones, J. B. & P. J. Mulholland (eds), Streams and Ground Waters. Academic Press, San Diego, 405–417.

Strayer, D. & E. Bannon-O'Donnell, 1988. Aquatic Microannelids (Oligochaeta and Aphanoneura) of undergroundwaters of Southeastern New-York. The American Midland Naturalist 119: 327–335.

Triska, F. J., J. H. Duff & R. J. Avanzino, 1990. Influence of exchange flow between the channel and hyporheic zone on $NO_3^-$ production in a small mountain stream. Canadian Journal of Fisheries and Aquatic Sciences 11: 2099–2111.

Valett, H. M., C. C. Hakenkamp & A. J. Boulton, 1993. Perspectives on the hyporheic zone: integrating hydrology and biology. Introduction. Journal of the North American Benthological Society 12: 40–43.

Vervier, P., J. Gibert, P. Marmonier & M. J. Dole-Olivier, 1992. A perspective on the permeability of the surface freshwater-groundwater interface. Journal of the North American Benthological Society 11: 93–102.

Ward, J. V., 1989. The four-dimensional nature of lotic ecosystems. Journal of the North American Benthological Society 8: 2–8.

Ward, J. V. & M. A. Palmer, 1994. Distribution patterns of interstitial freshwater meiofauna over a range of spatial scales, with emphasis on alluvial river-aquifer systems. Hydrobiologia 287: 147–156.

Weigelhoffer, G. & J. Waringer, 2003. Vertical distribution of benthic macroinvertebrates in riffles versus deep runs with differing contents of fine sediments (Weidlingbach, Austria). International Revue of Hydrobiology 88: 304–313.

White, D. S., 1993. Perspectives on defining and delineating hyporheic zones. Journal of the North American Benthological Society 12: 61–69.

Williams, N. V. & H. M. Taylor, 1968. The effect of *Psychoda alternata* (Say.) (Diptera) and *Lumbricillus rivalis* (Levinsen) (Enchytraeidae) on the efficiency of sewage treatment in percolating filters. Water Research 2: 139–150.

Hydrobiologia (2006) 564:183–193
P.F.M. Verdonschot, H. Wang, A. Pinder & R. Nijboer (eds), Aquatic Oligochaete Biology IX
DOI 10.1007/s10750-005-1718-8


# Surface and hyporheic oligochaete assemblages in a French suburban stream

Michel Lafont[1,*], Anne Vivier[2], Sylvie Nogueira[1], Philippe Namour[3] & Pascal Breil[4]
[1]*Biology Research Unit, Cemagref, 3 bis Quai Chauveau, C.P. 220, F-69336, Lyon Cedex 09, France*
[2]*SHU Laboratory, ENGEES, 1 Quai Koch, 67000, Strasbourg, France*
[3]*Water Quality Research Unit, Cemagref, Lyon France*
[4]*Hydrology-Hydraulic Research Unit, Cemagref, Lyon, France*
*(*Author for correspondence: E-mail: lafont@lyon.cemagref.fr)*

*Key words:* suburban watercourses, urban inputs, coarse sediments, hyporheic zone, Oligochaeta

**Abstract**

The Chaudanne stream received urban inputs discharged through combined sewer overflows (CSOs). The water quality was not severely impaired, with pollution mainly of organic origin. Oligochaete assemblages were studied in the coarse surface sediments and the hyporheic zone and sampled at four sites on seven occasions during 2000 and 2001. The seasonal distribution of oligochaete assemblages was analyzed by a PCA, the oligochaete species being assigned to functional traits (FTrs). Site 1, located upstream of the CSOs, was characterized by FTrs 1 and 2 (species indicating permeability and those intolerant to water pollution). Below the CSOs, high densities of oligochaetes occurred in the benthic layer of sites 3 and 4, with a predominance of FTr3 (species with tolerance to water pollution). At site 4, FTr4 species (indicative of sludge conditions) constantly predominated in the hyporheic system, but predominated in the benthos only during low stream discharges associated with peaks in CSOs. FTr3 was related to amounts of the oxidized forms of nitrogen, high stream discharges and probably to groundwater upwellings and the sludge tolerant species group (FTr4) was associated with high $NH_4^+$ contents. We are now testing the relevance and generalization of this new approach.

## Introduction

The assessment of the ecological impacts of combined sewer overflows (CSOs) has became a priority in urban aquatic systems (Cyr et al., 1998; Paul & Meyer, 2001; Rogers et al., 2002). The alteration of urban watercourses complicates prediction of the effects of CSOs, which are strongly related to rainfall patterns. In addition, the pollution originating from storm-water runoff is generally accompanied by other pollution sources, including permanent inputs and polluted overflows from impervious urban surfaces. These multiple sources of pollution are a common feature in urban aquatic areas, and are generally poorly known (Faulkner et al., 2000). The recent EC Water Framework Directive (UE, 2000) emphasized the conservation or restoration of good ecological condition in all aquatic ecosystems, including those located in urbanized areas.

In this study, we tried to assess the ecological effects of a CSO on the Chaudanne stream, a suburban watercourse of Lyon's urban area (France). We focused on the surface coarse sediments and hyporheic system which largely predominate in the studied stream. Invertebrate assemblages were mainly composed of oligochaetes and crustaceans, and aquatic insect larvae were not abundant (Bernoud, 1998). This was related to the urbanization of the surrounding landscape which had altered the physical habitats. The preserved area located near the source of the stream was poor in aquatic insects, however it was naturally rich in oligochaetes and crustaceans (Bernoud, 1998). Consequently, we focused on benthic and hyporheic oligochaetes as (i) they were the most abundant interstitial invertebrates in the study site and (ii) the present study was supported by data sets considering the oligochaete assemblages in the surface coarse sediments and hyporheic system of polluted and unpolluted watercourses (Lafont, 1989; Lafont & Durbec, 1990; Lafont et al., 1992, 1996; Lafont & Malard, 2001; Malard et al., 2001).

It was necessary, in this preliminary work, to assess the relevance of the information value of oligochaetes, in particular to see if gradients of alterations might be evident in a relatively unimpaired situation such as the Chaudanne stream. In addition, the choice of a suburban site was dictated (1) by the fact that catchment management is easier to control than in urban watercourses; (2) that the imperviousness of the urbanized surrounding landscape is not constant; and (3) that relatively preserved areas can still be found. If alteration gradients could be illustrated by oligochaete worms in a situation of intermediate disturbance, the generalization to similar or more altered situations should be possible in the future.

## Study site

The Chaudanne stream is a tributary of the Yzeron river (Bernoud, 1998). The Yzeron river enters the Rhône river at Oullins, a city located in the southwestern area of the Lyon's urban area. The length of the Chaudanne stream is 3843 m and it flows southeast, from 443 m elevation to its entry into the river Yzeron at 306 m (average gradient = 3.6%). The basin area is 2.83 $km^2$. Its headwater is located in a rural area (2.22 $km^2$). The length of the studied reach is about 2.5 km. Four sites were studied (Fig. 1). Site 1 is located near the spring, uninfluenced by agricultural, road or urban inflows. It is therefore considered as a reference site. Site 2 receives the first polluted input, from the drainage ditch of a factory. Site 3 is located 150 m below the main polluted input (Combined Sewer Overflow CSO of the 'Pont de la Barge'). Site 4 is 500 m downstream of the CSO and receives a third input from the car park of a commercial zone (Nogueira, 2001, Fig. 1). The imperviousness of the surrounding urbanized landscape of sites 2–4 does not exceed 15% (Bernoud, 1998). The streambed is composed of stones, gravel and sand. The current velocity of surface waters is about 30 cm $s^{-1}$ in the riffles where the biological and chemical samples were collected. The surface water velocity decreases to less than 10 cm $s^{-1}$ during low stream discharges (August, October 2000). The temperature of surface waters varies from 6.5 °C in winter (February 2000) to 18 °C in summer (August 2000). The temperature of hyporheic waters (−20 to −30 cm depth) varies from 6.4 °C in winter to 17 °C in summer. The pH

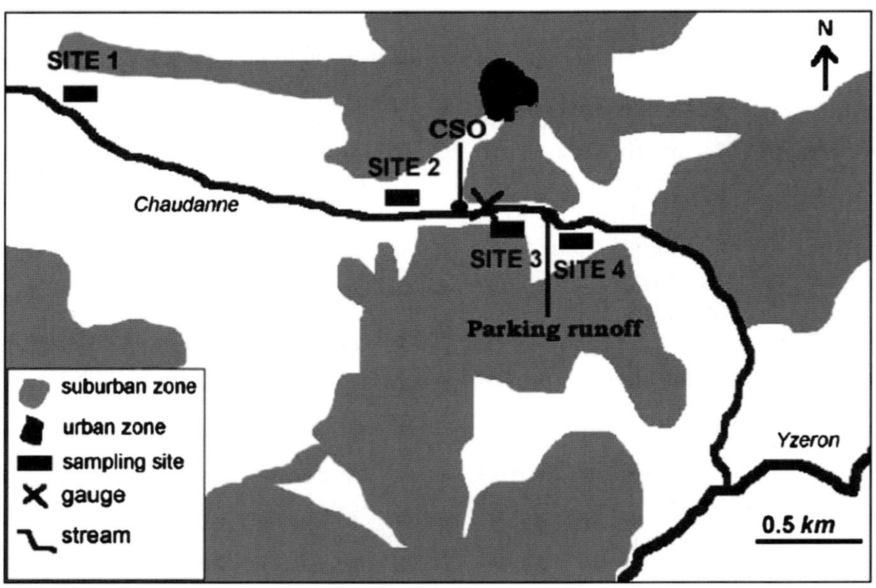


*Figure 1.* Sampling site locations in the stream Chaudanne.

and conductivity are similar in surface and hyporheic waters and, respectively, vary from 6.5 to 8.3 and from 170 $\mu S\ cm^{-1}$ (site 1) to 532 $\mu S\ cm^{-1}$ (site 4) (Namour, unpublished data). Daily stream discharge measurements were available from a gauge located at site 2.

## Material and methods

Benthic and hyporheic oligochaetes were sampled at the four sites on seven occasions (April, June, August, October and December 2000, February and April 2001). Benthic sediments were collected using a Surber-type net (400 $cm^2$ aperture; net mesh-size: 0.160 mm). In the hyporheic system, 10 l of material (sediments and interstitial water) were pumped from a depth of 20–30 cm using a Bou–Rouch pump (Bou & Rouch, 1967). At each site, three replicates were collected (a sample each 10 m). The three benthic replicates were pooled at each occasion. The three hyporheic replicates were kept separate. Samples were preserved in the field with 4% formaldehyde. In the laboratory, the mineral particles (stones, gravel, sand) of benthic and hyporheic samples were separated from the organic fractions (organic fragments and invertebrates) by decanting, and the organic fractions were washed through a 0.160 mm sieve. The residue from the sieving of each sample was poured into a gridded subsampling box. One hundred oligochaete specimens were sorted by hand under a binocular microscope, if necessary from subsamples taken randomly with a pipette from the squares of the subsampling box. Worm specimens were mounted on slides in a mixture of lactic acid and glycerin (50% glycerin, 50% lactic acid), covered by a cover-glass, and identified to species when possible under a microscope. Stream discharge was very low during August 2000, allowing benthic sampling of sites 2 and 3 only, the others being dry. The hyporheic sediments of the control site (Site 1) were sometimes clogged by silt, and the hyporheic system was dry during August 2000 at Site 4. Hyporheic samples with total oligochaete densities in the three replicates less than 10 specimens were not retained, as the relative abundances of oligochaete species were mainly considered in the statistical analyses. To allow a comparison with pooled benthic collections, the data of the hyporheic replicates were pooled together, each pooled sample being characterized by the mean values (calculated from the three replicates) of the biological variables. Consequently, 18 hyporheic and 26 benthic samples were retained for analyses. The surface replicates were pooled in the field to allow a comparison with previous data sets. Conversely, the hyporheic replicates were kept separately and pooled only in the multivariate analysis. The data of each hyporheic replicate remain available and will be considered in other papers.

### *Data analysis*

The seasonal distribution patterns of oligochaete assemblages at the four sites were analyzed by a standardized Principal Component Analysis (PCA, software: ADE-4, Thioulouse et al., 1997). The benthic system (26 samples) was separated from the hyporheic system (18 samples). The oligochaetes were gathered into four ecological groups (Table 1), called 'functional traits' (FTrs, Lafont et al., 2000). The FTrs were derived from the ecological knowledge of indicator oligochaete species that reveals the ecological functioning of the functional units 'coarse surface sediments' and 'hyporheic system' (FU3 & 4, Lafont, 2001). Two other variables were considered: the total number of oligochaete species and the densities per sample, $Log_{10}(n+1)$ transformed for normality (NOSP and EFBO, respectively). These two last variables enabled incorporation of the species whose ecological preferences are not known or not considered as ecologically distinct.

The PCA reduces the number of input variables, and was here expected to describe meaningful ecological gradients and spatio-temporal patterns of benthic and hyporheic assemblages. Because relationships among variables could be slightly different between the hyporheic and the benthic systems and as the sampling devices were different, we performed a PCA for each system.

## Results

A total of 46 taxa was found in the surface sediments and hyporheic system. The immature Tub-

*Table 1.* Functional traits (FTrs) of the surface and hyporheic sediments

| Functional traits FTrs | Oligochaete species characterizing each FTr |
|---|---|
| FTr1: Percentages of oligochaete species which indicate active hydrologic exchanges between surface and ground water (AED species) (Gaschignard, 1984; Juget, 1984; Juget & Dumnicka, 1986; Lafont, 1989; Lafont & Durbec, 1990; Lafont et al., 1992; Juget & Lafont, 1994; Lafont et al., 1996). | *Trichodrilus strandi, Stylodrilus heringianus, S. parvus, Rhyacodrilus ardierae, R. coccineus, R. falciformis, R. subterraneus, Haber speciosus, Pristina aequiseta, P. jenkinae, P. osborni, Cernosvitoviella atrata, Achaeta vesiculata, Marionina argentea, Haplotaxis gordioides* |
| FTr2: Percentages of oligochaete species which are intolerant to water pollution (Lafont, 1989; Lafont & Juget, 1993; Lafont et al., 1996). | *R. ardierae, R. falciformis, R. subterraneus, C. atrata, A. vesiculata, M. argentea, Eiseniella tetraedra* |
| FTr3: Percentages of oligochaete species which are tolerant to water pollution in coarse sediments (op. cited; Giani, 1984). | *Nais elinguis, P. jenkinae, Dero digitata, Marionina riparia* |
| FTr4: Percentages of taxa which indicate the presence of polluted sludge within coarse sediment interstices (Brinkhurst, 1965; Lafont, 1989; Lafont et al., 1996; Datry et al., 2003). | Immatures of Tubificidae with and without hair setae, *Tubifex ignotus, T. tubifex, Limnodrilus hoffmeisteri, Bothrioneurum* sp., *Lumbricillus* spp. (in bacterial sewage-treatment beds, Solbé, 1975). |
| *Other species* (less than 5% of oligochaete assemblages in the Chaudanne stream, except *Chaetogaster diastrophus* & *Nais communis*) are represented by the species richness and the absolute abundances of oligochaete assemblages per sample. | |

ificidae (with and without hair setae) and Lumbriculidae were not counted as individual taxa because adults of them were present. The unidentified Enchytraeidae accounted for 10 taxa. A notable record was *Rhyacodrilus ardierae*, which was first discovered near the spring of the river Ardières, in a headwater reach similar in size to the Chaudanne stream. It is considered a pollution-intolerant species (Lafont & Juget, 1993).

### *Distribution patterns of oligochaete assemblages*

For the benthic samples (Fig. 2a), the axis F1 (53.6% of the total variance) of a PCA of oligochaete assemblages separated species rich samples, dominated by the species considered as indicators of hydrological exchanges between surface and ground water (AED species) and by pollution-intolerant ones (NSPO, FTr1, FTr2), that typically reflected the assemblage features of site 1, from assemblages characteristic of sites 2–4 (Fig. 2b). These were characterized by high densities of oligochaetes (EFBO) and a predominance of pollution-tolerant species (FTr3). The second axis F2 (26.8%) discriminated the samples rich in Tubificidae (FTr4) (site 2 October 2000, site 3, August and October 2000, site 4 June and October 2000). In the surface sediments, the factor F1 of the PCA was considered as characterizing organically polluted waters. The factor F2 characterized another type of pollution, with a predominance of Tubificidae worms indicating the presence of sludge in the sediment interstices. The reference site (1) was clearly separated from the remaining sites by an increasing dominance of FTr3 + 4 species, indicating a downstream pollution gradient, although

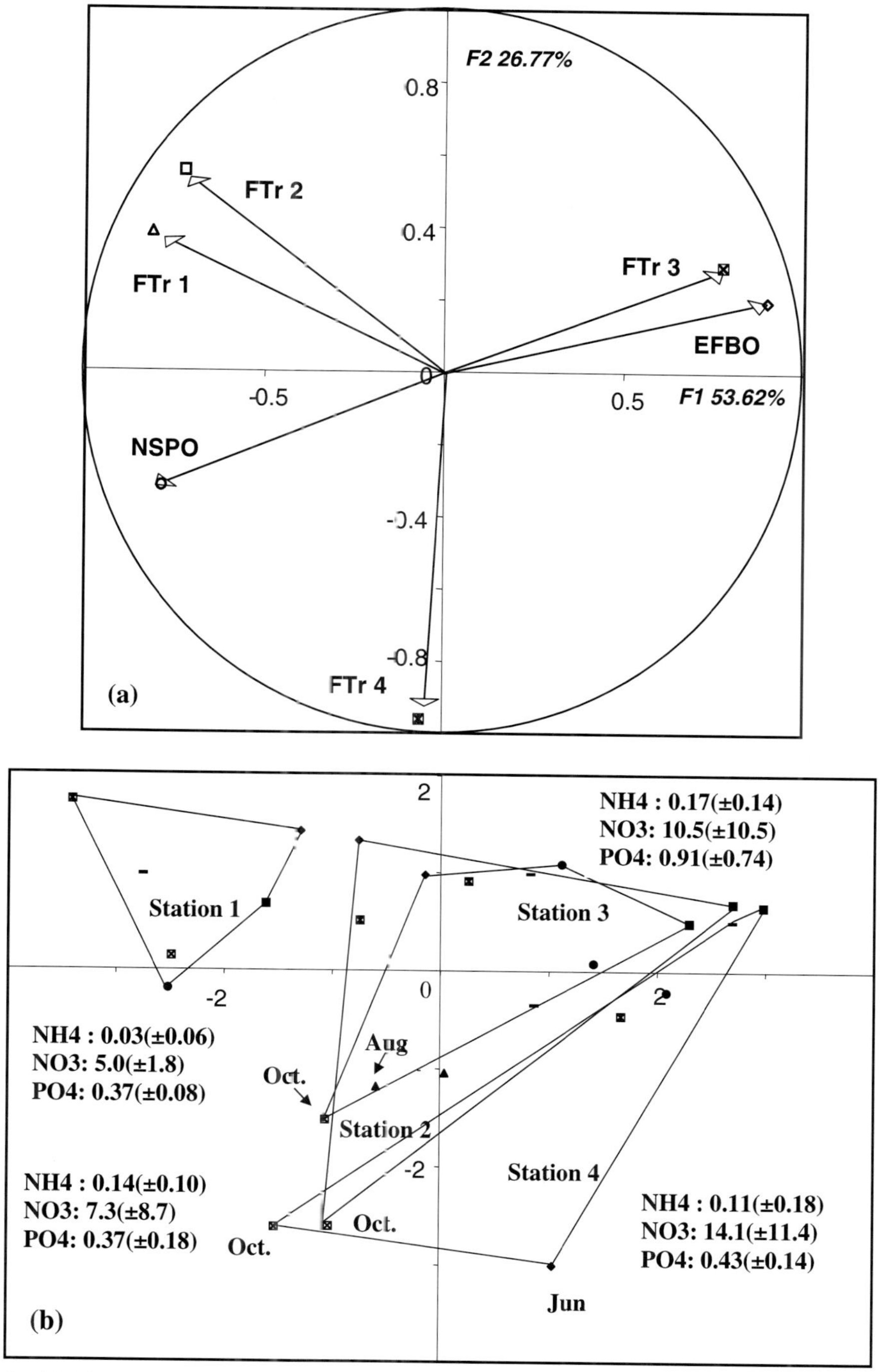


*Figure 2.* Standardized PCA on benthic oligochaete assemblages (26 rows, 6 columns); (a) correlation circle (F1–F2 plane); (b) plots and dispersion polygons of biological samples from the four sites on the first plane; NSPO: number of oligochaete species; EFBO: densities per 0.1 $m^2$ of oligochaete assemblages; FTr1, FTr2: functional traits permeability and intolerance to water pollution; FTr3, FTr4: functional traits tolerance to water pollution and sludge effect; the chemical variables are expressed as mg $l^{-1}$; the standard deviations are put in brackets; NH4: ammonium salts; NO3: nitrates; PO4: orthophosphates.

the site located just below the combined sewer overflow (3) was apparently no more impaired than the site immediately above the outflow (2).

In the hyporheic zone (Fig. 3a), axis F1 accounted for 44.4% of the total variance. It illustrated the difference between the global predominance of the Tubificidae (FTr4) at site 4 (Fig. 3b), from an association of AED and pollution-intolerant species (FTr1 & 2), particularly at sites 1 and 3. Axis F2 (24.5% of the total variance) was mainly correlated with the absolute densities of oligochaetes (EFBO), and characterized the sites 3 (December 2000) and 4 (June and October 2000). The F1–F2 plane discriminated site 1 from all the others. This site was frequently clogged by mineral particles and no oligochaetes were collected. But when silt was absent or reduced, the assemblages of the hyporheic zone were dominated by AED (FTr1) and pollution-intolerant species (FTr2).

### *Correlations with chemical features*

In the benthic system, the distribution pattern of oligochaetes on the F1–F2 plane was associated with an increasing gradient of nitrates (Fig. 2b), particularly from below the CSO (sites 3 and 4). Site 1 had good chemical quality compared to the three other ones, except for phosphorus. But the high standard deviation of phosphorus concentrations at the other sites reflected the phosphorus (and ammonium) peaks, characteristic of CSOs in urbanized environments (Paul & Meyer, 2001; Walsh et al., 2001). In the hyporheic system (Fig. 3b), site 1 exhibited relatively high values of $NH_4^+$ and $PO_4^{3-}$, probably related to mineral clogging at this site. The chemical alteration was marked at site 4 when considering ammonium salt contents (1.44 mg $l^{-1}$, Fig. 3b), but not at the first site below the CSO (site 3). The nitrate contents also declined at site 4, implying reducing conditions, in contrast to sites 2 and 3, where high nitrate contents indicated oxidized situations.

### *Relations between stream discharge and assemblage features*

The sludge effect, characterized by pollution-tolerant Tubificidae and *Lumbricillus* spp., was observed in the benthic system at all sites (Fig. 4) when the streambed discharges were very low with high CSO peaks (August and October 2000), but was least pronounced at the reference site (site 1). The sludge effect was also strongly pronounced at site 4 during June, when stream discharges were falling. These results suggested that rainfall led to overflows from the surrounding environment and brought sediments into the stream, even at the reference site, and that the purification capacity of the stream decreased (low discharges). The sludge effect (FTr4) remained most pronounced at the sites below the sources of pollution (2–4). When stream discharges increased ($> 14$ l $s^{-1}$), the sludge effect was much less marked in the benthic layer, but it remained strong in the hyporheic layer of site 4 (Fig. 4).

## Discussion

The species richness of the Chaudanne stream was high (46 taxa), probably because we considered both the surface and subsurface layers. The reference site (site 1) was predominated by pollution-intolerant species, which also are AED species, together with a greater species richness. We found similar ecological traits in a glacial unpolluted river (river Roseg, Lafont & Malard, 2001; Malard et al., 2001), where the surface and hyporheic habitats were colonized by pollution-intolerant AED species, particularly where upwellings of groundwater occurred. Site 3, located just below the main CSO, exhibited a distribution of epigean oligochaete assemblages similar to that of the site 2 above the CSO. In the hyporheic layer, the distribution of oligochaetes typified a rehabilitated situation compared to site 2 (Fig. 3b), with a predominance of pollution-intolerant AED species. This suggests that the effect of the CSO is not detrimental just below its inputs, and that benthic and hyporheic systems at site 3 were preserved by ground water upwellings. The high stream discharges had also a beneficial effect, probably by taking away the polluted sludge of the CSOs. Site 4 was the most vulnerable to a sludge effect, particularly in the hyporheic system that might be attributed to an attenuation of the slope of the stream and the predominance of downwellings of polluted surface waters. These last observations emphasize the fact that the hyporheic system can

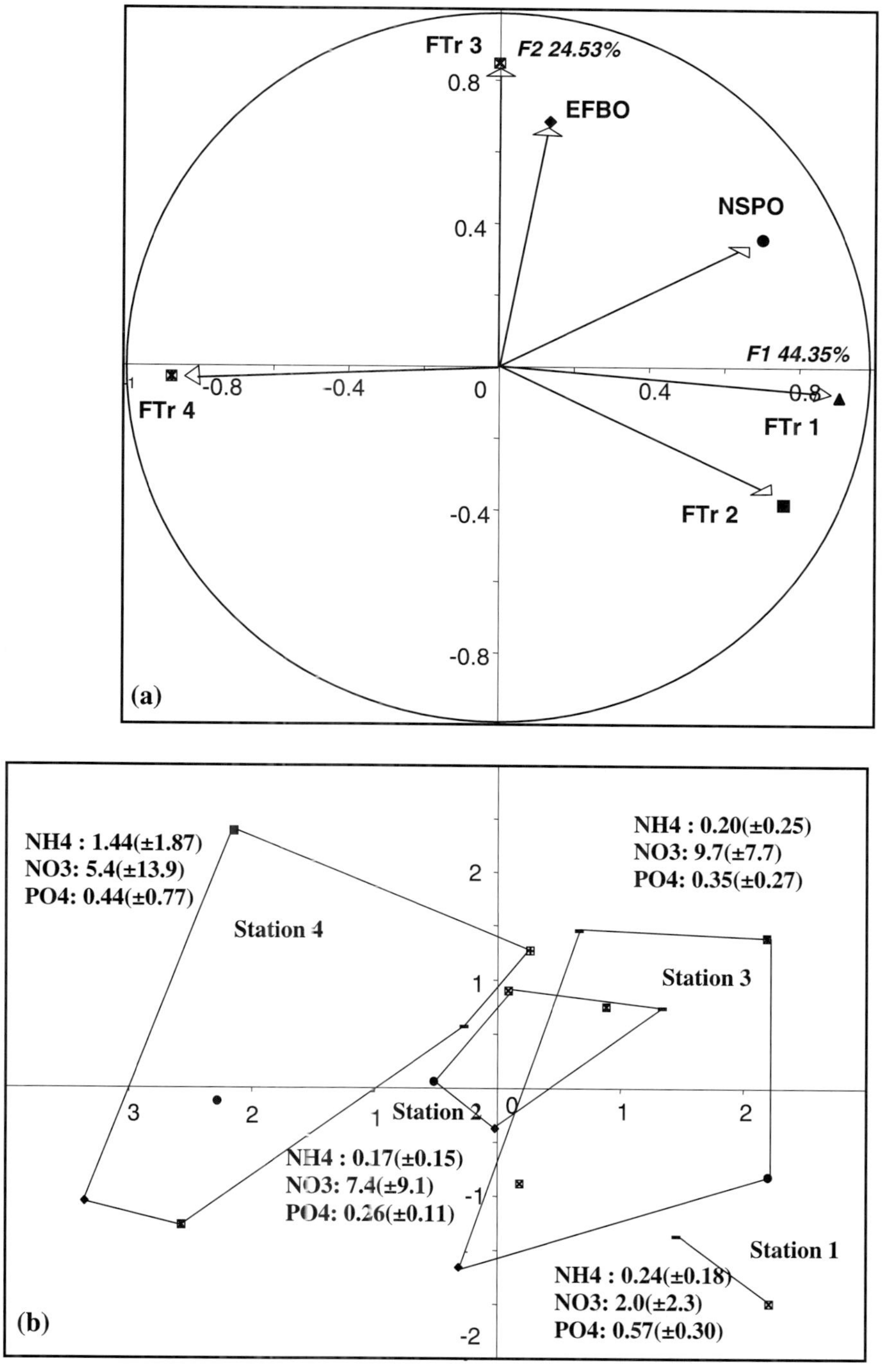


*Figure 3.* Standardized PCA on hyporheic oligochaete assemblages (18 rows, 6 columns); (a) correlation circle (F1–F2 plane); (b) plots and dispersion polygons of biological samples from the four sites on the first plane; NSPO: number of oligochaete species; EFBO: densities per 10 l of oligochaete assemblages; FTr1, FTr2: functional traits permeability and intolerance to water pollution; FTr3, FTr4: functional traits tolerance to water pollution and sludge effect; the chemical variables are expressed as mg $l^{-1}$; the standard deviations are put in brackets; NH4: ammonium salts; NO3: nitrates; PO4: orthophosphates.

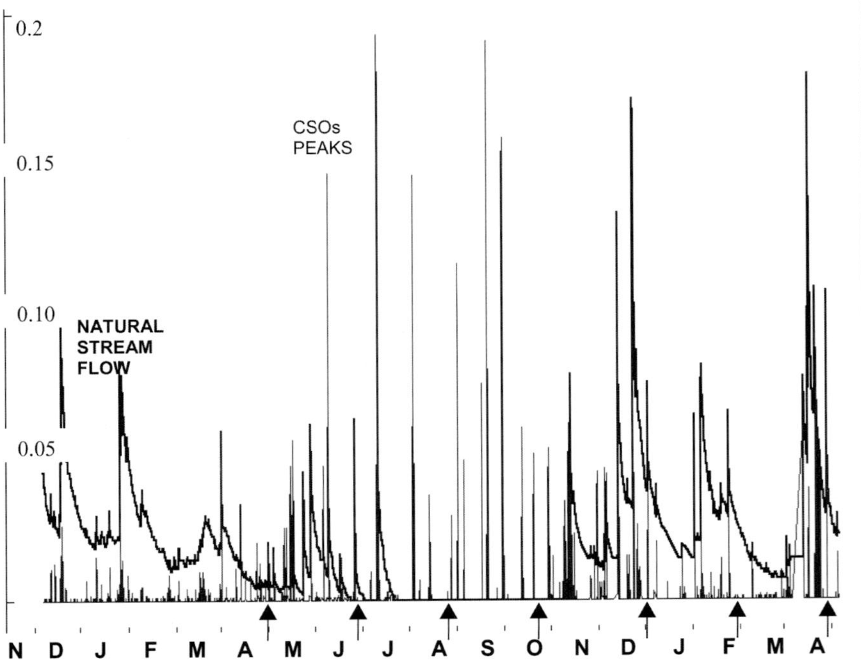


Surface sediments: percentages of the species indicating FTr4 (sludge effect)

| Sites | | | | | | | | |
|---|---|---|---|---|---|---|---|---|
| | **1** | 1% | 0% | ? | 11% | 0% | 1% | 7% |
| | **2** | 1% | 2% | 24% | 54% | 5% | 14% | 7% |
| | **3** | 0% | 0% | 30% | 33% | 0% | 1% | 0% |
| | **4** | 1% | 71% | ? | 50% | 22% | 3% | 16% |

Hyporheic zone: percentages of the species indicating FTr4 (sludge effect)

| Sites | | | | | | | | |
|---|---|---|---|---|---|---|---|---|
| | **1** | 0% | ? | ? | ? | ? | 0% | ? |
| | **2** | ? | 24% | ? | 17% | 7% | 2% | 9% |
| | **3** | ? | 7% | ? | 8% | 4% | 7% | 2% |
| | **4** | 40% | 87% | ? | 30% | 63% | 31% | 32% |

*Figure 4.* Comparison between the stream discharge (unbroken lines), the CSOs' peaks and the sludge effect (FTr4), at the four stations of the stream Chaudanne; vertical arrows: dates of biological samples; stream discharge and CSOs' peaks are expressed as $m^3\ s^{-1}$; ?: undetermined data.

accumulate nutrients and polluted inputs (Hynes, 1983; Jones & Mulholland, 2000) and retain residues of past and present pollution (Danielopol, 1989; Giere, 1993; Lafont et al., 1996).

The alteration of chemical quality is seen below the CSO, but with relatively low concentrations of solutes. The mean ammonia salt contents reached 1.44 mg $l^{-1}$ at site 4. These contents are considered indicative of a 'fair' quality in urban ecosystems (Faulkner et al., 2000). An increase in phosphorus and nitrogen loads was observed between site 2 (just above the CSO) and site 3 just below the CSO, but did not reach alarming values. The hyporheic zone of site 1 exhibited relatively high values of ammonia salts and phosphorus contents, potentially related to the clogging of interstices by mineral particles. This conflicts with the good biological quality of the site. We assume that the comparison between solute concentrations and oligochaetes will be better understood by considering the fluxes of solutes, rather than their instantaneous concentrations. Probably, nitrogen and phosphorus exhibited relatively high contents in the hyporheic zone at site 1 because they were not intensely assimilated by living organisms, which were scarce or absent.

The temperatures of surface and hyporheic waters are important factors, particularly for deriving patterns of downwelling surface water and upwelling ground water (Evans & Petts, 1997). However, temperature was not considered an important factor because temperature ranges did not reach extreme values in Chaudanne stream. Temperature, as well as other important factors like oxygen and organic matter (Strayer et al., 1997), will be considered in future studies.

The FTrs illustrated clear-cut gradients of oligochaete distribution in a situation not still severely polluted and where organic pollution predominated. The FTrs were empirically defined from literature data and our own experience and appear an attempt that by-passes regional and spatial heterogeneities and defines the functionalities of the studied habitats (Lafont et al., 2000). We are conscious that this approach is debatable. FTtr1 (permeability) seems relevant and supported by literature. FTr4 (sludge effect) is also relevant and supported by field experience on polluted coarse-grained habitats in running waters. The FTr2 (intolerance to water pollution) and FTr3 (tolerance to water pollution) are questionable. Moreover, the tolerance/intolerance to pollution of a given species is also strongly related to the habitat where it has been collected (Lafont, 1989).

The great majority of pollution studies regarding oligochaete communities considered fine sediments (see Chapman, 2001) and not coarse sediments or hyporheic systems. *Nais communis* is more tolerant of organic pollution in coarse than in fine sediments (Lafont, 1989). The pollution-tolerant species of fine sediment-dwelling Tubificidae are collected in coarse sediments only when polluted sludge is trapped within the interstices (Brinkhurst, 1965; Datry et al., 2003). Without polluted sludge, these species are not collected, even if the interstitial water is polluted. The ecological preferences of oligochaete species are also dependant upon the salinity of waters (Leland & Fend, 1998), habitat heterogeneity (Carter & Fend, 2001) and time (Verdonschot, 1999). But the hyporheic system is less sensitive to physical disturbances like floods than the benthic system, and its consideration in the ecological studies of watercourses has been recommended (Hynes, 1983; Boulton, 2000; Petts, 2000; Lafont, 2001; Boulton et al., 2003). The evaluation of the tolerance/intolerance to pollution of oligochaete species is probably more reliable in the hyporheic layer. Indeed, in the hyporheic layer, the species stay in more direct and constant exposition to pollutants than in the superficial zone, where flow regimes are prevailing factors (Bunn & Arthington, 2002) and reduce the probability of species exposure to pollutants.

Another issue is that several species occupy two FTrs while others occupy only one (e.g. *C. atrata* is characteristic of FTrs 1 & 2, the Tubificidae of FTr4). This means that the value of *C. atrata* is 0 to FTr4, and that the value of Tubificidae is 0 to FTr1 & 2, etc. This is an empiric simplification that needs several improvements. For example, *Chaetogaster diastrophus* is a common and abundant species in the Chaudanne stream, but it is above all a species which is tolerant to organic enrichment, but intolerant to severe organic pollution or toxic substances (Lafont, 1989). It did not fit any of the four FTrs considered here, but other FTrs will be investigated, and we are conscious that the functioning of coarse and hyporheic sediments cannot be described by only four FTrs.

This first general survey was a feasibility study. We now plan to conduct multidisciplinary research on the Chaudanne stream, including studies of crustaceans and hydraulics modeling. This research is expected to improve our present knowledge about the value of oligochaete species and present or future FTrs as functional descriptors (Lafont, 2001; Ruysschaert & Breil, 2004).

## Acknowledgements

We warmly thank Andrew Boulton, an anonymous reviewer and Adrian Pinder for their comments that enabled us to significantly improve our manuscript and the English language. We thank Christophe Lafont for his help in the field and the laboratory. This work was funded by the Observatoire de Terrain en Hydrologie Urbaine (OTHU), the Lyon's Urban Community, and the EU project URBEM (Urban River Basin Enhancement Methods, 5th PCRD).

## References


Bernoud S., 1998. Réponse écologique d'un ruisseau périurbain aux rejets de temps de pluie. TFE Mastère "Eau Potable et Assainissement", ENGEES, 55 pp.

Bou, C. & R. Rouch, 1967. Un nouveau champ de recherches sur la faune aquatique souterraine. Comptes Rendus de l'Académie des Sciences, Paris 265: 369–370.

Boulton, A. J., 2000. River ecosystem health down under: assessing ecological conditions in riverine groundwater zones in Australia. Ecosystem Health 6: 108–118.

Boulton, A. J., W. F. Humphreys & S. M. Eberhard, 2003. Imperilled subsurface waters in Australia: biodiversity, threatening processes and conservation. Aquatic Ecosystem Health and Management 6: 41–54.

Brinkhurst, R. O., 1965. Observation on the recovery of a British river from gross organic pollution. Hydrobiologia 25: 9–51.

Bunn, S. E. & A. H. Arthington, 2002. Basic principles and ecological consequences of altered flow regimes for aquatic biodiversity. Environmental Management 30: 492–507.

Carter, J. L. & S. V. Fend, 2001. Inter-annual changes in the benthic community structure of riffles and pools in reaches of contrasting gradient. Hydrobiologia 459: 187–200.

Chapman, P. M., 2001. Utility and relevance of aquatic oligochaetes in ecological risk assessment. Hydrobiologia 463: 149–169.

Cyr J. F., C. Marcoux, J. C. Deutsch & P. Lavalée, 1998. L'hydrologie urbaine : nouvelles problématiques, nouvelles approches de solutions. Revue des Sciences de l'Eau, no. spécial, 51–60.

Danielopol, D. L., 1989. Groundwater fauna associated with riverine aquifers. Journal of the North American Benthological Society 8: 18–35.

Datry, T, F. Malard, R. Niederreiter & J. Gibert, 2003. Video-logging for examining biogenic structures in deep heterogeneous subsurface sediments. Comptes Rendus Biologies 326: 589–597.

Evans, E. C. & G. E. Petts, 1997. Hyporheic temperature patterns within riffles. Hydrologic Sciences Journal des Sciences Hydrologiques 42: 199–213.

Faulkner, H., V. Edmonds-Brown & A. Green, 2000. Problems of quality designation in diffusely polluted urban streams – the case of Pymme's Brook, north London. Environmental Pollution 109: 91–107.

Gaschignard, O., 1984. Impact d'une crue sur les invertébrés benthiques d'un bras mort du Rhône. Verhandlungen Internationale Vereiningung für Theorische und Angewandte Limnologie 22: 1997–2001.

Giani, N., 1984. Le Riou Mort, affluent du Lot, pollué par des métaux lourds. IV. Les oligochètes. Annales de Limnologie 20: 167–181.

Giere, O., 1993. Meiobenthology. The Microscopic Fauna in Aquatic Sediments . Springer-Verlag, Berlin Heidelberg 328 pp.

Hynes, H. B. N., 1983. Groundwater and stream ecology. Hydrobiologia 100: 93–99.

Jones, J. B. & P. J. Mulholland, 2000. Streams and Ground Waters . Academic Press, San Diego 425 pp.

Juget, J., 1984. Oligochaeta of the epigean and underground fauna of the alluvial plain of the French upper Rhône (biotypological try). Hydrobiologia 115: 175–182.

Juget, J. & E. Dumnicka, 1986. Oligochaeta (incl. Aphanoneura) des eaux souterraines continentales. In Botosaneanu, L. (ed.) Stygofauna mundi. E. J. Brill, Leiden, 234–244.

Juget, J. & M. Lafont, 1994. Theoretical habitat templets, species traits, and species richness: aquatic oligochaetes in the Upper Rhône River and its floodplain. Freshwater Biology 31: 327–340.

Lafont, M., 1989. Contribution à la gestion des eaux continentales: utilisation des oligochètes comme descripteurs de l'état biologique et du degré de pollution des eaux et des sédiments. Thèse de Doctorat ès Sciences. Université Lyon I: 403 pp.

Lafont, M., 2001. A conceptual approach to the biomonitoring of freshwater: the ecological ambience system. Journal of Limnology 60(Supplementum 1): 17–24.

Lafont, M. & A. Durbec, 1990. Essai de description biologique des interactions entre eau de surface et eau souterraine: application à l'évaluation de la vulnérabilité d'un aquifère à la pollution d'un fleuve. Annales de Limnologie 26: 119–129.

Lafont, M. & J. Juget, 1993. Description de *Rhyacodrilus ardierae* n. sp. (Oligochaeta, Tubificidae) récoltée dans l'Ardières (affluent de la Saône, France). Bulletin de la Société Zoologique de France 118: 115–123.

Lafont, M. & F. Malard, 2001. Oligochaete communities in the hyporheic zone of a glacial river, the Roseg River, Switzerland. Hydrobiologia 463: 75–81.

Lafont, M., A. Durbec & C. Ille, 1992. Oligochaete worms as biological describers of the interaction between surface and groundwaters: a first synthesis. Regulated Rivers 7: 65–73.

Lafont, M., J. C. Camus & A. Rosso, 1996. Superficial and hyporheic oligochaete communities as describers of pollution and water exchanges in the River Moselle system (France). Hydrobiologia 334: 147–155.

Lafont, M., S. Bernoud, J. C. Camus, Ph. Namour, P. Breil & P. Le Pimpec, 2000. Etat écologique d'un ruisseau péri-urbain soumis à des rejets de temps de pluie ; premiers résultats et perspectives. Acta 2ème Symposium International Québec-Paris, *La réhabilitation et l'aménagement des cours d'eau en milieu urbain, 18–20 octobre 2000, Paris*, 11–22.

Leland, H. V. & S. V. Fend, 1998. Benthic invertebrate distributions in the San Joaquin River, California, in relation to physical and chemical factors. Canadian Journal of Fisheries and Aquatic Sciences 55: 1051–1067.

Malard, F., M. Lafont, P. Burgherr & J. V. Ward, 2001. A comparison of longitudinal patterns in hyporheic and benthic oligochaete assemblages in a glacial river. Arctic, Antarctic, and Alpine Research 33: 457–466.

Nogueira, S., 2001. Incidence des rejets urbains de temps de pluie sur les communautés d'invertébrés interstitiels (oligochètes et crustacés) d'un petit cours d'eau périurbain. *Cemagref/BELY*, 42 pp.

Paul, M. J. & J. L. Meyer, 2001. Streams in the urban landscape. Annual Review of Ecology and Systematics 32: 333–365.

Petts, G. E., 2000. A perspective on the abiotic processes sustaining the ecological integrity of running waters. Hydrobiologia 422/423: 15–27.

Rogers, C. E., D. J. Brabander, M. T. Barbour & H. F. Hemond, 2002. Use of physical, chemical, and biological indices to assess impacts of contaminants and physical habitat alteration in urban streams. Environmental Toxicology and Chemistry 21: 1156–1167.

Ruysschaert F. & P. Breil, 2004. Assessment of the hyporheic fluxes in a headwater stream exposed to combined sewer overflows. Acta Fifth International Symposium on Ecohydraulics. Aquatic Habitats: Analysis and Restoration, Madrid 2004: 293–299.

Solbé, J. F. & L. G. de, 1975. Annelida. In Curds, C. R. & H. A. Hawkes (eds), Ecological Aspect of Used Water Treatment . Academic Press, London, 305–335.

Strayer, D. L., S. E. May, P. Nielsen, W. Wollheim & S. Hausam, 1997. Oxygen, organic matter, and sediment granulometry as controls on hyporheic animal communities. Archiv für Hydrobiologie 140: 131–144.

Thioulouse, J., D. Chessel, S. Doledec & J.-M. Olivier, 1997. ADE-4: a multivariate analysis and graphical display software. Statistics and Computing 7: 75–83.

UE, 2000. Directive 2000/60/CE du Parlement Européen et du Conseil du 23 octobre 2000 établissant un cadre pour une politique communautaire dans le domaine de l'eau. Adoptée le 22/12/2000.

Verdonschot, P. F. M., 1999. Micro-distribution of oligochaetes in a soft-bottomed lowland stream (Elsbeek; The Netherlands). Hydrobiologia 406: 149–163.

Walsh, C. J., A. K. Sharpe, P. F. Breen & J. A. Sonneman, 2001. Effects of urbanization on streams of the Melbourne region, Victoria, Australia. I. Benthic macroinvertebrate communities. Freshwater Biology 46: 535–551.

# Part Four
# Sludge Studies

Hydrobiologia (2006) 564:197–211
P.F.M. Verdonschot, H. Wang, A. Pinder & R. Nijboer (eds), Aquatic Oligochaete Biology IX
DOI 10.1007/s10750-005-1719-7



# Sludge reduction by predatory activity of aquatic oligochaetes in wastewater treatment plants: science or fiction? A review

Christa H. Ratsak[1,*] & Jaap Verkuijlen[2]
[1]*Sub-department of Environmental Technology, Wageningen University and Research Centre, P.O. Box 8129, 6700 EV Wageningen, Bennekom, The Netherlands*
[2]*DHV BV, P.O. Box 1132, 3818 BC Amersfoort, The Netherlands*
*(*Author for correspondence: E-mail: christa.ratsak@hetnet.nl)*

*Key words:* oligochaetes, *Aeolosoma*, Naidids, Tubificidae, Lumbriculidae, review, wastewater treatment plant, predation, sludge reduction

## Abstract

Biological aerobic wastewater treatment plants (WWTPs) produce a lot of excess sludge. The costs for handling this residual product are increasing, so the search for alternative techniques to reduce the amount of sludge has to be continued. Activated sludge consists of inorganic and organic substances, bacteria, protozoa and metazoa. Due to incomplete biomass conversion, sludge consumption yields less oligochaete biomass. From a technological point of view, the application of aquatic oligochaetes to reduce the sludge production offers interesting perspectives. This paper aims to review the feasibility for the reduction of activated sludge in WWTPs by means of aquatic oligochaetes. Also the current techniques concerning sludge reduction are taken into account. Several of the WWTPs relevant parameters, which may influence predatory activity of aquatic oligochaetes, are discussed: particle size, organic content of substrate, bacteria preference, life cycle and population dynamics of aquatic oligochaetes, temperature, pH, oxygen and process conditions. From the literature it appeared that most research has been performed on laboratory scale. Only a few authors mention a significant reduction of the sludge production by 'sessile' species such as *Lumbriculus*. Vermicultures for the reduction of activated sludge are rather common in developing countries. Incidentally large annelid blooms have been noticed in WWTPs. It remains obscure which factors trigger the initiation of annelid blooms in WWTPs and which are of importance to maintain a stable annelid population in WWTPs. The influence of a considerable worm bloom on the waste sludge production is still under investigation.

## Introduction

Oligochaete worms are found in both fresh and saline continental water. Many river and lake biotopes are often dominated by oligochaetes, which may comprise 50–60% and even 100% of the benthic biomass.

Oligochaete worms dwelling in the profundal zone of lakes play an important role in the exchange of matter and are responsible to a notable degree for the rate of mud formation and mineralization of bottom sediments. Additionally, the presence of a large population of saprophytic species of oligochaetes is an important factor in the self-purification of polluted waters.

Many environmental variables including water temperature, oxygen concentration, presence of competitors or predators and substrate properties influence habitat selection and selective feeding by oligochaetes. A number of species of oligochaete worms are able to feed on waste organic materials such as sewage sludge and cattle excrement. These worms have subsequently been used as a feed for fish or farm animals, for example, 60–70% of the dry weight of *Eisenia foetida* is protein, high in essential amino acids. Other studies also indicate that enchytraeid worms are a highly suitable fish food (Kirk & Howell, 1972; Kirk, 1973, Bouguenec, 1992). In the tropical regions of India systems of fishes grown in treated domestic sewage has been investigated (Inakollu & Wanganeo, 2002).

Many different types of sessile (benthic species which are attached to carrier material in WWTPs) and free-swimming oligochaetes have frequently been found in low loaded (0.04–0.07 kg BOD/kg DW d) WWTPs (Ratsak, 1994, 2001; Janssen et al., 1998; Elissen et al., submitted for publication). A WWTP may be considered as an artificial ecosystem (Hawkes, 1963) that is determined by its influent characteristics, its design and treatment management. The concentration of activated sludge is controlled by the sludge wastage rate. As a result of high sludge wastage rate the sludge concentration will decrease.

Excess sludge production is affected by the sludge loading rate, temperature and the oxygen concentration. Raising the concentration of dissolved oxygen in the sludge and thus stimulating the microbial activity responsible for mineralization of the sludge enhances the reduction of the excess sludge production (Abbassi et al., 1999). The number of parameters affecting the bacteria in biological wastewater treatment plants is immense and is continuously varying in time. Occasionally, the appearance of oligochaetes seems to coincide with a considerable reduction of the waste sludge (Ratsak, 1994; Wei et al., 2003).

Since the introduction of legislation prohibiting the use of sludge as fertilizer in agriculture in the Netherlands, disposal of activated sludge from wastewater treatment plants has become very expensive. Generally, large amounts of bio-sludge are formed in biological wastewater treatment processes and the separation, dewatering, treatment and disposal of this sludge require large investments and form a major percentage of the operating costs for treatment. The disposal of excess sludge from WWTPs represents a rising challenge in activated sludge processes for modification in design and operation to minimize sludge.

Sludge minimization may be a result of reduced production of sludge and/or disintegration processes both in the wastewater treatment stage and in the sludge stage. In biological wastewater treatment processes there are principally three main strategies by which sludge reduction can be achieved, (a) by enhanced pre-treatment (reduced load), (b) by yield reduction and (c) by sludge disintegration processes by which a greater part of the sludge becomes more biodegradable so that more can be mineralized and assimilated by the biomass.

In the literature, various sludge disintegration technologies for sludge minimization are discussed (see Fig. 1), including mechanical methods (focusing on stirred ball-mill, high pressure homogenizer, ultrasonic disintegrator), electrical methods, chemical methods (focusing on the use of ozone and hydrogen peroxide), physical methods (focusing on thermal and thermal/chemical hydrolysis) and biological methods (focusing on enzymatic processes such as lyses-cryptic growth, decay, uncoupling metabolism, maintenance metabolism and anaerobic treatment). Finally, an alternative strategy is sludge reduction by predation. At each transfer to a higher trophic level of the food chain, a large proportion of the potential energy is lost as heat, due to maintenance processes, respiration and reproduction. Ultimately, a lower biomass remains. Vermiculture is considered as post-treatment of dewatered or thickened sludge and sludge reduction by aquatic oligochaetes is comprehensively dealt with in this review. Excellent reviews including pros and contras of the above mentioned methods are outlined by Ødegaard (1993), Mayhew & Stephenson (1997), van Loosdrecht &

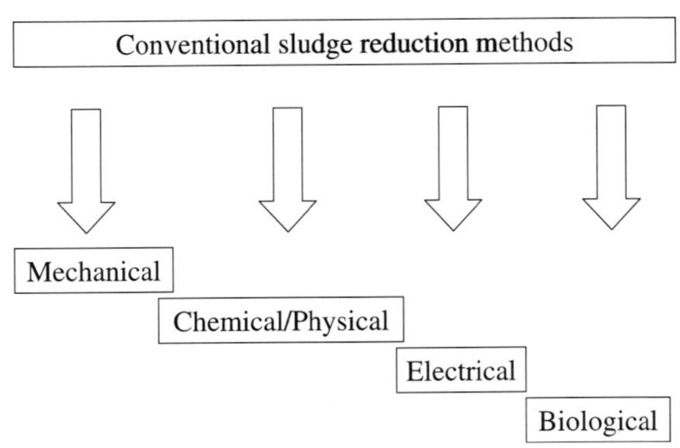


*Figure 1.* Sludge disintegration methods used for activated sludge minimization.

Henze (1999), Wei et al. (2003), WIRES (2004). The potential to manipulate the process to reduce yield is increasing and each individual methodology has its advantages depending on the process configuration.

The present discharge standards on nitrogen have lead to a decrease in both total amount of sludge and organic content of excess sludge, due to increase of retention time and therefore an ongoing mineralization. However, due to strengthened environmental standards, total number of WWTPs is increasing.

Especially in small WWTP (<100 000 population equivalents) without an anaerobic digester, it is interesting to find ways of designing and operating biological treatment to reduce biomass production. Oligochaetes may be introduced to the aeration process or the waste sludge may be separately treated by oligochaetes.

The oligochaetes used for treatment of excess activated sludge can be divided into two groups, firstly the large aquatic worms such as the Tubificidae, Lumbriculidae and the semi-aquatic or terrestrial Enchytraeidae and, secondly the small aquatic worms such as Naidids and Aeolosomatids. The usability of these groups is quite different because of the different niches they occupy. Whereas the former needs a substratum in which it can swirl around, the latter occurs in the mixed liquor or in the biofilm attached to a carrier material. Because it is difficult to create favorable conditions for their maturation and successful sexual reproduction, these small aquatic species are not suitable for stable long-term cultivation so far. This important difference has to be taken into account for the development of a predation stage in the wastewater treatment process.

The objective of this review is to provide insight in the feasibility of sludge reduction in WWTPs by using worms. To reach this insight a comparison was made between the information available on oligochaetes in their natural habitat, and in a laboratory environment, with the information available on worms in WWTPs. Their role in the latter is becoming more and more interesting from both sustainable and economical point of view. The review is not covering all the literature published about the subjects listed below. This would be quite a laborious exercise.

In the subsequent sections several important parameters, which are relevant for wastewater treatment plants and may influence the predatory activity of aquatic oligochaetes are discussed: particle size, organic content of substrate, bacteria preference, life cycle and population dynamics, temperature, pH, oxygen and process conditions.

## Particle size

Particle size is supposed to play a major role in the distribution and abundance of tubificids. This section deals with two distinct aspects of particle size selectivity: the influence of particle size on distribution (habitat selection) and selective feeding.

Sauter & Güde (1996) found the grain size fraction less than 63 $\mu$m was an important environmental factor affecting the presence of many of the tubificid species of Lake Constance. In substrates with a heterogeneous grain size where tubificids can be selective many species occur almost exclusively in their particle size preference range, which is highly species specific. Five species showed a preference for substrates with a high clay and silt content. These are *Tubifex tubifex*, *Potamothrix heuscheri*, *Potamothrix bedoti*, *Ilyodrilus templetoni* and *Aulodrilus pluriseta*. Three species *Pothamothrix moldaviensis*, *Limnodrilus claparedeianus* and *Tubifex ignotus* are more likely to occur in sandy substrates, having clay and silt content of less than 10% (see Table 1).

Juget (1979) reported by microscopic analysis, the maximum diameter of the biggest particle observed in the gut of the tubificid worm *Tubifex tubifex* was 130 $\mu$m. *Tubifex tubifex* mainly ingested particles <63 $\mu$m.

Rodriguez et al. (2001) measured the particle size distribution of fecal pellets produced by *T. tubifex*. The fecal material was composed of particles with a mean diameter under 63 $\mu$m. This suggested that this species actively selected the silt-clay fraction, avoiding the larger sand particles. A more detailed analysis showed that about 75%, by volume, was composed of particles with a mean diameter <25 $\mu$m, and the mode was <10 $\mu$m.

Only one reference has been found with respect to small aquatic oligochaetes and size selective feeding. Bowker et al. (1983) indicated the shape

*Table 1.* Substrate preference of species in relation to the clay and silt content of the sediment (after Sauter & Güde, 1996)

| Species | Clay and silt content (%) |
|---|---|
| *Tubifex tubifex* | >60 |
| *Potamothrix heuscheri* | >60 |
| *Potamothrix bedoti,* | >60 |
| *Ilyodrilus templetoni* | >60 |
| *Aulodrilus pluriseta* | >60 |
| *Stylodrilus heringianus* | 20–30 |
| *Limnodrilus claparedeianus* | <10 |
| *Pothamothrix moldaviensi* | <10 |
| *Tubifex ignotus* | <10 |
| *Psammoryctides barbatus* | <10 |
| *Limnodrilus udekemianus* | 50–70 |
| *Spirosperma ferox* | 50–70 |
| *Limnodrilus hoffmeisteri* | No preference |
| *Limnodrilus profundicola* | No preference |
| *Ptomothrix hammoniensis* | No preference |

of the pharynx from the study of *Nais elinguis* possibly determined the maximum size of food. The rate at which algae were ingested was negatively correlated with the lengths of the algae (Bowker et al., 1985a). Also unicellular algae were ingested at a greater rate than colonial and filamentous algae.

**Organic content of substrate**

The aspects distribution (habitat selection) and selective feeding also applies to the relationship between tubificid abundances and organic matter. Most available literature concerns the (semi-) aquatic worms Lumbriculidae and Tubificidae, especially *Tubifex tubifex*.

Substrate selection is a taxis as opposed to kinesis. It appears to be a taxic response to dispersing microbiota or molecules with molecular weight greater than 12,000–14,000 (McMurtry et al., 1983). This means that worms must be able to detect the presence and direction of a sediment component dispersing through the water. However individual variation in habitat selection behavior exists due to the considerable behavioral heterogeneity of individuals.

Studies of the influence of substrate properties on habitat selection by tubificids have produced conflicting results. Wachs (1967) found that *T. tubifex* preferred nutrient-rich sediment (i.e. high organic C and total N) regardless of particle size while Zahner (1967) found that this species chose coarse-grained nutrient-poor sediment (i.e. low in organic C, organic N, and organic P) to fine-grained nutrient-rich sediment.

Brinkhurst & Jamieson (1971) pointed out that few correlations between the variations of total organic matter and the distribution and abundance of worms have been demonstrated. The quality of organic matter in the sediment is therefore more important in determining tubificid preference than the quantity of organic matter. However, when the correlation of abundance of tubificids with total organic matter in sediments was investigated, the results were contradictory. Sometimes positive correlation has been observed (Brinkhurst, 1970), but in many instances tubificid abundance was not correlated or was even negatively correlated with the organic content of sediments (McMurtry et al., 1983; Robbins et al., 1979, 1989).

Total organic carbon is an unsatisfactory measure of the nutrients available to tubificids. Poor correlations occur because total organic carbon includes not only valuable food sources (bacteria or algae) but also substances resistant to decomposition and presumably unsuitable as a food source (lignin and humic materials) (Pasteris et al., 1994). All in all it can be concluded that the results are not always unambiguous.

According to Rodriguez et al. (2001), *T. tubifex* exhibits two levels of selectivity in its feeding behavior. Firstly, particle size and secondly the organic content of the particles. *T. tubifex* selectively fed the organic rich particles and this feeding was independent of particle size (within the fine (silt-clay) fraction) of the sediment.

Also other studies have shown selective feeding by oligochaetes on the organic component (living or non-living) of the sediment (Brinkhurst & Chua, 1969; Wavre & Brinkhurst, 1971; Brinkhurst et al., 1972; Brinkhurst & Austin, 1979; Matisoff et al., 1999). *Branchiura sowerbyi* was able to grow on activated sludge but only when this was mixed with sand or river mud (Aston & Milner, 1982; Aston et al., 1982).

Finogenova & Lobasheva (1987) studied the growth of *T. tubifex* on highly caloric activated sludge and poor in organic matter natural silt in glass vessels (10 cm in diameter). Selectivity in feeding of oligochaetes did not only depend on the type of ground inhabited by the worms but also on their age. The specific growth rate of worms on activated sludge was twice as high as that on natural silt. Furthermore the worms reached maturity more quickly. The number of eggs produced and values of fecundity also were obviously attributed to specific characteristics of the food involved. The average number of eggs/ind/100 days was respectively 2.5 and 5 times higher of the worms grown on activated sludge compared with the worms cultured on natural silt or fainted lettuce. The extreme tolerance to pollution and ability to accelerate growth and development allow *T. tubifex* and *L. hoffmeisteri* (ecologically closely related) to be predominant under conditions of great quantities of easily assimilated organic material, i.e. in waters affected by human activity such as sewage sludge.

Sewage sludge (50–70% organic carbon) originating from a low loaded conventional activated sludge plant can be used as a food source for oligochaetes (Rensink & Rulkens, 1997; Klapwijk et al., 2000). Pilot plant experiments have been carried out to study the mineralization of sludge with and without addition of tubificids in trickling filters filled with lava slag or synthetic carrier material. Sludge reduction through predation varied between 18 and 67%. Unfortunately however, it was not possible to reproduce the experiments. This could be ascribed to the fact that an unknown amount of sludge accumulates within the filters.

Buijs et al. (submitted for publication) investigated the effect of predation by *Lumbriculus variegatus* (Family Lumbriculidae) on the stabilization of activated sludge and found an enhanced stabilization of activated sludge by predation of the sludge in aerobic/anoxic batch cultures. An additional sludge reduction up to 30% was found compared with a reference setup without *L. variegatus*.

There is not much literature available about the small aquatic worms and activated sludge with respect to this item. Ratsak et al. (1993) developed a mathematical model for the growth of *Nais elinguis* on activated sludge.

**Bacteria preference**

Coler et al. (1968) demonstrated selective feeding on bacterial communities for the first time. The importance of bacteria in the diet of tubificids is now well established (Brinkhurst & Chua, 1969; Wavre & Brinkhurst, 1971; Brinkhurst et al., 1972). Further, Coler et al. (1968) have shown that mixed populations of tubificids can discriminate between different species of bacteria. It is assumed that while tubificids can respond to many different environmental characteristics, the relative importance of these differs and only a few of them exert primary influence on the preference response in the natural environment (Cummins & Lauff, 1968).

Also Brinkhurst (1974) and Milbrink (1993) investigated the existence of selective feeding on bacterial communities associated with sediment particles or with organic matter in the sediment. The role of potential food sources such as anaerobic bacteria, protozoa and large organic molecules in substrate selection is largely unknown. McMurtry et al. (1983) concluded that the microbial constituents of the sediment are more important than its physical and chemical properties in influencing substrate selection by Tubificids. There was a significant correlation between the abundance of heterotrophic aerobic bacteria in sediment and tubificid preference.

The substrate preference can be caused by selective colonization of different bacteria/algae species in the substrate. However, the occurrence of a certain species is not determined by a single environmental factor. If individual tubificid worms were exposed to more heterogeneous substrates, worms were more selective in the range of grain size in which they occurred.

Oligochaete species have been shown to differ in their ability to digest different bacterial species. Wavre & Brinkhurst (1971) found most of the heterotrophic bacteria (about 70%) were killed on passage through the gut, and the freshly produced feces became rich in monospecific bacteria. Different theories about the origin of bacterial concentrations on feces say that bacteria surviving gut passage may multiply rapidly due to relaxed competition, factors in the gut itself may stimulate bacterial growth, or selection may favor rapidly growing strains of microorganisms (Juniper, 1981).

According to Brinkhurst (1974) tubificids as a group ingest particles selectively and three tubificid species fed the same food source egests different proportions of at least one representative fraction of the bacterial flora present, the heterotrophic aerobes. The feces of one worm species contain a high proportion of bacteria that are the preferred food of another species in the association and thus worms may profit from each other (mutualism). These findings are supported by the research of Milbrink (1987). He found an increased growth rate of 50% and sexual maturity was reached earlier in mixed cultures due to each species selectively grazing concentrations of bacteria associated with the feces of another species in the community.

Selective feeding therefore permits more tubificid species to coexist, thereby lowering the respiration rate and increasing growth due to higher assimilation efficiency by feeding on available food. Solitary species may exhaust their preferred food and may actively burrow in search of preferred food hence raising their respiration loss, reducing time available for feeding and lowering their assimilation efficiency by feeding on available food.

There is ample evidence that polychaetes, oligochaetes, chironomid larvae, crustaceans, etc., are selectively attracted to colonies of bacteria, and it has been argued 'microbial stripping' of particles like fecal pellets can enhance microbial turnover and biomass of ingested particles (Milbrink, 1993). Mermillod-Blondin et al. (2003) found the combination of *Limnodrilus* and *Tubifex* worms produced higher sediment redistribution and a higher stimulation of bacterial numbers in the top 5 cm of the sediment than tested separately. Furthermore, the combination of the two genera produced lower oxygen consumption than the two genera tested separately in the deep layers of the sediment.

The fact that many deposit-feeders produce fecal pellets enriched in organic matter relative to available sediment indicates selective ingestion (Hargrave, 1976; Juniper, 1981; Cammen, 1982). Brinkhurst et al. (1971) found oligochaete feces had a higher percentage of organic matter, energy and nitrogen than the surrounding sediments. It has been argued, however, that the degree of enrichment is often too low to meet energy demands. Therefore most deposit feeders would also require nonliving food for subsistence, i.e. microorganisms largely for proteins, and undifferentiated detritus for carbohydrates (Lopez & Levington, 1987).

Enchytraeids had no detectable effect on microbial biomass, but the microbial respiration increased by 35% in the surface horizon. Because of the enhanced microbial activity in this horizon, Cole et al. (2000) suggested enchytraeids indirectly drive processes of decomposition and nutrient mineralization in organic upland soils. There were no effects on the release of inorganic N or P, although the release of ammonium and phosphate was correlated with the number of enchytraeids. Enchytraeids almost doubled the availability of organic carbon.

Selective feeding by *Nais elinguis* and *Nais variabilis* is investigated by Harper et al. (1981a, b), Bowker et al. (1983, 1985a, b). No conclusive evidence was obtained indicating a preferential ingestion of certain kinds of bacteria, that the gut environment was selectively hostile to the particular kinds of bacteria studied or that worms had a specialized gut micro-flora. On the contrary, *N. elinguis* actively selected diatoms in preference to other algae species (Bowker et al., 1985a, b). The worms were able to discriminate between filamentous algae and/or diatoms and unicellular or colonial cholorphycean algae. Diatoms were selected in preference to latex beads but the worms could not discriminate between living filaments and nylon monofilaments. A thigmotactic response to filaments and a chemo tactic response to diatoms may be implicated.

The (mostly small) metazoa frequently found in the activated sludge and biofilm systems are the oligochaetes *Aeolosoma hemprichi*, *Pristina longiseta*, *N. variabilis* and *N. elinguis*. Additionally, Inamori et al. (1983) also encountered *Dero* sp. (especially on biofilm), *Chaetogaster* sp., *Paranais* sp., *Limnodrilus* sp. and *Tubifex* sp.

Only two authors investigated food selectivity in small aquatic oligochaetes (Inamori et al., 1983, 1987; Kuniyasu et al., 1997). *Nais* sp., *Aeolosoma hemprichi*, *Pristina* sp. and *Dero* sp. were cultured on sterilized activated sludge in order to investigate maximum specific growth rates of the worms fed different diets and at different temperatures. *A. hemprichi* was able to monoxenically grow on *Pseudomonas putida*, *Escherichia coli*, *Acinetobacter calcoaceticus*, *Klebsiella pneumoniae*, *Flavobacterium luteus*, *Flavobacterium suaveolens* en *Streptococcus acidominius*. *P. longiseta* and *N. variabilis* could also grow on these species and on *Micrococcus luteus*. Neither of the tested oligochaetes was able to grow on *Bacillus cereus*, *Bacillus subtilis* and *Achromobacter cycloclastus* (Kuniyasu et al., 1997; Inamori et al., 1987, 1990). Perhaps the worms had a preference for Gram-negative bacteria. This aspect however, is not mentioned in the discussion. Especially the bacterial species was an important factor for the growth of *A. hemprichi* and a still unknown sludge component was important for the growth of *Nais* sp. (Inamori et al., 1983).

## Life cycle and population dynamics

The structure and persistence of biological communities is strongly influenced by biological interactions between component species: few would deny the importance of competition and degradation in ecological theory development and in natural communities. The available literature about this item is innumerable. Only some remarks are placed with respect to wastewater treatment in advance of the discussion section.

The reproduction modes of the oligochaetes are very important for manipulation of the population growth. The formation of paratomic budding is genetically determined and limited (Hämmerling, 1924). A great variety of asexual reproductive modes are known among aquatic oligochaetes.

Several intrinsic density regulation mechanisms exist; for example, a decrease in population abundance (e.g. by predation) causes both the percentage maturation and the specific fecundity to increase, so leading to an overall numerical growth. This new situation in its turn will bring out, through the decreased maturation and fecundity, a decline in the population size (Adreani et al., 1984). From population dynamics modeling, density dependence appeared to be extremely important as far as regulation of population abundance of the tubificids was concerned (Bonacina et al., 1989b). Generation time and fecundity were lower in 'eutrophic' than in 'oligotrophic' species. Also a positive correlation between minimum generation time and the embryonic developmental time was shown. Maturation time is revealed as density dependent and depends on the individual growth rate. Density affects negatively the percentage of the population that reaches maturity and the specific fecundity (Adreani et al., 1984).

Population density controls the fecundity, growth, maturation and ovigeration rates of *Tubifex tubifex* (Bonacina et al., 1989a, b). Growth rates of *T. tubifex* and mean number of eggs laid/ovigerous are inversely related to density, but generation time appears to be directly related. Very low or very high initial densities display efficient density controls.

Some associations are the results of species-specific feeding on bacteria colonizing the fecal pellets of the component species (Milbrink, 1987, 1993). Each species is supposed to grow significantly better when fed upon the fecal pellets of the other species (see also previous section). The results suggest facultative mutualism may be more common than generally believed. The mutualistic interaction is likely to be 'indirect' via the species-specific intake by the worms of bacteria associated with the fecal pellets of the component species.

Christensen found in dense mass cultures (with densities above 400), *Enchytreaeus bigeminus*, reproduced solely by fragmentation and subsequent regeneration. At densities less than 300–350 (in 200 ml glass jars and 75 $cm^3$ wet coarse sand substratum), sexual reproduction was not suppressed. The ecological importance of this reproductive behavior in dense populations of *E. bigeminus* is based upon suppression of sexual reproduction in some potential competitors that are obligatory sexual breeders (Christensen, 1973).

Although some authors claim Naididae and Aeolosomatidae are characterized by a regular, mostly seasonal alternating of sexual and asexual reproduction (Christensen, 1984; Timm, 1984), we did not find any evidence for sexual reproduction of Aeolosomatidae and Naididae in wastewater treatment systems. The sexual generation appears in natural waters, as a rule, before winter or drought and survive by diapausing cocoons only. Old worms always die after the egg-laying period.

According to Loden (1981), asexual reproduction is employed by species of Naididae during favorable conditions. Sexual reproduction occurs rarely and sporadically. His observations indicated a pattern to the occurrence of sexual and asexual phases in Naididae of the eastern and Midwestern United States. These occurrences are related to habitat, geographic distribution and certain species adaptations. In general, there has been conflicting evidence concerning where, why, by what stimulus, and how sexual maturity occurs among species of Naididae. The cocoon is a device to survive adverse conditions. Species of Naididae in unstable situations appear adapted for rapid breeding and cocoon deposition. Loden found four patterns to the abundance of naidids in unstable habitats: those that thrive in winter and spring, the ('winter species'); those present in summer and autumn, ('summer species'); those present throughout the year, but with a population maximum in spring; and those present year-round, but with a late summer or autumnal abundance.

## Temperature, pH and oxygen

Indications from field data suggest that both dissolved oxygen and temperature regimes of lakes are critical and highly species-specific variables for freshwater invertebrates and specifically aquatic oligochaetes thereby affecting species composition and abundance. Lakes liable to develop anoxic conditions had fewer species (Reynoldson, 1987).

The optimal temperatures for both somatic growth and reproduction for *Tubifex tubifex* usually is in the range between 20 and 25 °C (Reynoldson et al., 1996). However, genotypic or phenotypic variations exist. *T. tubifex* and *Limnodrilus hoffmeisteri* in the English Lake District have similar narrow temperature ranges over which growth occurs, from 10 to 13 °C and both show a considerable tolerance to extended periods of anoxia, of at least 16 and 10 weeks respectively. However no growth was observed under anoxic conditions. A third common species *Ilyodrilus templetoni* is sensitive to anoxia and could not survive more than 4 weeks. There was no evidence of differences in substrate quality affecting growth rates of either *T. tubifex* or *L. hoffmeisteri* (Reynoldson, 1987). *L. hoffmeisteri* has a higher egg production at higher temperatures (20–25 °C) and *T. tubifex* at lower temperatures 8 °C. Famme & Knudsen (1985) however, have been found evidence for anoxic survival, growth and reproduction by *Tubifex* sp. in an anoxic chemostat. *Tubifex* was able to survive more than 10 months of anoxia and they concluded that adult *Tubifex* might be considered as obligate anaerobe.

Aston et al. (1982) studied the effect of temperature on the culture of *Branchiura sowerbyi* (Oligochaeta, Tubificidae) on activated sludge. Highest reproduction and growth rates occurred between 21 and 29 °C with a population doubling time of 1.48 week at 25 °C (the optimum temperature). In an other study he found the optimum temperature for cocoon laying in mature worms, also growth in sexually immature worms, was near 25 °C. The optimum temperature for growth in sexually mature worms, however, was lower (10 °C in the Avon population and 15 °C in the Thames population). Probably the growth of sexually mature worms at 20–24 °C was depressed by the high rate of cocoon production (Aston, 1968).

Dumnicka & Pasternak (1978) found the occurrence of *N. elinguis* was positively correlated with temperature in the range from 0.3 to 20.6 °C. This species finds the best conditions of development in moderately oxygenated and fertile waters, but not in strongly polluted ones ($BOD_5$ > 9.24 mg/l). They were more numerous in water when the $BOD_5$ values of the water were lower and the oxygen content is higher. Thus *N. elinguis* is a very sensitive indicator of any slight pollution or even eutrophication of water. This species did not show any allegiance to specific types of habitat, since it is equally numerous in stony, sandy, and slimy bottoms.

Lochhead & Learner (1983) and Juget et al. (1989) investigated the effect of temperature on asexual population growth of *N. variabilis*, *N. elinguis* and *P. equiseta*. The mean population doubling time was about 3 days for *N. variabilis* and *P. aequiseta* at 20 °C and the lowest was 22 days for *P. aequiseta* at 8 °C. The values for *N. elinguis* were somewhat lower than expected probably due to less suitable experimental conditions.

Research on the growth of small aquatic oligochaetes such as *Aeolosoma hemprichi*, *Pristina longiseta* and *Nais variabilis* on activated sludge has especially been investigated by Japanese researchers. Temperature, pH and oxygen concentration in conventional domestic wastewater treatment systems are summarized in Table 2.

There has been very little recent research published by other authors. Kuniyasu et al. (1997) and Inamori et al. (1983) measured the effects of temperature, pH, phosphate buffer concentration, shake stress endurance and food concentration on the specific population growth rate of *A. hemprichi*, *P. longiseta* and *N. variabilis* (Table 3) and *Nais* sp., *A. hemprichi*, *Pristina* sp. and *Dero* sp. (Table 4), respectively, monoxenic cultivated on sterilized activated sludge. The relationship between the specific population growth rate ($\mu$) and feed concentration was described according to the Monod formula, $\mu = \mu_{max}\{S/(K_s + S)\} \cdot \mu_{max}$ = maximum specific population growth rate, $K_s$ is the saturation constant where $\mu = 0.5\mu_{max}$ and $S$ is the feed concentration. The activation energy was estimated with the Arrhenius formula, $\log \mu = -\Delta E/2.303RT + b$; where $\Delta E$ is the activation energy (cal/mole), $R$ is the gas constant, $T$ is the absolute temperature and b is a constant. The Arrhenius formula describes how physiological processes depend on temperature. The activation energy of *A. hemprichi* was lowest. This means that *A. hemprichi* was most uninfluenced by temperature. It was concluded that the feed concentration affected growth of the oligochaetes very much (Kuniyasu et al. 1997).

*Table 2.* Some abiotic parameters in conventional wastewater treatment systems

| Zones in WWTPs | $T$ (°C) | pH | Oxygen (mg/l) |
|---|---|---|---|
| Anaerobic | 10–20 | 7–8 | <0 |
| Anoxic | 10–20 | 7–8 | $0 < O_2 < 0.5$ |
| Oxic | 10–20 | 7–8 | $0.5 < O_2 < 2$ |

*Table 3.* Optimum growth parameters of *A. hemprichi*, *P. longiseta* and *N. variabilis* (after Kuniyasu et al., 1997)

| Parameter | *A. hemprichi* | *P. longiseta* | *N. variabilis* |
|---|---|---|---|
| Phosphate buffer (M) | 1/375 | 1/150 | 1/150 |
| pH | 5–9 | 5–9 | 5–9 |
| Shake stress | High | Low | Low |
| $K_s$* (mg/l) | 120 | 100 | 120 |
| Activation energy (cal mol$^{-1}$) | 12 500 | 18 500 | 18 000 |

*Growth on sterilized activated sludge.

## Process conditions

In wastewater treatment plants the following parameters are involved: process design, organic loading rate, sludge and hydraulic retention time, type of electron acceptor (for instance oxygen, nitrate/nitrite or sulfur compounds), wastewater composition, sludge volume index (svi), oxygen concentration, pH and temperature. The svi is the volume in milliliters occupied by 1 g of a suspension after 30 min settling. SVI typically is used to monitor settling characteristics of activated sludge and other biological suspensions. Although svi is not supported theoretically, it is an integral part of routine process control.

The abiotic factors pH, temperature and oxygen concentration already are discussed in the previous section. This section focuses on the remaining parameters mentioned above.

The occurrence of the free-swimming species strongly depends on the process conditions of the

*Table 4.* Optimum growth parameters of *A. hemprichi*, *Nais* sp., *Pristina* sp. and *Dero* sp. (after Inamori et al., 1983)

| Parameter | *A. hemprichi* | *Nais* sp. | *Pristina.* sp. | *Dero* sp. |
|---|---|---|---|---|
| Temperature (°C) | 33 | 25 | 30 | n.d. |
| Specific population growth rate[1] ($\mu$) (d$^{-1}$) | 0.35–0.43 | 0.10–0.13 | n.d. | n.d. |
| Phosphate buffer (M) | 1/375 | 1/150 | 1/150 | 1/7500 |
| pH | 6–8 | 6–8 | 6–8 | 6–8 |
| Shake stress | High | Low | Low | Nil |
| Max. pop. Density[2] (number/ml) (batch culture) | 100 | 60 | 60 | n.d. |
| Max. pop. density (continuous culture) | 1000 | 200 | 200 | n.d. |

[1]Sludge concentration >500 mg/l.
[2]The maximal density is measured under favorable growth conditions.

plant. Elissen et al. (submitted for publication) monitored the presence of several free-swimming oligochaetes in 4 wastewater treatment plants and investigated the relationship between the amount of worms in the plant and the prevailing process conditions. There was no relationship between the temperature and the presence of the worms. The worms seemed to influence the sludge volume index positively. A statistical analysis to quantify the relationship between some selected operational parameters and the presence of the worms is now in progress.

Generally, free-swimming worms only occur in (ultra) low loaded systems with enough oxygen available for growth and reproduction. The sludge age has to be not too short in order to prevent wash out of the worms. Also other still obscure factors can determine whether or not worms are present in WWTPs (see Elissen et al., submitted for publication).

Another new appearance in wastewater treatment is the membrane bioreactor (MBR). In the MBR the separation of biomass and the liquor takes place in the membrane module instead of a settling tank. Boran Zhang and Kazuo Yamamoto studied the role of predators on sludge mineralization in a MBR (Thesis, 1997). Wei et al. (2003) compared the performances of a submerged MBR and a conventional activated sludge process (CAS) on sludge reduction induced by oligochaetes. Only two types of worms were found in the MBR, *Aeolosoma hemprichi* and *Nais elinguis*. Worm presence and absence alternated in the MBR. The worms in the CAS reactor maintained throughout the operating period. In the CAS reactor besides *A. hemprichi* and *N. elinguis*, also *Pristina aequiseta* was found. The latter only occasionally appeared in the reactor. The worm growth in the MBR did not lead to a lower sludge production but worm bloom in the CAS reactor greatly decreased sludge yield and improved sludge settling characteristics at high density. It appeared that *Nais* had more potential for sludge reduction than *Aeolosoma*.

A final type of WWTP which will 'become extinct' by tightened regulations, is the trickling filter, a low loaded activated sludge systems with carrier material to encourage bacterial and metazoan growth. Trickling filters are the oldest WWTPs. In 1893–1894, Joseph Corbett in Salford, England developed the first trickling filter (Koot, 1980). Nowadays the trickling filters are more and more replaced by conventional activated sludge systems. Trickling filters examined in the United Kingdom contained *Eisenia foetida*. *E. foetida* is capable of rapidly converting large quantities of activated sludge in earthworm biomass (Hartenstein et al., 1984).

## Vermiculture or vermicomposting

Vermiculture or vermicomposting involves the use of semi-aquatic earthworms as novel bioreactors for bioprocessing of agricultural and non-toxic city and industrial wastes for the production of biofertilizers and primary proteins. During the process organic material is fed to a variety of worm species for the purpose of converting the organic material into increased worm biomass and vermicast. The worm bodies represent a valuable source of protein that could be utilized for livestock (pig, fish and chickens) or even human consumption. Vermiculture is the latest aspect of technology, fast developing in advanced countries like the UK, the USA and Japan.

The use of vermiculture to stabilize sewage sludge and other organic wastes has been mooted for the past 50 years. Considerable work has been done on the stabilization process and value of the vermicast byproduct (Lotzof, 1999). Vermicast is worm excreta that can be used as a plant growth medium and soil conditioner. The biomass of the worms themselves has been sold for bait, animal feed and domestic and small composting systems.

For application of the worms several critical elements are mentioned (Lotzof, 1999):

- Preparation of the sludge before feeding it to the worms.
- Controlled application of the feed to the worm beds.
- Raised cage bed structures.
- Environmental control systems for managing moisture, temperature and wind.
- Worm biomass management.
- Harvesting of the vermicast.
- Post-processing of the vermicast to sale.
- Leachate control.

- Sampling and analyses systems for sludge and vermicast.
- Information management systems.

A variety of vermicomposting systems have been investigated in recent decades, treating both cakes and liquid sludge and results have proved promising. For maximum potential to be realized the worm beds should be carefully designed and operated (i.e. with due consideration to the temperature, moisture content, drainage and other physiochemical requirements of the worms). Careful selection of the worm species is also required. The epigeic worm *Eisena foetida* has been the focus of much of the research done on using worms to treat sewage sludges (Clark, 1997).

According to Vigueros & Camperos (2002), a determinant factor in the survival of *Eisenia foetida* is humidity. The most adequate value for this parameter was in the range of 70–88%, which should be maintained throughout the process. The mixture with the greatest biomass production measured in numbers of cocoons per kilogram of vermicompost was that of 70% sewage sludge + 30% water hyacinth, for which its use is recommended, with the option of adding backyard waste. The reduction in the TVS/TS ratio pointed out the important role that worms play in the degradation of waste. Also Kaplan et al. (1980) mentioned a humidity percentage between 70 and 80% the sludge-conditioning material mixture should have to assure an adequate process and this humidity should be maintained during the entire process.

Athanasopouluos (1993) treated anaerobically stabilized effluents of dried vine fruit industry in earthworm filters. He used the species *Lumbricus rubellus*. COD removal was 95% for loadings of 0.10 and 0.15 kg COD/$m^2$·d. The earthworm biomass seems to be in its upper bearing capacity of approximately 2 kg/$m^2$ and did not increase seriously with time.

Elvira et al. (1996) investigated the bioconversion of solid paper-pulp mill sludge by the epigeic earthworm *Eisenia andrei*. Regardless of the presence of earthworms, degradation occurred during the bioconversion period, but the presence of earthworms accelerated the mineralization of organic matter, favored the breakdown of structural polysaccharides and increased the humification rate. Consequently, the C/N ration and the degree of extractability of heavy metals were lower in the worm-worked end product. The Bioverm worms are of terrestic origin, but with some relatively small adaptations of the culture operation design, the positive experiences with vermicultures opens new perspectives for application of aquatic oligochaetes to be of usage for aerobic sludge stabilization.

Worm species such as *Allobophora chloritica* and *Dendrobena venteta* also offer potential in sewage sludge systems (Clark, 1997). The African nightcrawler *Eudrilus eugeniae* can grow twice as fast as *E. foetida*. Once fully populated worms can eat between one half and their own body weight a day. To prevent deterioration of the worm activity composting of the material or anaerobic sites has to be circumvented. It is therefore essential that the quantity fed matches the daily quantity consumed by the worms.

## Discussion and conclusion

This paper reviews the possibility to put forward a new sustainable technique for handling and reduction of the excess sludge through predation of the sludge by aquatic oligochaetes. Another benefit of this approach is the application of annelid biomass as a resource for protein production, fish feed, etc.

Only four scientific research groups have been performed research on oligochaetes in activated sludge: two Japanese groups (MBR), a Chinese group (MBR) and the Dutch group (sludge reduction). Incidentally other researchers are interested in this item but until now no articles have been published. There is evidence that also commercial groups are working on this item, but for policy reasons access to their results is not possible. The limited data available so far are sometimes incomplete, contradictory or difficult to interpret. Furthermore both the experimental conditions and the reactor configurations differ. Therefore future laboratory experiments should be set up according a general concept, which facilitates comparison between other experiments and scale up from laboratory to applied research.

Until now, the experiments have mainly been performed on small laboratory scale and relevant data about oligochaetes is limited. Therefore oligochaete knowledge collected from field data or small-scale laboratory experiments is necessary in order to understand and to be able to predict the behavior of oligochaetes in WWTPs. Also deducible measures can be taken in order to guarantee optimal growth conditions for the oligochaetes.

In general, interpretation of data concerning oxygen demand at different sludge concentrations and temperatures is difficult because other factors like micro-turbulence and mechanical aeration determine the prevailing oxygen concentration during the experiments.

There is a wide variation in the estimates of specific sludge production in aerobic wastewater treatment processes, varying from 35 to 85 g dry solids per population equivalent per day. The total European sludge production is put at 6 million dry solid tonnes per annum (EC, 1991) and is expected to rise by the year 2006 to 10 million tonnes or even 17 million tonnes according to Saabye & Schwinning (1994). About 60% of the total operation costs is accounted for by the cost of sludge storage, transport, digestion and disposal. It is clear therefore that the major part of the cost in aerobic wastewater treatment is associated with the treatment and disposal of sewage sludge.

The benefits of sludge application in agriculture is limited by its potential health hazards, i.e., transmission of pathogenic organisms, accumulation of nitrates in the groundwater, soil and water contamination by eventual highly toxic organic constituents of the sludge, and accumulation of toxic heavy metals in soil, water and crops. The danger of transmission of heavy metals and pathogenic organisms appears to be the major element of concern.

Preliminary short time (7 days) experiments (unpublished data) with oligochaetes fed contaminated sludge have been showed the heavy metals remained in the sludge although the organic part of the sludge was completely digested by the worms. Whether or not heavy metals will accumulate in worm tissue depends on the heavy metal concentration in the sludge, the exposure time, the organic content of sludge and the worm species themselves.

Most annelid worms are able to grow on bacterial biomass. The worms tolerate temperatures between 2 and 8 °C. The optimum temperature for growth and reproduction ranges between 20 and 25 °C. The prevailing operating conditions in conventional wastewater treatment plants are thus suitable for the worms. However one has to prevent that species out compete themselves and that the food source does not become depleted.

Since small free-swimming oligochaetes are able to survive under conditions of less than 5 °C these oligochaetes would contribute to wastewater treatment in biological wastewater treatment plants except at extremely low temperatures. The application of these worms for reduction of bacterial biomass is, however, despite all research on this topic, hampered by the fact that this process is difficult to control.

Within Oligochaeta, asexual reproduction is a regular phenomenon in the lifecycles of Naidids and Aeolosomatids whereas it is of sporadic occurrence or is unknown within the other families. Only five species of Enchytraeids are known to have adopted this reproductive method (Christensen, 1973). In some species sexual reproduction was not correlated with density but with temperature. Sexually mature individuals are mainly recorded in the colder season, which coincides with the period when enchytraeid populations usually attain maximal densities. Finally one has to keep in mind that the factors responsible for sexual reproduction in naidids are different from those for example in *E. bigeminus*. Since reproductive data available in the literature still show conflicting results, the predation suitability of a certain type of sludge has to be measured under standardized conditions. Only if the worms are able to grow and reproduce (either sexually and/or asexually) on the sludge, application of these worms in sludge treatment may be technically feasible.

From an economical point of view the main item to solve is how to reduce the required area for the application of oligochaetes in WWTPs. Also it is necessary to design a reliable process that can cope with changes in food supply and environmental factors. Before using the worms for fish food, an environmental risk assessment for the public health should be carried out in order to allow consumption of the fish.

In conclusion, it is clear that the total amount of sludge can be reduced by the introduction of oligochaetes in WWTPs. For the free-swimming oligochaetes future research should focus on the factors that trigger the initiation of annelid blooms in WWTPs and how to maintain a stable annelid population in WWTPs. Use of these worm species is not an option unless knowledge about these subjects becomes available.

Over recent years there has been a growing realization that the treatment and disposal of sludge presents a major challenge in wastewater management. Consequently rising sludge production and costs due to strengthened environmental standards demand a new approach of sludge treatment. Now, the challenge is to introduce a sustainable technology, which also yields a product that will satisfy defined quality standards. These facilities require a large investment of capital and to be cost effective they need to have long and useful lives. There is therefore great pressure in the wastewater management to maximize the returns on existing investments by improving and refurbishing existing capital plants to meet the new demands. The search after new sustainable technologies justifies an accelerated acceptance for the application of sludge predation by aquatic oligochaetes in the wastewater treatment process.

## Acknowledgements

The authors are grateful to Gary Yanover for his kind assistance to read this manuscript and if needed improve the English grammar.

## References

Abbassi, B., S. Dullstein & N. Räbiger, 1999. Minimization of excess sludge production by increase of oxygen concentration in activated sludge flocs; experimental and theoretical approach. Water Research 34: 139–146, SRA II.

Adreani, L., C. Bonacina, G. Bonomi & C. Monti, 1984. Cohort cultures of *Psammoryctides barbatus* (Grube) and *Spirosperma ferox* Eisen: a tool for a better understanding of demographic strategies in Tubificidae. Hydrobiologia 115: 113–119.

Aston, R. J., 1968. The effect of temperature on the life cycle, growth and fecundity of *Branchiua sowerbyi* (Oligochaeta: Tubificidae). Journal of Zoology 154: 29–40.

Aston, R. J. & A. G. P. Milner, 1982. Conditions required for the culture of *Branchiura sowerbyi* (Oligochaeta: Tubificidae) in activated sludge. Aquaculture 26: 155–160.

Aston, R. J., K. Sadler & A. G. P. Milner, 1982. The effects of temperature on the culture of *Branchiura sowerbyi* (Oligochaeta, Tubificidae) on activated sludge. Aquaculture 29: 137–145.

Athanasopouluos, N., 1993. Use of earthworm biotechnology for the management of anaerobically stabilized effluents of dried vine fruit industry. Biotechnology letters 15: 1281–1286.

Bonacina, C., G. Bonomi & C. Monti, 1989a. Population analysis in mass cultures of *Tubifex tubifex*. Hydrobiologia 180: 127–134.

Bonacina, C., G. Bonomi & C. Monti, 1989b. Density-dependent processes in cohorts of *Tubifex tubifex*, with special emphasis on the control of fecundity. Hydrobiologia 180: 135–141.

Bouguenec, V., 1992. Oligochaetes (Tubificidae and Enchytraeidae) as food in fish rearing: a review and preliminary tests. Aquaculture 102: 201–217.

Bowker, D. W., M. T. Wareham & M. A. Learner, 1983. The selection and ingestion of epilithic algae by *Nais elinguis* (Oligochaeta: Naididae). Hydrobiologia 98: 171–178.

Bowker, D. W., M. T. Wareham & M. A. Learner, 1985a. A choice chamber experiment on the selection of algae as food and substrate by (Oligochaeta: Naididae). Freshwater Biology 15: 547–557.

Bowker, D. W., M. T. Wareham & M. A. Learner, 1985b. The ingestion and assimilation of algae by *Nais elinguis* (Oligochaeta: Naididae). Oecologia 67: 282–285.

Brinkhurst, R. O. & K. E. Chua, 1969. Preliminary investigation of the exploitation of some potential nutritional resources by three sympatric tubificid oligochaetes. Journal Fisheries Research Board of Canada 26: 2659–2668.

Brinkhurst, R. O., 1970. Distribution and abundance of tubificid (Oligochaeta) species in Toronto Harbour, Lake Ontario. Journal Fisheries Research Board of Canada 27: 1961–1969.

Brinkhurst, R. O. & B. G. M. Jamieson (eds), 1971. Distribution and Ecology. From: Aquatic Oligochaeta of the World. Edinburgh, Oliver and Boyd, 122–147.

Brinkhurst, R. O., K. E. Chua & N. K. Kaushik, 1972. Interspecific interactions and selective feeding by tubificid oligochaetes. Limnology and Oceanography 17: 122–133.

Brinkhurst, R. O., 1974. Factors mediating interspecific aggregation of tubificid oligochaetes. Journal Fisheries Research Board of Canada 31: 460–462.

Brinkhurst, R. O. & M. J. Austin, 1979. Assimilation by aquatic Oligochaeta. International Review of Hydrobiology 64: 245–250.

Cammen, L. M., 1982. Effect of particle size on organic content and microbial abundance within four marine sediments. Marine Ecology-Progress Series 9: 273–280.

Christensen, B., 1973. Density dependence of sexual reproduction in *Enchytraeus bigeminus* (Enchytraeidae). Oikos 24: 287–294.

Christensen, B., 1984. Asexual propagation and reproductive strategies in aquatic Oligochaeta. Hydrobiologia 115: 91–95.

Clark, P., 1997. Vermistabilisation as a viable rural works treatment technology. Technical report.

Cole, L., R. D. Bardgett & P. Ineson, 2000. Enchytraeid worms (Oligochaeta) enhance mineralization of carbon in organic upland soils. European Journal of Soil Science 51: 185–192.

Coler, R. A., H. B. Gunner & B. M. Zuckermann, 1968. Selective feeding of tubificids on bacteria. Nature 216: 1143–1144.

Cummins, K. W. & G. H. Lauff, 1968. The influence of substrate particle size on the microdistribution of stream benthos. Hydrobiologia 34: 145–181.

Dumnicka, E. & K. Pasternak, 1978. The influence of physicochemical properties of water and bottom sediments in the River Nida on the distribution and numbers of Oligochaeta. Acta Hydrobiologica 20: 215–232.

EC, 1991. Commission of European Communities, Survey of Sludge Production, Treatment, Quality and disposal in European Union, Report EC 3646.

Elvira, C., M. Goicoechea, L. Sampedro, S. Mato & R. Nogales, 1996. Bioconversion of solid paper-pulp mill sludge by earthworms. Bioresource Technology 57: 173–177.

Famme, P. & K. Knudsent, 1985. Anoxic survival, growth and reproduction by the freshwater annelid, *Tubifex* sp., demonstrated using a new simple anoxic chemostat. Comparative Biochemistry and Physiology 81A: 251–253.

Finogenova, N. P. & T. M. Lobasheva, 1987. Growth of *Tubifex tubifex* Mueller (Oligochaeta, Tubificidae) under various trophic conditions. International Review of Hydrobiology 72: 709–726.

Hämmerling, J., 1924. Die ungeschlechtliche Fortplanzung und Regeneration bei *Aeolosoma hemprichi*. Zoologische Jahrbucher-Abteilung für Allgemeine Zoologie und Physiologie der Tiere 41: 581–656.

Harper, R. M., J. C. Fry & M. A. Learner, 1981a. A bacteriological investigation to elucidate the feeding biology of *Nais variabilis* (Oligochaeta: Naididae). Freshwater Biology 11: 227–236.

Harper, R. M., J. C. Fry & M. A. Learner, 1981b. Digestion of bacteria by *Nais variabilis* (Oligochaeta) as established by autoradiography. Oikos 36: 211–218.

Hargrave, B. T., 1976. The central role of invertebrate feces in sediment decomposition. In Anderson, J. M. & A. Macfadyen (eds), The Role of Terrestrial and Aquatic Organisms in Decomposition Processes. Blackwell, Oxford: 301–321. 17th Symposium on British Ecology Society.

Hartenstein, R., D. L. Kaplan & E.F. Neuhauser, 1984. Earthworms and trickling filters. Journal Water Pollution Control Federation 56: 294–298.

Hawkes, H. A., 1963. The Ecology of Wastewater Treatment. Permagon Press, London.

Inakollu, S. V. & A. Wanganeo, 2002. Environmental assessment of fishes grown in treated domestic sewage. Journal of Environmental Systems 29: 55–70.

Inamori, Y., R. Suzuki & R. Sudo, 1983. Mass culture of small aquatic oligochaeta. Research Report from the National Institute for Environmental Studies 47: 125–137 (in Japanese).

Inamori, Y., Y. Kuniyasu & R. Sudo, 1987. Role of smaller metazoa in water purification and sludge reduction. Japanese Journal of Water Treatment Biology 23: 15–23 (in Japanese).

Inamori, Y., Y. Kuniyasu, N. Hayashi, H. Ohtake & R. Sudo, 1990. Monoxenic and mixed cultures of the small metazoa *Philodina erythrophthalma* and *Aeolosoma hemprichi* isolated from a waste-water treatment process. Applied Microbial Biotechnology 34: 404–407.

Janssen, P. M. J., W. H. Rulkens, J. H. Rensink, & H. F. van der Roest, 1998. The potential for metazoa in biological wastewater treatment. Water Quality International: 25–27.

Juget, J., 1979. La texture granulometrique des sediments et le regime alimentaire des oligochètes limnicoles. Hydrobiologia 65: 145–154.

Juget, J., V. Goubier & D. Barthélémy, 1989. Intrinsic and extrinsic variables controlling the productivity of asexual populations of *Nais* spp. (Naididae, Oligochaeta). Hydrobiologia 180: 177–184.

Juniper, S. K., 1981. Stimulation of bacterial activity by a deposit feeder in two New Zealand intertidal inlets. Bulletin of Marine Science 31: 691–701.

Kaplan, D. L., R. Hartenstein & E. F. Newhauser, 1980. Coprophagic relations among the earthworms *Eisena foetida*, *Eudrilus eugenia* and Amynthae spp. Pedobiologia 20: 74–84.

Kirk, R. G., 1973. Growth of sole (*Solea solea* L.) on Enchytraeid womrd diets. International Council of Exploration of the sea. Fisheries Improvement Committee C. M. 197/E: 18.

Kirk, R. G. & B. R. Howell, 1972. Growth rates and food conversion in young plaice fed on artificial diets. Aquaculture 1: 29–34.

Klapwijk, A., V. Blom, C. H. Ratsak, B. R. Buijs & H. E. H. Elissen, 2000. Reduction of sludge production in an activated sludge system with the use of metazoa and other higher organisms. CREST Workshop on Integrated Water Quality Management – The 4th Japan-Netherlands Workshop, Hokkaido Kiroro Resort, Japan, 2000: 369–375.

Koot, A. C. J., 1980. Behandeling van afvalwater. Uitgeverij Waltman. Delft. ISBN 90 212 3172 7.

Kuniyasu, K., N. Hayashi, Y. Inamori & R. Sudo, 1997. Effect of environmental factors on growth characteristics of oligochaeta. Japanese Journal of Water Treatment Biology 33: 207–214 (in Japanese).

Learner, M. A., 1972. Laboratory studies on the life histories of four enchytraeid worms (Oligochaeta) which inhabit sewage percolating filters. Annals of Applied Biology 70: 251–266.

Lee, N. M., 1996. Parameters affecting microorganisms and the process performance in biological wastewater treatment. Doctoral dissertation. Department of biotechnology, Center for Chemistry and Chemical Engineering, Lund University, Sweden.

Lochhead, G. & M. A. Learner, 1983. The effect of temperature on asexual population growth of three species of Naididae (Oligochaeta). Hydrobiologia 98: 107–112.

Loden, M. S., 1981. Reproductive ecology of Naididae (Oligochaeta). Hydrobiologia 83: 115–123.

van Loosdrecht, M. C. M. & M. Henze, 1999. Maintenance, endogenous respiration, lysis, decay and predation. Water Science and Technology 39: 107–117.

Lopez, G. R. & J. S. Levinton, 1987. Ecology of deposit feeding animals in marine sediments. Quarterly Review of Biology 62: 235–260.

Lotzof, M, 1999. The wonder of worms for sludge stabilisation. Journal Australian Water and Wastewater Association 26: 38–42 SRM II.

Matisoff, G., X. S. Wang & P. L. McCall, 1999. Biological redistribution of lake sediments by tubificid oligochaetes:

*Branchiura sowerbyi* and *Limnodrilus hoffmeisteri/Tubifex tubifex*. Journal of Great Lakes Research 25: 205–219.

Mayhew, M. & T. Stephenson, 1997. Low biomass yield activated sludge: a review. Environmental Technology 18: 883–892.

Mermillod-Blondin, F., M. Creuzé des Châtelliers & M. Gérino, 2003. Effects of the interaction between tubificid worms on the functioning of hyporheic sediments: an experimental study in sediment columns. Archiv für Hydrobiology 156: 203–223.

Milbrink, G., 1987. Mutualistic relationships between cohabiting tubificid species. Hydrobiologia 155: 193.

Milbrink, G., 1993. Evidence for mutualistic interactions in freshwater oligochaete communities. Oikos 68: 317–322.

McMurtry, M. J., D. J. Rapport & K. E. Chua, 1983. Substrate selection by tubificid oligochaetes. Canadian Journal of Fisheries and Aquatic Sciences 40: 1639–1646.

Ødegaard, H., 2003. Sludge minimization technologies – an overview, 2003. Proceedings of the International Water Association Specialist Conference. Biosolids 2003. Wastewater Sludge as a Resource, 23–25 June, 2003. Norwegian University of Science and Technology, Trondheim, Norway.

Pasteris, A., C. Bonacina & G. Bonomi, 1994. Observations on cohorts of *Tubifex tubifex* cultured at different food levels, using cellulose substrate. Hydrobiologia 278: 315–320.

Ratsak, C. H., S. A. L. M. Kooijman & B. W. Kooi, 1993. Modelling the growth of an oligochaete on activated sludge. Water Research 27: 737–749.

Ratsak, C. H., 1994, Grazer induced sludge reduction in wastewater treatment. Thesis, Vrije Universiteit Amsterdam.

Ratsak, C. H., 2001. Effects of *Nais elinguis* on the performance of an activated sludge plant . Hydrobiologia 463: 217–222.

Rensink, J. H. & W. H. Rulkens, 1997. Using metazoa to reduce sludge production. Water Science and Technology 36: 171–179.

Reynoldson, T. B., 1987. The role of environmental factors in the ecology of tubificid oligochaetes – an experimental study. Holarctic Ecology 10: 241–248.

Reynoldson, T. B., P. Rodriguez & M. A. Martinez Madrid, 1996. Comparison of reproduction, growth and acute toxicity in two populations of *Tubifex tubifex* (Müller, 1774) from the North America Great Lakes and northern Spain. Hydrobiologia 334: 199–206.

Robbins, J. A., P. L. McCall, J. B. Fisher & J. R. Krezoski, 1979. Effects of deposit-feeders on migration of $^{137}$Cs in lake sediments. Earth Planet Science Letters 42: 277–287.

Robbins, J. A., T. J. Keilty, D. S. White & D. N. Edgington, 1989. Relationships among tubificid abundances, sediment composition, and accumulation rates at Lake Erie. Canadian Journal of Fisheries and Aquatic Sciences 46: 223–231.

Rodriquez, P., M. Martinez-Madrid, J. A. Arrate & E. Navarro, 2001. Selective feeding by the oligochaete *Tubifex tubifex* (Tubificidae, Clitellata). Hydrobiologia 463: 133–140.

Saabye, A. & H.-G. Schwinning, 1994. Treatment and beneficial use of sewage sludge in the European Union. ISWA Times 3: 1–6.

Sauter, G. & H. Güde, 1996. Influence of grain size on the distribution of tubificid oligochaete species. Hydrobiologia 334: 97–101.

Timm, T., 1984. Potential age of aquatic Oligochaeta. Hydrobiologia 115: 101–104.

Vigueros, L. C. & E. R. Camperos, 2002. Vermicomposting of sewage sludge: a new technology for Mexico. Water Science and Technology 46: 153–158.

Wachs, B., 1967. Die Oligochaeten-Fauna der Fliessgewässer unter besonderer Berücksichtigung der Beziehungen zwischen der Tubificiden-Besiedlung und dem Substrat. Archiv für Hydrobiology 63: 310–386.

Wavre, M. & R. O. Brinkhurst, 1971. Interactions between some tubificid oligochaetes and bacteria found in the sediments of Toronto Harbour, Ontario. Journal Fisheries Research Board of Canada 28: 335–341.

Wei, Y., R. T. van Houten, A. R. Borger, D. H. Eikelboom & Y. B. Fan, 2003. Comparison performances of membrane bioreactor and conventional activated sludge processes on sludge reduction induced by oligochaete. Environmental Science and Technology 37: 3171–3180.

Wei, Y., R. T. van Houten, A. R. Borger, D. H. Eikelboom & Y. B. Fan, 2003. Minimization of excess sludge production for biological wastewater treatment. Water Research 37: 4453–4467.

WIRES, 2004. EU Project ID nr EVK1-CT-2000-00050.Ways of innovation for the reduction of excess sludge.

Zahner, R., 1967. Refinement in empirical functions for realistic soil-moisture regimes under forest cover. In Sopper, W. E. & H. W. Lull (eds), Forest Hydrology. Pergamon Press, New York, NY, 261–272.

Zhang, B. & K. Yamamoto, 1997. The role of predators on sludge mineralization in membrane separation bioreactor. Thesis Zhang, Boran (Prof. Yamamoto): A Study on Microbial Activities and The Role of Predators in Membrane Separation Activated Sludge Process. Dept. of Urban Engineering, the University of Tokyo.

Hydrobiologia (2006) 564:213–222
P.F.M. Verdonschot, H. Wang, A. Pinder & R. Nijboer (eds), Aquatic Oligochaete Biology IX
DOI 10.1007/s10750-005-1720-1


# Sludge reduction with a novel combined worm-reactor

Yuansong Wei* & Junxin Liu
*Department of Water Pollution Control Technology, Research Center for Eco-Environmental Sciences, Chinese Academy of Sciences, P.O. Box 2871, 100085, Beijing, P.R. China*
*(*Author for correspondence: Tel.: +86-10-62923543; Fax: +86-10-62849108; E-mail: ys_wei@yahoo.com)*

*Key words:* combined worm-reactor, excess sludge, sludge reduction, Tubificidae

## Abstract

In China, as a result of economic, environmental and regulation factors, excess sludge treatment and disposal represents a rising challenge for small and mid-scale municipal wastewater treatment plants (WWTPs). Although the presence of worms in aerobic wastewater treatment may lead to substantial sludge reduction, the practical application is still uncontrollable because of unstable worm growth. To overcome unstable worm growth in activated sludge process, a combined worm-reactor, consisting of sections for both free-swimming and sessile worms, was developed to enhance worm growth. This plug flow type of worm-reactor has the sessile worm growing section filled with plastic carriers, and recirculation of sludge from the section for sludge settling was carried out in order to avoid worm washing out. Sludge reduction of treating the discharged excess sludge using this worm-reactor was investigated during 53 days. In the start-up phase the combined worm-reactor was inoculated with Tubificidae. Tubificidae mainly occurred in the carriers and on the bottom of the worm-reactor, however, neither *Aeolosoma* nor *Nais* was found in the worm-reactor. Results clearly showed that the average sludge reduction was 48%, and the COD removal was low, about 8%. High $NH_4^+$-N or $NO_2^-$-N concentrations or their combination may limit the growth of Tubificidae. Few $PO_4^{3-}$-P release into the effluent was observed in the worm-reactor. However, results obtained from this study cannot unequivocally be attributed to the presence of worms because it did not include a control system, and thus much more work will be needed in this concept in the further research.

## Introduction

The activated sludge process is the most widely used biological wastewater treatment for both domestic and industrial plants in the world. One of the drawbacks of conventional activated sludge (CAS) processes is high sludge production. In activated sludge plants the sludge yield coefficient is typically 0.4 g volatile suspended solids (VSS)/g biodegradable chemical oxygen demand (bCOD) (Metcalf & Eddy, 2001). Treatment and disposal of sewage sludge from wastewater treatment plants (WWTPs) accounts for about 50%, even up to 60% of the total costs of wastewater treatment (Davis & Hall, 1997; Spellman, 1997). In China, about 4.22 billion tonnes municipal wastewater was treated in more than 150 municipal wastewater treatment plants, which produced about 0.55–1.06 million tonnes of dry sewage sludge in 2001 (National Environmental Protection Agency of China, 2001). It is estimated that sludge production in China is increasing annually at about 20%. However, most of WWTPs in China lack full process facilities for sewage sludge treatment such as anaerobic sludge digestion so that large amounts of sewage sludge produced are directly used for agriculture without any pre-treatment (Wei et al., 2000). This practice may pose a potential threat to both the environment and human health. Sewage sludge treatment and disposal is, therefore, becoming one of the most serious challenges in WWTPs as a result of legal constraints, rising costs and public sensitivity. However, these factors also provide considerable impetus to explore and develop strategies and technologies for sludge reduction. Different engineering strategies are being developed for sludge reduction using the following approaches: lysis-cryptic growth, uncoupling metabolism, maintenance metabolism, and predation on bacteria (Wei et al., 2003a).

It is well known that presence of protozoa and metazoa in aerobic wastewater treatment processes plays an important role in keeping effluent clear by consuming dispersed bacteria. In the past, protozoa and metazoa were usually used as important indicators of process performance and efficiency in biological wastewater treatment processes. Recently, many researchers have focused on sludge reduction induced by grazing on bacteria (Ratsak et al., 1993, 1994, 1996; Ratsak, 1994, 2001; Welander & Lee, 1994; Lee & Welander, 1996a, b; Rensink et al., 1996; Rensink & Rulkens, 1997; Janssen et al., 1998; Ghyoot, 1998; Ghyoot & Verstraete, 2000; Zhang, 2000; Luxmy et al., 2002; Lapinski & Tunnacliffe, 2003; Wei et al., 2003b). Worms are the largest organisms observed during the microscopic investigation of activated sludge (Eikelboom, 2000), and are easy to be monitored than protozoa. The performance of oligochaetes on sludge reduction in biological wastewater treatment has been paid more attention than that of protozoa (Ratsak et al., 1993; Ratsak, 1994, 2001; Welander & Lee, 1994; Rensink et al., 1996; Rensink & Rulkens, 1997; Janssen et al., 1998; Zhang, 2000; Wei et al., 2003b). Principal taxa of worms present in activated sludge systems and trickling filters are Naididae, Aeolosomatidae and Tubificidae. The Naididae and Aeolosomatidae are free-swimming worms. Whereas the Tubificidae is sessile worms that do not normally occur in activated sludge suspensions, but are sometimes present in sludge 'bankets' on the bottom of the basin (Ratsak et al., 1993; Rensink & Rulkens, 1997). Different worms were found in their observations such as *Nais elinguis*, *Pristina* sp. and *Aeolosoma hemprichicii*, but either *Nais* or *Aeolosoma* was predominant (Ratsak, 1994, 2001; Zhang, 2000; Wei et al., 2003b). Although the presence of worms in the aerobic wastewater treatment may lead to substantial sludge reduction (Table 1), the practical application of worms for sludge reduction is still uncontrollable in biological wastewater treatment processes because of unstable worm growth, e.g. alternative bloom and absence of worms, high variation of worm density. In these mentioned researches, worms grow together with bacteria in the same bioreactor of

*Table 1.* Literature data of sludge reduction with Oligaochaeta in aerobic wastewater treatment processes

| Operation conditions | Type of worms | Sludge yield ($kgSS/kgCOD_{removed}$) | Sludge reduction (%) | References |
|---|---|---|---|---|
| Two pilot tricking filters (filled with lava slags and plastic media, respectively); 18–23 °C; excess sludge | Tubificidae | – | 10–50 (lava slags)<br>10–45 (plastic media) | Rensink et al. (1996); Rensink & Rulkens (1997); Janssen et al. (1998) |
| Pilot activated sludge system; 18–23 °C; domestic wastewater | Tubificidae | 0.15 | – | Rensink et al. (1996); Rensink & Rulkens (1997); Janssen et al. (1998) |
| Pilot oxidation ditch; 18–23 °C; domestic wastewater | Tubificidae | 0.17 | – | Rensink et al. (1996); Rensink & Rulkens (1997); Janssen et al. (1998) |
| Pilot suction submerged MBR; 20 °C; municipal wastewater | Aeolosomatidae, Naididae | 0.00–0.12 | – | Zhang (2000) |
| Pilot gravitational submerged MBR; 20 °C; municipal wastewater | Aeolosomatidae, Naididae | 0.10–0.15 | - | Zhang (2000) |
| Pilot activated sludge system; 20 °C; domestic wastewater | Aeolosomatidae, aididae | 0.17 | – | Wei et al. (2003b) |

aerobic wastewater treatment processes. But bioreactors such as aeration tank in activated sludge process are designed for bacterial growth rather than for worm growth. A suitable habitat may therefore be essential for stable growth and retention of worms. On the basis of worms occurring in aerobic wastewater treatment processes, a novel worm-reactor was developed to enhance worm growth, and consists of three sections: (i) a section for free-swimming worms, (ii) a section for sessile worms and (iii) a section for sludge settling (Wei & Liu, 2002). This worm-reactor can be used to treat discharged excess sludge from the activated sludge process, or be integrated into the activated sludge process through treating returned sludge. In this preliminary study, the purpose was to investigate sludge reduction using this new worm-reactor treating discharged excess sludge. The impact of worm growth on nutrients release was also discussed.

## Materials and methods

### *Experiment facility*

Figure 1 shows the flow sheet for the worm-reactor to treat the discharged excess sludge. In this study, a bench scale worm-reactor was used to treat the discharged excess sludge. The reactor is made of Lucite, and the working volumes of three sections are 3.4 l (free-swimming worms), 11.0 l (sessile worms) and 0.5 l (sludge settling), respectively. This is a plug flow type of worm-reactor, with the section for sessile worm growth filled with filar polypropylene carriers. The thread diameter of carriers is 1–2 mm, and its density and porosity are 0.9–0.92 g/cm$^3$ and 90–95%, respectively.

### *Operating conditions*

During 53 days, this combined worm-reactor was continuously fed with the discharged excess sludge (Table 2) from a pilot scale activated sludge system using a peristaltic pump (BT00-100M, China), and a part of the processed sludge from the sludge settling section of the reactor was returned to the section for free-swimming worms using a second peristaltic pump in case of worms' washing out with the effluent sludge. No sludge from the worm-reactor was wasted throughout the operation period. The worm-reactor was alternatively aerated with two air compressors controlled by a timer. In this pilot scale activated sludge system, municipal wastewater was treated from the beginning to Day 21, and Day 35 to Day 53, respectively, and the mixed wastewater of municipal sewage and the landfill leachate was treated from Day 21 to Day 35 due to lack municipal wastewater during this period.

Worm growth is highly dependent on, and sensitive to temperature (Ratsak, 1994, 2001; Janssen et al., 1998). On the basis of literature results (Table 1), a temperature of 20 °C may be suitable for worm growth in aerobic wastewater treatment systems. In this study, the temperature

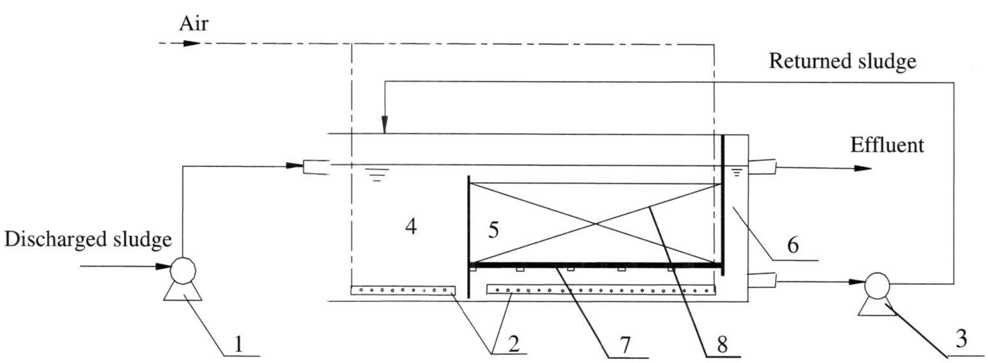


*Figure 1.* The diagram of the worm-reactor treating the discharged excess sludge: (1) the influent pump; (2) the perforated pipe; (3) the recycle pump; (4) the section for free-swimming worms; (5) the section for sessile worms; (6) the section for sludge settling; (7) the perforated plate; (8) the plastic carriers.

*Table 2.* Characteristics of the fed excess sludge in the worm-reactor

| | Average ± SD | Range |
|---|---|---|
| TSS (mg/l) | 5.42 ± 3.25 | 0.63–11.52 |
| VSS (mg/l) | 3.23 ± 2.04 | 0.24–5.86 |
| VSS/TSS | 0.72 ± 0.12 | 0.38–0.80 |
| COD (mg/l) | 246.30 ± 137.24 | 117.96–526.33 |
| $NH_4^+$-N (mg/l) | 26.81 ± 30.65 | 3.77–95.96 |
| $NO_3^-$-N (mg/l) | 61.46 ± 25.15 | 31.79–102.05 |
| $NO_2^-$-N (mg/l) | 49.39 ± 36.06 | 5.92–101.61 |
| $PO_4^{3-}$-P (mg/l) | 4.29 ± 2.34 | 1.27–7.84 |

in the worm-reactor was thus controlled at 20 °C (Table 3) with a heater. After the worm-reactor was in the steady state of running, the total amount of 200 ml Tubificidae bought from a local fish-shop were used to initially inoculate into both sections of free-swimming and sessile worms. The same amount of Tubificidae inoculation was repeated again on Day 29 because Tubificidae disappeared on Day 26. The dissolved oxygen (DO) concentration (YSI 52CE, USA), pH (PHS-3C, China) and temperature in both sections for free-swimming and sessile worms were monitored daily. The pH values in both sections were stable at 7.21 and 7.44, respectively, and the average DO concentrations in both sections were kept at or above 2.0 mg/l. The average hydraulic retention times (HRT) in both sections were 6.03 and 17.34 h, respectively (Table 3).

### *Analysis*

Sludge samples were taken from the worm-reactor twice a week in the morning. All concentrations of chemical oxygen demand (COD), $NO_2^-$-N, $NO_3^-$-N and $PO_4^{3-}$-P were measured with the supernatant of samples in influent sludge and effluent sludge. The concentrations of COD were determined with a COD analyser (CTL-12, China). The concentrations of $NO_2^-$-N, $NO_3^-$-N and $PO_4^{3-}$-P were determined with the filtrate through 0.45 $\mu$m membrane by a 4500i Ion Chromatograph (Dionex Corp., USA). The concentrations of total suspended solids (TSS) and volatile suspended solids (VSS) were determined at 104 °C (4 h) and 600 °C (2 h), respectively. The sizes of sludge flocs were determined by a laser particle size analyser (Mastersizer 2000, UK). Due to the difficulty of determining TSS concentration in the section containing plastic carriers, sludge reduction was estimated as the difference of TSS concentrations between the influent sludge and the effluent sludge (Equation 1)

$$SR\% = (1 - TSS_{out}/TSS_{in}) \times 100 \quad (1)$$

*Table 3.* Operation conditions of the worm-reactor

| | Section for free-swimming worms Average ± SD (Range) | Section for sessile worms Average ± SD (Range) |
|---|---|---|
| T (°C) | 20.32 ± 1.23 (18.10–22.90) | 18.85 ± 0.97 (16.40–19.90) |
| pH | 7.21 ± 0.51 (5.61–8.20) | 7.44 ± 0.55 (6.30–8.47) |
| DO (mg/l) | 3.91 ± 2.60 (0.10–7.80) | 4.69 ± 2.15 (0.50–8.00) |
| HRT (h) | 6.03 ± 1.43 (3.37–8.29) | 17.34 ± 4.11 (9.68–23.84) |
| TSS (g/l) | 5.01 ± 2.09 (2.07–8.86) | 5.25 ± 2.54 (2.14–12.96)* |
| VSS (g/l) | 3.62 ± 1.53 (1.03–6.46) | 3.91 ± 1.81 (1.86–9.48)* |
| VSS/TSS | 0.72 ± 0.06 (0.50–0.77) | 0.75 ± 0.04 (0.71–0.87)* |
| Recycle ratio | 1.41 ± 0.48 (0.70–2.28) | |

*These values are of the returned sludge.

where SR% is the percent of sludge reduction, % (weight); $TSS_{in}$ and $TSS_{out}$ are the sludge concentrations of influent sludge and effluent sludge, respectively, g/l.

## Results

### *Sludge reduction*

Figure 2 shows differences of TSS concentrations in the influent sludge and the effluent sludge in the worm-reactor. The average sludge reduction and the TSS concentration of the effluent sludge throughout the operation period were 48.36 ± 44.76% and 2.04 ± 1.35 g/l, respectively (Fig. 2). It is noted that the average sludge reduction (29 ± 80%) from Day 24 to Day 35 was much lower than those (53 ± 29 and 59 ± 30%) from Day 4 to Day 21 and from Day 38 to Day 51, respectively. In addition, the average TSS concentrations in the free-swimming worm growth section and the returned sludge were 5.01 and 5.25 g/l, respectively. The average ratios of VSS and TSS in the section for free-swimming worms and the returned sludge were 0.72 and 0.75, respectively (Table 2), which showed little inert solids accumulated in biomass. No apparent sludge accumulation occurred in the worm-reactor based on changes of TSS concentrations (Fig. 2).

### *Sludge settling characteristics*

The presence of Tubificidae did not contribute to improvement of sludge settling characteristics (Fig. 3). The average sludge volume index (SVI) of influent sludge, effluent sludge, sludge in aeration tank, and the returned sludge were 188 ± 48, 201 ± 78, 178 ± 48, and 184 ± 57 ml/g, respectively. Compared to average SVI (< 80 ml/g) in a pilot conventional activated sludge system (Wei et al., 2003b), these values of SVI were very higher. The microscopic investigation clearly showed that large

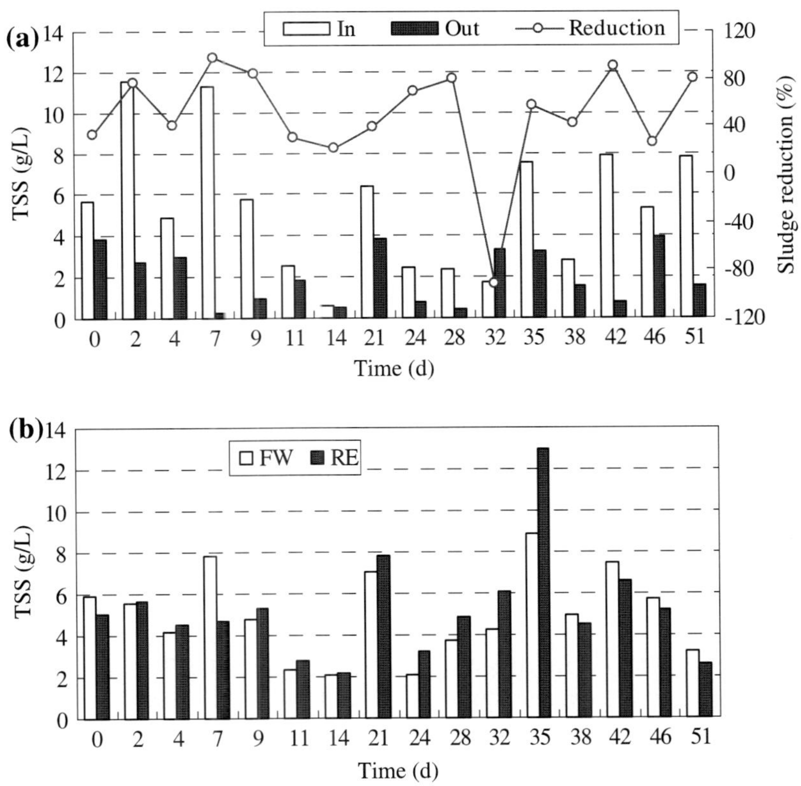


*Figure 2.* TSS concentrations (bars) and percent sludge reduction (line) in the worm-reactor (In-Influent, Out-Effluent, FW-Section for free-swimming worms, RE-Returned sludge).

*Table 4.* Characteristics of influent sludge and effluent sludge in the worm-reactor

| Parameter | Day 0–51 | | Day 21–35 | |
|---|---|---|---|---|
| | Influent Average ± SD | Effluent Average ± SD | Influent Average ± SD | Effluent Average ± SD |
| TSS (mg/l) | 5.42 ± 3.25 | 2.04 ± 1.35 | 4.10 ± 2.65 | 2.33 ± 1.58 |
| VSS (mg/l) | 3.23 ± 2.04 | 1.47 ± 0.97 | 3.05 ± 1.92 | 1.71 ± 1.11 |
| VSS/TSS | 0.72 ± 0.12 | 0.75 ± 0.04 | 0.75 ± 0.02 | 0.75 ± 0.04 |
| SVI (ml/g) | 188.07 ± 99.08 | 200.76 ± 77.60 | 176.75 ± 79.21 | 189.63 ± 46.40 |
| COD (mg/l) | 246.30 ± 137.24 | 236.38 ± 168.36 | 405.70 ± 100.99 | 399.68 ± 168.79 |
| $NH_4^+$-N (mg/l) | 26.81 ± 30.65 | 18.44 ± 29.30 | 55.39 ± 34.41 | 46.23 ± 35.85 |
| $NO_2^-$-N (mg/l) | 49.39 ± 36.06 | 34.73 ± 37.95 | 64.54 ± 40.26 | 61.00 ± 38.01 |
| $NO_3^-$-N (mg/l) | 61.46 ± 25.15 | 59.83 ± 28.76 | 68.54 ± 10.82 | 67.94 ± 34.49 |
| TIN* (mg/l) | 115.63 ± 66.89 | 89.56 ± 73.54 | 188.48 ± 41.79 | 149.38 ± 92.68 |
| $PO_4^{3-}$-P (mg/l) | 4.29 ± 2.34 | 4.85 ± 2.86 | 2.92 ± 0.34 | 3.38 ± 0.60 |

*TIN (total inorganic nitrogen) = $NH_4^+$-N + $NO_2^-$-N + $NO_3^-$-N.

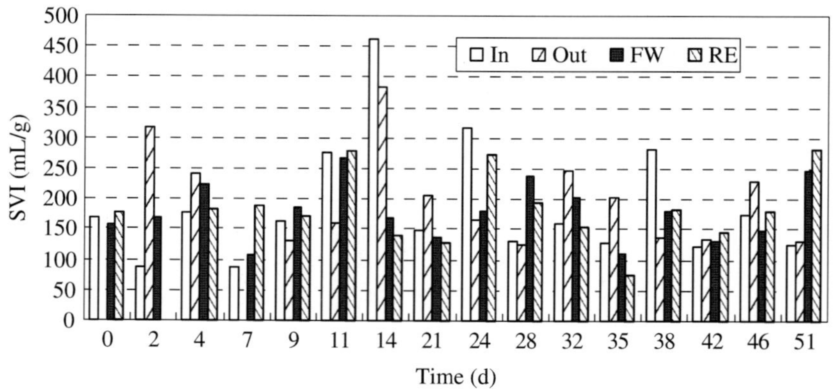


*Figure 3.* SVI changes in the worm-reactor (In-Influent, Out-Effluent, FW-Section for free-swimming worms, RE-Returned sludge).

amounts of straight filaments were found in activated sludge. The size of the sludge flocs changed little as the *d*(0.5) values of influent sludge, effluent sludge, sludge in section of free-swimming worms, and the returned sludge were 90.976, 89.840, 89.073, and 90.901 $\mu$m, respectively.

*COD and nutrients removal*

Obviously, few differences in COD were observed between the influent sludge and the effluent sludge (Fig. 4). Throughout the running period, the average COD and $NH_4^+$-N removals were 10 ± 24 and 46 ± 29%, respectively, and the average COD and $NH_4^+$-N concentrations of the effluent sludge were 232 ± 163 and 27 ± 31 mg/l, respectively (Table 3). No increase of total inorganic nitrogen and $NO_3^-$-N concentrations in the effluent sludge occurred (Table 3 and Fig. 5). No nitrite-nitrogen in both the influent sludge and the effluent sludge was determined before the 21st day. However, $NO_2^-$-N concentrations in both the influent sludge and the effluent sludge increased correspondingly with high $NH_4^+$-N concentrations of the influent sludge from Day 21 to Day 35 (Fig. 5). High $NH_4^+$-N or $NO_2^-$-N concentrations or their combination therefore may limit Tubificidae growth in the worm-reactor, and then resulted in the decreased sludge reduction from Day 21 to Day 35 (Fig. 6). In contrast to total inorganic nitrogen, $PO_4^{3-}$-P

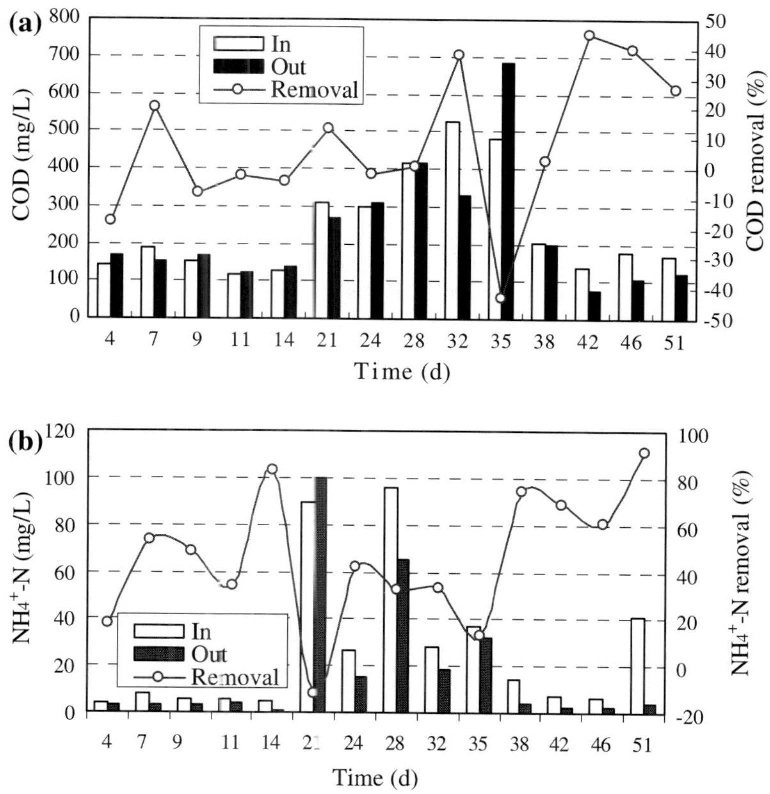


*Figure 4.* COD and $NH_4^+$-N changes in influent sludge and effluent sludge.

release into the effluent sludge occurred, i.e. the average $PO_4^{3-}$-P concentration (4.85 ± 2.86 mg/l) in the effluent sludge was more than that (4.29 ± 2.34 mg/l) in the influent sludge. However, such phosphate increase in the effluent sludge was not substantial.

## Discussion

In their researches (Rensink et al., 1996; Janssen et al., 1998; Rensink & Rulkens, 1997), performances of Tubificidae on sludge reduction in different aerobic wastewater treatment processes equipped with carriers were compared with those without Tubificidae. Their results clearly showed that the presence of Tubificidae could lead to sludge reduction (Table 1). In this study, the average sludge reduction using a worm-reactor treating the discharged sludge was similar to literature data of sludge reduction with Oligaochaeta in aerobic wastewater treatment processes. Aggregations of Tubificidae were visible, and appeared to increase in the carriers and on the bottom of the reactor several days after the worm inoculation. Different from results that *Aeolosoma hemprichicii* and *Nais elinguis* naturally occurred in activated sludge systems (Ratsak, 1994, 2001; Zhang, 2000; Wei et al., 2003b), no free-swimming worms such as *Aeolosoma hemprichicii* and *Nais elinguis* were found in both the worm-reactor and the discharged excess sludge in this study. This result suggested that little or no free-swimming worm reproduction occurred in the pilot activated sludge system.

Few differences of COD were observed between the influent sludge and the effluent sludge in the worm-reactor. The reason was that COD residue in the discharged excess sludge from the pilot activated sludge system was recalcitrant to be biodegraded. High $NH_4^+$-N concentration in the fed wastewater from Day 21 to Day 35 and the

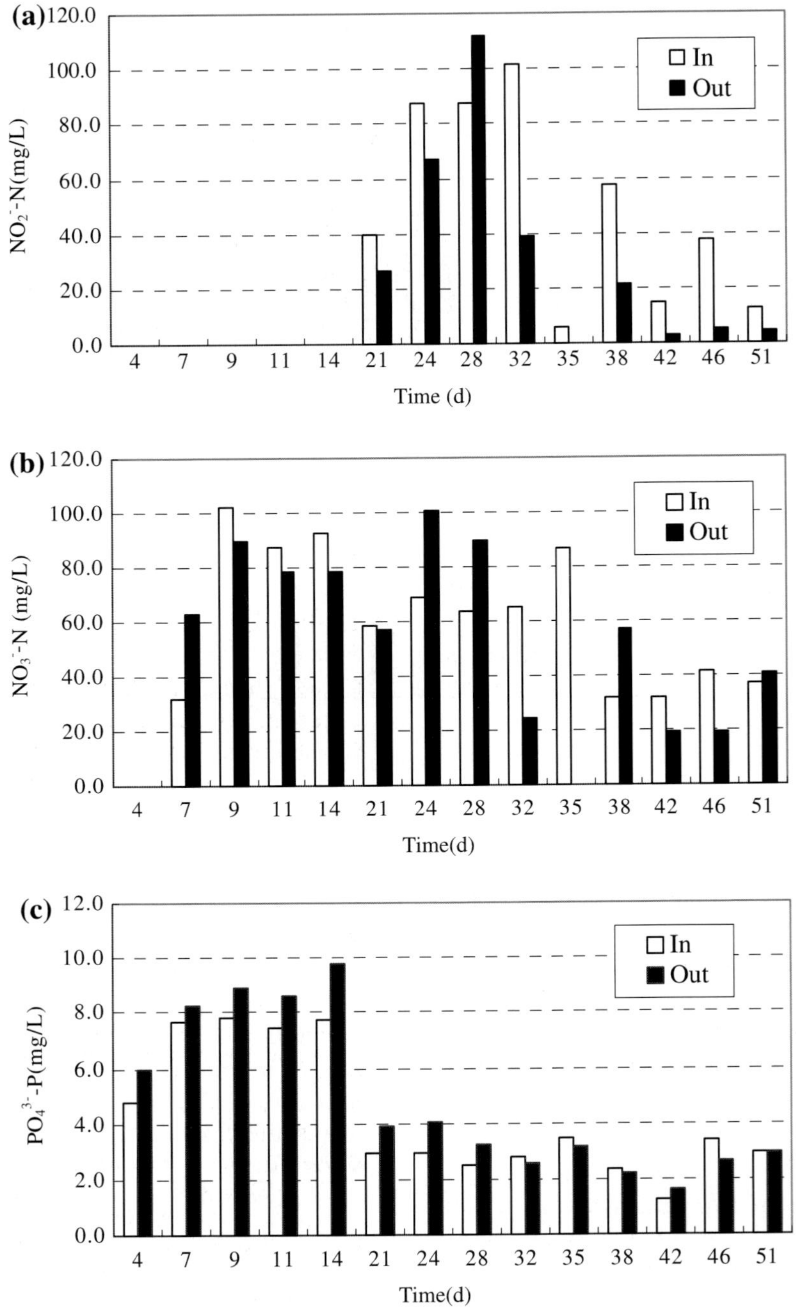


*Figure 5*. Changes of $NO_2^-$-N, $NO_3^-$-N and $PO_4^{3-}$-P in influent sludge and effluent sludge.

operation at low temperature (in the winter) could affect nitrification process, and then resulted in high $NO_2^-$-N accumulation in the pilot activated sludge system. Therefore, high $NH_4^+$-N or $NO_2^-$-N concentrations should be avoided in case of inhibiting worm growth. From the point view of sludge mineralization, an increase of phosphorus, nitrogen (nitrate), $CO_2$ and even dissolved COD hardly seems avoidable. The nutrient concentrations in the effluent would be expected to

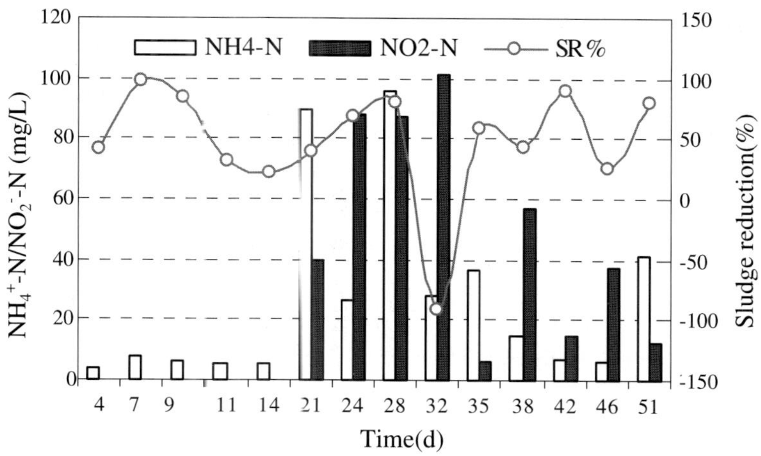


*Figure 6.* Relationships of $NH_4^+$-N, $NO_2^-$-N and sludge reduction.

be equal to or higher than those in the influent for sludge reduction. In this study, the average total inorganic nitrogen concentrations in both the influent sludge and the effluent sludge were similar, and the increase of $PO_4^{3-}$-P in the effluent sludge was not great.

## Conclusions

In this paper, we have presented some preliminary results. However, further research is needed to (i) understand the effects of free-swimming and sessile worms on sludge reduction, including use of reactors with and without oligochaetes to better assess the contribution of worms to the process, and (ii) optimise the performance of the worm-reactor. From the results to date, we made the following conclusions:

(1) The new worm-reactor, consisting of sections suitable for both free-swimming and sessile worms, achieved reduction of sludge volume by an average of 48%. This reduction was most likely a result of the presence of tubificid oligochaetes.
(2) No free-swimming worms were found in the worm-reactor. High $NH_4^+$-N or $NO_2^-$-N concentrations or their combination may have limited the growth of such worms.
(3) $PO_4^{3-}$-P increase of effluent was not substantial.

## Acknowledgements

This work is financially supported by the Scientific Research Foundation for the Returned Overseas Chinese Scholars, State Education Ministry of China, the Youth Fund Project of National Hi-Tech Development Plan of China (2002AA649260), and the Key Project of National Natural Science Foundation of China (50238050).

## References

Davis, R. D. & J. E. Hall, 1997. Production, treatment and disposal of wastewater sludge in Europe from a UK perspective. European Water Pollution Control 7(2): 9–17.

Eikelboom, D. H., 2000. Process Control of Activated Sludge Plants by Microscopic Investigation . IWA Publishing, UK, pp 85–102.

Ghyoot W., 1998. Membrane-Assisted Bioreactor for Anaerobic Sludge Digestion and Low Sludge Production in Aerobic Wastewater treatment. Ph.D thesis. University Gent, Belgium.

Ghyoot, W. & W. Verstraete, 2000. Reduced sludge production in a two-stage membrane-assisted bioreactor. Water Research 34: 205–215.

Janssen, P. M. J., W. H. Rulkens, J. H. Rensink & H. F. van der Poest, 1998. The potential for metazoa in biological wastewater treatment. Water Quality International (WQI), Sep./Oct., 25–27.

Lapinski, J. & A. Tunnacliffe, 2003. Reduction of suspended biomass in municipal wastewater using bdelloid rotifers. Water Research 37: 2027–2034.

Lee, N. M. & T. Welander, 1996a. Reducing sludge production in aerobic wastewater treatment through manipulation of the ecosystem. Water Research 30: 1781–1790.

Lee, N. M. & T. Welander, 1996b. Use of protozoa and metazoa for decreasing sludge production in aerobic wastewater treatment. Biotechnology Letters 18: 429–434.

Luxmy, B. S., T. Kubo & K. Yamamoto, 2002. Sludge reduction potential of metazoa in membrane bioreactors. Water Science and Technology 44(10): 197–202.

Metcalf, & Eddy, 2001. Wastewater Engineering: Treatment, Disposal and Reuse, 4th edn. McGraw-Hill, Inc., Chapter 8.

National Environmental Protection Agency of China, 2001. Chinese Environmental Statistical Gazette of 2001.

Ratsak C. H., 1994. Grazer induced sludge reduction in wastewater treatment. Ph.D. thesis. Vrije University, the Netherlands.

Ratsak, C. H., 2001. Effects of *Nais elinguis* on the performance of an activated sludge plant. Hydrobiologia 463: 217–222.

Ratsak, C. H., S. A. L. M. Kooijman & B. W. Kooi, 1993. Modelling the growth of an oligochaete on activated sludge. Water Research 27: 739–747.

Ratsak, C. H., B. W. Kooi & H. W. van Verseveld, 1994. Biomass reduction and mineralization increase due to the ciliate *Tetrahymena Pyriformis* grazing on the bacterium *Pseudomonas Fluorescens*. Water Science and Technology 29(7): 119–128.

Ratsak C. H., K. A. Maarsen and S. A. L. Kooijman, 1996. Effects of protozoa on carbon mineralization in activated sludge. Water Research 30(1): 1–12.

Rensink, J. H. & W. H. Rulkens, 1997. Using metazoa to reduce sludge production. Water Science and Technology 36(11): 171–179.

Rensink, J. H., R. Corstanje, J. H. van der Pal, 1996. A new approach to sludge reduction by metazoa. In 10th European Sewage and Reuse Symposium, IFAT 1996, Munchen, pp. 339–364.

Spellman, F. R., 1997. Wastewater Biosolids to Compost . Technomic Publishing Company, Inc, Lancaster, Pennsylvania, USA, pp 223–235.

Wei, Y. S., J. X. Liu, 2002. China invention patent application No. 02159048.6.

Wei, Y. S., Y. B. Fan, M. J. Wang & J. S. Wang, 2000. Composting and compost application in China. Resource, Conservation & Recycling 30: 277–300.

Wei, Y. S., R. T. van Houten, A. R. Borger, D. H. Eikelboom & Y. B. Fan, 2003a. Minimization of excess sludge production for biological wastewater treatment. Water Research 37: 4453–4467.

Wei, Y. S., R. T. van Houten, A. R. Borger, D. H. Eikelboom & Y. B. Fan, 2003b. Comparison performances of membrane bioreactor (MBR) and conventional activated sludge (CAS) processes on sludge reduction induced by Oligochaete. Environmental Science & Technology 37: 3171–3180.

Welander, T., N. M. Lee, 1994. Minimization of sludge production in aerobic treatment by use of predators. The Second International Symposium on Environmental Biotechnology, 4–6 July 1994, Brighton, UK.

Zhang S. Y., 2000. Polluted water treatment by the combining processes of membrane separation and biodegradation. Ph.D thesis. Research Centre for Eco-Environmental Sciences, Chinese Academy of Sciences, China.